Environmental Earth Sciences

Series editor

James W. LaMoreaux, Tuscaloosa, USA

More information about this series at http://www.springer.com/series/8394

Maria Luisa Calvache · Carlos Duque
David Pulido-Velazquez
Editors

Groundwater and Global Change in the Western Mediterranean Area

Springer

Editors
Maria Luisa Calvache
Department of Geodynamics
University of Granada
Granada
Spain

David Pulido-Velazquez
Spanish Geological Survey (IGME)
Granada
Spain

Carlos Duque
Department of Geoscience
Aarhus University
Aarhus
Denmark

ISSN 2199-9155 ISSN 2199-9163 (electronic)
Environmental Earth Sciences
ISBN 978-3-319-69355-2 ISBN 978-3-319-69356-9 (eBook)
https://doi.org/10.1007/978-3-319-69356-9

Library of Congress Control Number: 2017955361

Printed on acid-free paper

This Springer imprint is published by Springer Nature
The registered company is Springer International Publishing AG
The registered company address is: Gewerbestrasse 11, 6330 Cham, Switzerland

Preface

The Western Mediterranean area is a very sensitive area that suffers frequents droughts due to climate conditions, significant anthropogenic impacts on land use (urbanization or increment of cultivated areas) and high seasonality both in precipitation and water resources demands. This effect will be exacerbated in the future due to the global change and specially in the coastal areas with growing urban development and intensive agriculture. In this framework, groundwater plays an important role in the definition of management alternatives of Water Resources systems. An exhaustive knowledge of those systems and their problems is required to identify appropriate sustainable decision, which is one of the most important "challenges of our society".

This book aims to contribute to the dissemination of the knowledge about impacts of global change on water resources systems in the Western Mediterranean area, with special emphasis on groundwater. It is a compilation of works carried out by researchers from Algeria, France, Italy, Morocco, Portugal, Tunisia and Spain. Although most global change investigation is focused on surface water, the number of research papers dealing with global change and groundwater has grown fast in recent years, as shown in recent review papers. This compilation coversa particularly interesting area, the Western Mediterranean countries, from the perspective of the water resources with frequent scarcity periods and societies highly dependent on groundwater. It includes work on this Mediterranean area of both, southern Europe and North Africa, where important impacts are expected on the sustainability, quantity, quality, and management of water resources. This volume is composed by a selection of contributions presented in the Conference "Groundwater and Global Change in the Western Mediterranean" (Granada, November 2017). It covers a wide range of aspects linking global change to groundwater, from monitoring and modeling historical and future impacts to adaption strategies. This provides an overview of methods, study areas and case studies in multiple countries essential for facing future challenges.

Granada, Spain — Maria Luisa Calvache
Granada, Spain — Carlos Duque
Aarhus, Denmark — David Pulido-Velazquez

Contents

Contributors

H. Abahous Laboratory of Applied Geology and Geo-Environment, Ibn Zohr University, Agadir, Morocco

Y. Ait Brahim Laboratory of Applied Geology and Geo-Environment, Ibn Zohr University, Agadir, Morocco

I. Alhama Departamento de Ingeniería Civil, Universidad Politécnica de Cartagena, Cartagena, Spain

B. Andreo Department of Geology and Centre of Hydrogeology, University of Málaga (CEHIUMA), Málaga, Spain

Y. Arjdal Applied Geology and Geo-Environment Laboratory, University Ibn Zohr, Agadir, Morocco

L. Baena Ruiz Spanish Geological Survey (IGME), Urb. Alcázar Del Genil, Granada, Spain

M. Bahir Laboratoire de Géosciences et Environnement, Département de Géologie, Ecole Normale Supérieure, Université Cadi Ayyad, Marrakech, Maroc

J. A. Barberá-Fornell Department of Geology and Centre of Hydrogeology, University of Málaga (CEHIUMA), Málaga, Spain

S. Benabdallah Centre de Recherches et des Technologies des Eaux, Soliman, Tunisia

J. Benavente University of Granada, Granada, Spain

E. Beraaouz Laboratory of Applied Geology and Geo-Environment, Ibn Zohr University, Agadir, Morocco

J. Berbel University of Córdoba, Córdoba, Spain

L. Bouchaou Applied Geology and Geo-Environment Laboratory, Faculty of Sciences, Ibn Zohr University, Agadir, Morocco

A. Bouderbala Department of Earth Sciences, University of Khemis Miliana, Khemis Miliana, Algeria

R. Bouhlila Laboratory of Modeling in Hydraulics and Environment (LMHE), National Engineering School of Tunis, University of Tunis El Manar, Tunis, Tunisia

G. Braca ISPRA—Italian National Institute for Environmental Protection and Research, Rome, Italy

N. Brini Faculty of Sciences of Tunis, University of Tunis El Manar, Tunis, Tunisia

M. L. Calvache Department of Gedynamics, University of Granada, Granada, Spain

P. M. Carreira Centro de Ciências e Tecnologias Nucleares (C2TN), Instituto Superior Técnico, Universidade de Lisboa, Bobadela, LRS, Portugal; Campus Tecnológico e Nuclear, Instituto Superior Técnico, Universidade Técnica de Lisboa, Lisboa, Portugal

R. C. Carvalho TARH Lda, Sacavém, Portugal

T. Carvalho TARH Lda, Sacavém, Portugal

S. Chargui Georesources Laboratory, Water Research and Technology Centre, Soliman, Tunisia

B. Chebli Laboratory of Mechanics, Processes Energy and Environment, National School of Applied Sciences, Agadir, Morocco

A. Chehbouni CESBIO, Rabat Agdal, Morocco

A. Chekirbane Georesources Laboratory, Water Research and Technology Centre, Soliman, Tunisia

H. Cherkaoui Dekkaki Laboratory of Research and Development in Engineering Sciences (LRDSI), Department of Geology, Faculty of Sciences and Techniques of al Hoceima, University Mohammed First, Al Hoceima, Morocco

R. Choukr-allah Laboratory of Salinity and Plant Nutrition, Hassan II Institute of Agronomy and Veterinary Medicine, Agadir, Morocco

A. J. Collados-Lara Spanish Geological Survey (IGME), Urb. Alcázar Del Genil, Granada, Spain; Unidad del IGME en Granada, Granada, Spain

E. Custodio Royal Academy of Sciences of Spain, Civil and Environmental Engineering Department, Technical University of Catalonia (UPC), Barcelona, Spain

S. Danni Laboratory of Applied Geology and Geo-Environment, Ibn Zohr University, Agadir, Morocco

D. Ducci Department of Civil, Architectural and Environmental Engineering, University of Naples Federico II, Naples, Italy

C. Duque Department of Geoscience, Aarhus University, Aarhus, Denmark; Department of Geological Sciences, University of Delaware, Newark, DE, USA

S. M. El Kanti Geosciences Laboratory of Natural Resources, Faculty of Sciences, University Ibn Tofail, Kenitra, Morocco

B. El Mansouri Geosciences Laboratory of Natural Resources, Faculty of Sciences, University Ibn Tofail, Kenitra, Morocco

M. El Otmani Laboratory of Salinity and Plant Nutrition, Hassan II Institute of Agronomy and Veterinary Medicine, Agadir, Morocco

A. Elmouden Applied Geology and Geo-Environment Laboratory, Faculty of Sciences, Ibn Zohr University, Agadir, Morocco

S. Engelhardt UMS AG, Munich, Germany

P. Engesgaard Department of Geosciences and Natural Resources Management, University of Copenhagen, Copenhagen, Denmark

M. Erostate UMR CNRS SPE 6134, Hydrogeology Department, University of Corsica, Corte, France

A. Expósito University of Sevilla, Sevilla, Spain

A. Fernández-Ayuso University Pablo de Olavide, Seville, Spain

G. García Departamento de Ingeniería Civil, Universidad Politécnica de Cartagena, Cartagena, Spain

E. Garel UMR CNRS SPE 6134, Hydrogeology Department, University of Corsica, Corte, France

S. Gil Câmara Municipal de Monchique, Monchique, Portugal

J. V. Giráldez Departamento de Agronomía, Universidad de Córdoba, Córdoba, Spain

J. M. Gómez-Fontalva Aljibe Consultores S.L., Granada, Spain

F. Hamzaoui-Azaza Research Unit of Geochemistry and Environmental Geology, Department of Geology, Faculty of Mathematical, Physical and Natural Sciences, University of Tunis El Manar, Tunis, Tunisia

L. Hanich Laboratory of Geo-Resources, Faculty of Science and Technology, Cadi Ayyad University, Marrakech, Morocco

M. Hassani Zerrouk Laboratory of Research and Development in Engineering Sciences (LRDSI), Department of Geology, Faculty of Sciences and Techniques of al Hoceima, University Mohammed First, Al Hoceima, Morocco

I. Hassen Laboratory of Modeling in Hydraulics and Environment (LMHE), National Engineering School of Tunis, University of Tunis El Manar, Tunis, Tunisia

F. M. Hummel Ludwig–Maximilians–University, Munich, Germany

F. Huneau UMR CNRS SPE 6134, Hydrogeology Department, University of Corsica, Corte, France

F. Jarraya Horriche Georesources Laboratory, Centre of Research and Water Technologies, Technopole Borj Cedria, Soliman, Tunisia

L. Kenny IAV Hassan II, Ait Melloul, Morocco

C. Kohfahl Instituto Geológico y Minero de España, Sevilla, Spain

T. Kouz Laboratory of Research and Development in Engineering Sciences (LRDSI), Department of Geology, Faculty of Sciences and Techniques of al Hoceima, University Mohammed First, Al Hoceima, Morocco

F. Lachaal Georesources Laboratory, Water Research and Technology Centre, Soliman, Tunisia

A. Larabi Laboratory LIMEN, Mohammadia School of Engineers, University Mohammed V of Rabat, Rabat, Morocco

C. Leduc IRD, UMR G-EAU, Montpellier, France

C. Llopis-Albert Universitat Politècnica de València, Valencia, Spain

M. López-Chicano Departamento de Geodinámica, Facultad de Ciencias, Universidad de Granada, Granada, Spain

H. Mairech Centre de Recherches et des Technologies des Eaux, Soliman, Tunisia

M. Malki Applied Geology and Geo-Environment Laboratory, Faculty of Sciences, Ibn Zohr University, Agadir, Morocco

I. Mansir Laboratory of Mechanics, Processes Energy and Environment, National School of Applied Sciences, Agadir, Morocco

S. Mansour Laboratory of Research and Development in Engineering Sciences (LRDSI), Department of Geology, Faculty of Sciences and Techniques of al Hoceima, University Mohammed First, Al Hoceima, Morocco

J. M. Marques Centro de Recursos Naturais e Ambiente (CERENA), Instituto Superior Técnico, Universidade de Lisboa, Lisbon, Portugal

J. Martín-Arias Departamento de Geología y Centro de Hidrogeología de la Universidad de Málaga, Facultad de Ciencias, Universidad de Málaga, Málaga, Spain

C. Martín-Montañés Instituto Geológico y Minero de España (IGME), Granada, Spain

D. J. Martínez Suárez Instituto Geológico y Minero de España, Sevilla, Spain

P. Martínez-Santos Departamento de Geodinámica, Facultad de Ciencias Geológicas. Universidad Complutense de Madrid, Ciudad Universitaria, Madrid, Spain

J. Mas-Pla Institut Català de Recerca de l'Aigua (ICRA), Girona, Spain

S. Massuel IRD, UMR G-EAU, Montpellier, France

L. Mateos CSIC-IAS, Córdoba, Spain

A. Menció Grup de Geologia Aplicada i Ambiental (GAiA-Geocamb), Deptartament de Ciències Ambientals, Campus de Montilivi, Universitat de Girona, Girona, Spain

N. Mgaidi Faculty of Sciences of Tunis, University of Tunis El Manar, Tunis, Tunisia

J. L. Michelot UMR IDES 8148 CNRS—Université Pari-Sud, Orsay Cedex, France

A. Mlayah Georesources Laboratory, Water Research and Technology Centre, Soliman, Tunisia

L. Molano-Leno Instituto Geológico y Minero de España, Sevilla, Spain

I. Morell Universitat Jaume I, Castelló, Spain

T. Mourabit Laboratory of Research and Development in Engineering Sciences (LRDSI), Department of Geology, Faculty of Sciences and Techniques of al Hoceima, University Mohammed First, Al Hoceima, Morocco

J. M. Murillo Instituto Geológico y Minero de España, Madrid, Spain

A. N. Martínez Sánchez de la Nieta Instituto Geológico y Minero de España, Sevilla, Spain

J. M. Nieto-López Department of Geology and Centre of Hydrogeology, University of Málaga (CEHIUMA), Málaga, Spain

B. N'da Applied Geology and Geo-Environment Laboratory, Faculty of Sciences, Ibn Zohr University, Agadir, Morocco

J. T. Olsen Norwegian Road Administration, Bodø, Norway

S. Ouhamdouch Laboratoire de Géosciences et Environnement, Département de Géologie, Ecole Normale Supérieure, Université Cadi Ayyad, Marrakech, Maroc

S. Oumarou Danni Applied Geology and Geo-Environment Laboratory, Faculty of Sciences, Ibn Zohr University, Agadir, Morocco

E. Pardo-Igúzquiza Instituto Geológico y Minero de España, Madrid, Spain

V. Pasqualini UMR CNRS SPE 6134, Hydrogeology Department, University of Corsica, Corte, France

M. Polemio Research Institute for Geo-Hydrological Protection (CNR-IRPI), Bari, Italy

L. Portell Institut Català de Recerca de l'Aigua (ICRA), Girona, Spain

D. Pulido-Velazquez Spanish Geological Survey (IGME), Urb. Alcázar Del Genil, Granada, Spain; Unidad del IGME en Granada, Granada, Spain

R. Purtschert Climate and Environmental Physics, University of Bern, Bern, Switzerland

J. M. Ramírez-González Andalusian Regional Agency of Environment and Water, Málaga, Spain

B. Reichert Steinmann-Institute, Geology, Bonn University, Bonn, Germany

A. Renau-Pruñonosa Universitat Jaume I, Castelló, Spain

M. Rendón-Martos Andalusian Regional Office of Environment and Land Management, Fuente de Piedra, Spain

M. Rodríguez-Rodríguez University Pablo de Olavide, Seville, Spain

T. Rodríguez Departamento de Ingeniería Minera, Geológica y Cartográfica, Universidad Politécnica de Cartagena, Cartagena, Spain

J. Ronchail Université Paris 7-Denis Diderot, UMR LOCEAN, Paris, France

F. Ruiz Bermudo Instituto Geológico y Minero de España, Sevilla, Spain

S. Santoni Hydrogeology Department, University of Corsica, CNRS UMR 6134 SPE, Corte, France

J. Senent-Aparicio Spanish Geological Survey (IGME), Urb. Alcázar Del Genil, Granada, Spain

A. Sifeddine IRD-Sorbonne Université (IRD, UPMC, CNRS MNHN) UMR LOCEAN, Bondy Cedex, France

F. R. Sousa TARH Lda, Sacavém, Portugal

J. P. Sánchez-Úbeda Departamento de Geodinámica, Facultad de Ciencias, Universidad de Granada, Granada, Spain

J. Sültenfuß Institute of Environmental Physics, University of Bremen, Bremen, Germany

M. Tsujimura Faculty of Life and Environmental Sciences, University of Tsukuba, Tsukuba, Ibaraki, Japan

K. Vanderlinden IFAPA, Centro Las Torres-Tomejil, Seville, Spain

M. Zammouri Faculty of Sciences of Tunis, University of Tunis El Manar, Tunis, Tunisia

K. Zouari Laboratoire de Radio-Analyse et Environnement, Ecole Nationale d'Ingénieurs de Sfax, Université de Sfax, Sfax, Tunisie

Introduction

The Western Mediterranean areas present a series of specific characteristics due to climate, anthropogenic pressure over natural environments, hydrological conditions and water demands that make it very sensitive to the effects of the global change. This has a direct impact over the quality and quantity of groundwater resources, essential for the water supply of population and the maintenance of agriculture. Valuable ecosystems connected with groundwater are also affected as for example has been shown in wetlands in the South of Spain in Málaga (Nieto-López et al. 2017) and in the sand dune ponds in the Doñana National Park (Fernandez-Ayuso et al. 2017) or in the Biguglia Lagoon in Corsica (Erostate et al. 2017). Agriculture is the base of the economy in multiple regions of the Western Mediterranean and it can be seriously damaged due to the decrease in water resources connected to global change. For example, the cereal in the region of Souss-Massa in Morocco is strongly affected by the precipitation regime (Abahous et al. 2017) and requires a sustainable development as indicated by Mansir et al. (2017). Other potential impacts have been also described for the nitrate occurrence as shown in the Catalonian Inner Basins (Mas Pla 2017a).

Climate change is one of the major concerns when referring to global change. Climate presents a variability that can affect groundwater resources as has been researched in North Algeria (Bouderbala 2017) or in Southern Italy (Ducci et al. 2017). Other case studies describe climatic changes impacting groundwater in areas of Morocco (Ouhamdouch et al. 2017) and in Tunisia (Benabdallah et al. 2017). Successive climate change reports analyzing climate change issues can be summarized as Mas Pla et al. (2017b) did for the Catalonia region in Northwestern Spain.

Potential future hydrological impacts of climatic change can be assessed by propagating future climate scenarios by simulation with hydrological models. This has been done for the groundwater in the Mijas mountains in Spain (Martin-Arias et al. 2017) or for the dynamics of snowpack in Sierra Nevada (Pardo-Iguzquiza et al. 2017). For the correct evaluation of the trends in future scenarios a full understanding of aquifer systems is required. The use of numerical models is a common tool for the assessment of the water budget as shown in Torrevieja aquifer

(Duque et al. 2017a) or for the management of water resources as has been exposed in Dakhla Bay in Southern Morocco (El Kanti et al. 2017).

In some cases, the changes observed in the quality of groundwater can be related to global changes or a natural evolution of the system. The differentiation between both is possible with the application of numerical models but demands a good knowledge of the current and the historical conditions of the aquifers. This principle has been applied for understanding the salinity changes of the Lower Sado aquifer in Portugal in the last 20000 years (Carreira and Marques 2017), or the Motril-Salobreña aquifer in Spain for the last 6000 years (Duque et al. 2017b). In these cases the use of environmental tracers can determine the age of groundwater (Sanchez-Úbeda et al. 2017a) that can be eventually used for the integration into numerical models to answer hydrogeological questions (Sanchez-Úbeda et al. 2017b).

In addition to the models, the prevention of the impact of global change requires the use of other methods for the research of aquifer systems. The application of hydrogeochemical and isotopic tracers is a useful tool as demonstrated in the Plaine of Kasserine (Hassen et al. 2017), the Moroccan High Atlas (N'Da et al. 2017), the Massa catchment (Oumarou Danni et al. 2017) or with a multitracer approach in Corsica (Santoni et al. 2017).

In this global change framework, coastal areas are critical for the groundwater resources as they are exposed to the additional risk of saltwater intrusion. For this reason, it is important to develop methods to summarize impacts as for example an index-based method (Baena Ruiz et al. 2017) or for evaluating the vulnerability as in the Ghiss-Nekor aquifer (Kouz et al. 2017). The Mediterranean coast of Spain has an elevated number of aquifers with variable circumstances and different sea-water intrusion processes (Custodio 2017b) but the risk can be extended to almost all the coastal areas in the western Mediterranean as for example is seen in North East Tunisia (Lachaal et al. 2017).

One of the methods to adapt to global change impacts and to optimize water resources is the artificial recharge. This technique requires a good knowledge of the land infiltration capacity. For this purpose, high precision lysimeters in South Spain are being used (Molano-Leno et al. 2017). In Tunisia, Horriche et al. (2017) propose to apply an integrated model as Wetpass and Zammouri et al. (2017) the assessment of the efficiency against groundwater stress.

Since global change is a phenomenon that probably will continue in the future, there are investigations presenting adaptions to future changes. In this sense, Alhama et al. (2017) propose the improvement of the management through numerical models in Calasparra aquifer, Braca and Ducci (2017) evaluate groundwater resources with GIS tools in Italy and Berbel et al. (2017) suggest better regulations measurements in Llanos de la Puebla aquifer. Climate change adaptation requires a special effort for managing aquifer recharge as proposed for the Monchique mountain (Carvalho et al. 2017) and also it is important to use a clear and homogeneous terminology when referring to groundwater resources (Custodio 2017a).

A solid knowledge about the groundwater resources, a variety of tested methods for the research and understanding of aquifers and reliable potential future scenarios as well as science-based adaptations will be essential to confront the challenges that global change will bring to the Western Mediterranean countries.

References

Abahous H, Ronchail J, Sifeddine A, Kenny L, Bouchaou L (2017) Impacts of the water resources variability on cereal yields in the region of Souss-Massa Southern Morocco. In: Groundwater and global change in the western mediterranean area. Environmental earth sciences. ISBN: 978-3-319-69355-2

Alhama I, García G, Rodríguez T (2017) Groundwater as an useful resource in the adaptation to the climate change: the case of the synclinal de Calasparra aquifer (Murcia, SE Spain). In: Groundwater and global change in the western mediterranean area. Environmental earth sciences. ISBN: 978-3-319-69355-2

Baena Ruiz L, Pulido-Velazquez D, Renau-Pruñonosa A, Morell I, Llopis-Albert C, Collados-Lara AJ, Senent-Aparicio J (2017) An index-based method to assess impacts of global change on seawater intrusion problems. In: Groundwater and global change in the western mediterranean area. Environmental earth sciences. ISBN: 978-3-319-69355-2

Benabdallah S, Mairech H, HummelF M (2017) Assessing the impacts of climatic change on groundwater recharge for the Chiba Basin in Tunisia. In: Groundwater and global change in the western mediterranean area. Environmental earth sciences. ISBN: 978-3-319-69355-2

Berbel J, Expósito A, Mateos L (2017) The importance of the groundwater governance in the global change context: a proposal for a Mediterranean aquifer (Llanos de la Puebla, Spain). In: Groundwater and global change in the western mediterranean area. Environmental earth sciences. ISBN: 978-3-319-69355-2

Bouderbala A (2017) Effects of climate variability on groundwater resources in coastal aquifers (case of Mitidja Plain in the North Algeria) In: Groundwater and global change in the western mediterranean area. Environmental earth sciences. ISBN: 978-3-319-69355-2

Braca G, Ducci D (2017) Development of a GIS based procedure (BIGBANG 1.0) for evaluating groundwater balances at National scale and comparison with groundwater resources evaluation at local scale. In: Groundwater and global change in the western mediterranean area. Environmental earth sciences. ISBN: 978-3-319-69355-2

Carreira PM, Marques JM (2017) Groundwater salinity and environmental change over the last 20,000 years: isotopic evidences in the Lower Sado aquifer recharge, Portugal. In: Groundwater and global change in the western mediterranean area. Environmental earth sciences. ISBN: 978-3-319-69355-2

Carvalho RC, Carvalho T, Sousa FR, Gil S, Agostinho R (2017) Sowing water in Monchique mountain: a multidisciplinary MAR project for climate change adaptation. In: Groundwater and global change in the western mediterranean area. Environmental earth sciences. ISBN: 978-3-319-69355-2

Custodio E (2017) Concepts on groundwater resources. In: Groundwater and global change in the western mediterranean area. Environmental earth sciences. ISBN: 978-3-319-69355-2

Custodio E (2017) Consequences of seawater intrusion in Mediterranean Spain. Project SASMIE. In: Groundwater and global change in the western mediterranean area. Environmental earth sciences. ISBN: 978-3-319-69355-2

Ducci D, Polemio M (2017) Quantitative impact of climate variations on groundwater in southern Italy. In: Groundwater and global change in the western mediterranean area. Environmental earth sciences. ISBN: 978-3-319-69355-2

Duque C, Gomez-Fontalva JM, Murillo JM, Calvache ML (2017a) Assessing the uncertainties in the water budget of the Torrevieja aquifer (Southeast Spain). In: Groundwater and global change in the western mediterranean area. Environmental earth sciences. ISBN: 978-3-319-69355-2

Duque C, Thu Olsen J, Sanchez-Ubeda JP, Calvache ML (2017b) Paleohydrogeological model of the groundwater salinity in the Motril-Salobreña aquifer (Southeast Spain). In: Groundwater and global change in the western mediterranean area. Environmental earth sciences. ISBN: 978-3-319-69355-2

El Kanti SM, El Mansouri B, Arjdal Y, Larabi A (2017) Numerical groundwater modelling as an effective tool for management of the deep groundwater at the Dakhla Bay (South Morocco). In: Groundwater and global change in the western mediterranean area. Environmental earth sciences. ISBN: 978-3-319-69355-2

Erostate M, Garel E, Huneau F, Pasqualini V (2017) Delineating the aquifer role in the anthropogenic fingerprint on the groundwater-dependent ecosystem of the Biguglia Lagoon (Corsica, France). In: Groundwater and global change in the western mediterranean area. Environmental earth sciences. ISBN: 978-3-319-69355-2

Fernández-Ayuso A, Rodríguez-Rodríguez M, Benavente J (2017) Change in the hydrological functioning of sand dune ponds in Donaña National Park (Southern Andalusia, Spain). In: Groundwater and global change in the western mediterranean area. Environmental earth sciences. ISBN: 978-3-319-69355-2

Hassen I, Hamzaoui-Azaza F, Bouhlila R (2017) Hydrogeochemical and isotopic investigations for evaluation of the impact of climate change on groundwater quality, a case study of the Plaine of Kasserine (Central Tunisia). In: Groundwater and global change in the western mediterranean area. Environmental earth sciences. ISBN: 978-3-319-69355-2

Horriche FJ, Mgaidi N, Ghouili N, Zammouri M (2017) Groundwater recharge assessment using Wetpass: case study of the Sidi Marzoug-Sbiba aquifer (Tunisia). In: Groundwater and global change in the western mediterranean area. Environmental earth sciences. ISBN: 978-3-319-69355-2

Kouz T, Cherkaoui Dekkaki H, Mansour S, Hassani Zerrouk M, Mourabite T (2017) Application of GALDIT index to assess the intrinsic vulnerability of coastal aquifer to seawater intrusion: Case of the Ghiss-Nekor aquifer (North East of Morocco). In: Groundwater and global change in the western mediterranean area. Environmental earth sciences. ISBN: 978-3-319-69355-2

Lachaal F, Chargui S, Ben Messaoud R, Chekirbane A, Tsujimura M, Mlayah A, Massuel S, Leduc C (2017) Impacts of global changes on groundwater resources in North-East Tunisia: the case of the Grombalia phreatic aquifer. In: Groundwater and global change in the western mediterranean area. Environmental earth sciences. ISBN: 978-3-319-69355-2

Mansir I, Bouchaou L, Choukr-allah R, Chebli B, El Otmani M (2017)Groundwater resources scarcity in Souss-Massa region and alternative solutions for sustainable agricultural development. In: Groundwater and global change in the western mediterranean area. Environmental earth sciences. ISBN: 978-3-319-69355-2

Martín-Arias J, Martínez-Santos P, Andreo B (2017) Hydrogeological model of Mijas mountain aquifers under different climate conditions (Malaga, Spain). In: Groundwater and global change in the western mediterranean area. Environmental earth sciences. ISBN: 978-3-319-69355-2

Mas-Pla J (2017a) Global change and groundwater in Catalonia: contributions from the three "Climate change reports" (2005, 2010, 2016). In: Groundwater and global change in the western mediterranean area. Environmental earth sciences. ISBN: 978-3-319-69355-2

Mas-Pla J, Menció A, Portell L (2017b) Qualitative evaluation of climate change effects on nitrate occurrence at several aquifers in the Catalonia Inner Basins. In: Groundwater and global change in the western mediterranean area. Environmental earth sciences. ISBN: 978-3-319-69355-2

Molano Leno L, Kohfahl C, Martinez Suarez J, Ruiz Bermudo F, Martinez Sanchez de la Nieta AN, Mediavilla C, Engelhardt S, Vanderlinden K, Giraldez JV (2017) Investigating the impact of climate change on groundwater recharge using a high precision meteo lysimeter in a dune belt of the Doñana National Park. In: Groundwater and global change in the western mediterranean area. Environmental earth sciences. ISBN: 978-3-319-69355-2

N'Da B, Bouchaou L, Reichert B, Hanich L, Danni S, Ait Brahim Y, Chehbouni A, Beraaouz E, Michelot J (2017) Chemical tracers and stable isotopes mixing models for the study of groundwater quality and recharge in the Moroccan high Atlas mountains. In: Groundwater and global change in the western mediterranean area. Environmental earth sciences. ISBN: 978-3-319-69355-2

Nieto-Lopez JM, Andreo-Navarro B, Barbera Fornell JA, Ramirez Gonzalez JM, Rendon Martos M (2017) Preliminary study of the impact of Guadalhorce river mouth channeling (Malaga, Spain) on groundwater related wetlands. In: Groundwater and global change in the western mediterranean area. Environmental earth sciences. ISBN: 978-3-319-69355-2

Ouhamdouch S, Bahir M, Carreira PM, Zouari K (2017) Groundwater responses to climate change in a coastal semi-arid area from Morocco; case of Essaouira Basin. In: Groundwater and global change in the western mediterranean area. Environmental earth sciences. ISBN: 978-3-319-69355-2

Oumarou Danni S, Bouchaou L, Elmouden A, N'da B, Malki M (2017) Isotopic and chemical tracers for the sustainable management of water resources in a semi-arid area: case of Massa catchment, Morocco. In: Groundwater and global change in the western mediterranean area. Environmental earth sciences. ISBN: 978-3-319-69355-2

Pardo-Iguzquiza E, Collados Lara AJ, Pulido Velazquez D (2017) Future effects of climate change on the dynamics of the Sierra Nevada snowpack: conclusions from cellular automata models. In: Groundwater and global change in the western mediterranean area. Environmental earth sciences. ISBN: 978-3-319-69355-2

Sánchez-Úbeda JP, Calvache ML, Engesgaard P, Duque C, López-Chicano M, Purtschert R (2017a) Numerical modeling of age distribution in Motril-Salobrena coastal aquifer (SE Spain). In: Groundwater and global change in the western mediterranean area. Environmental earth sciences. ISBN: 978-3-319-69355-2

Sánchez-Úbeda JP, López-Chicano M, Calvache ML, Purtschert R, Engesgaard P, Martin-Montanes C, Sültenfuβ J, Duque C (2017b) Groundwater age dating in Motril-Salobrena coastal aquifer with environmental tracers ($\delta^{18}O/\delta^{2}H$, $^{3}H/^{3}He$, ^{4}He, ^{85}Kr, and ^{39}Ar). In: Groundwater and global change in the western mediterranean area. Environmental earth sciences. ISBN: 978-3-319-69355-2

Santoni S, Garel E, Huneau F (2017) Multi-tracers strategy to define a conceptual model for the coastal aquifers of Mediterranean islands, case study of the Bonifacio aquifer (Corsica, France). In: Groundwater and global change in the western mediterranean area. Environmental earth sciences. ISBN: 978-3-319-69355-2

Zammouri M, Brini N, Jarraya Horriche F (2017) Assessment of artificial recharge efficiency against groundwater stress in El Khairat aquifer. In: Groundwater and global change in the western mediterranean area. Environmental earth sciences. ISBN: 978-3-319-69355-2

Impacts of the Water Resources Variability on Cereal Yields in the Region of Souss-Massa Southern Morocco

H. Abahous, J. Ronchail, A. Sifeddine, L. Kenny and L. Bouchaou

1 Introduction

The demographic growth and the emergent position of Souss Massa region as competitive economic pole in Morocco represent the main pressures to the regional water resources. The climate conditions influences the availability of water in this semi-arid area and at the same time allows a developed agriculture in the region. Since the second half of the last century, an intensive agriculture is practiced. Yet, the region has experienced in the past, periods of prosperous agriculture.

Recent works show a decline of surface water discharge in Souss Massa region since the early 1970s, directly linked to decrease in annual precipitation during 1970–2007 (Brahim et al. 2016). An alarming decrease in groundwater levels is also shown by Bouchaou et al. (2011) due to intensive agriculture. In this context, questions like food security and/or sustainable development are of great importance for local populations.

H. Abahous (✉) · L. Bouchaou
Laboratory of Applied Geology and Geo-Environment, Ibn Zohr University, Agadir, Morocco
e-mail: houriaabahous@gmail.com

J. Ronchail
Université Paris 7-Denis Diderot, UMR LOCEAN, Paris, France

A. Sifeddine
IRD-Sorbonne Université (IRD, UPMC, CNRS MNHN) UMR LOCEAN, Centre IRD France Nord. 32, Avenue Henri varagant, 93143 Bondy Cedex, France

L. Kenny
IAV Hassan II, Ait Melloul, Morocco

M. L. Calvache et al. (eds.), *Groundwater and Global Change in the Western Mediterranean Area*, Environmental Earth Sciences,
https://doi.org/10.1007/978-3-319-69356-9_1

The region of Souss Massa knows a continuous improvement of techniques of irrigation for a sustainable management of water resources. To better understand the impacts of inter-annual precipitations variability on rainfed and irrigated crop yields, we analyze the correlations between cereals yields and seasonal precipitations.

2 Data and Methods

2.1 Rainfall Data

A dataset of monthly precipitations records are provided by the Hydraulic Agency of Souss-Massa-Draa Basin (ABHSMD). The stations were chosen to cover rainfed and irrigated perimeters during the corresponding period of available data of barley, hard and soft wheat yields. Figure 1 represents topography and locations of the stations used in this work.

2.2 Cereals Yields Data

The crop data collected from the Office of ORMVASM are superficies in hectares and annual productions in hundredweight from 1973 to 2014. The main species of crop cultivated in the region are barley, "hard" and "soft" wheat. The spatial distribution of rainfed and irrigated perimeters are shown in Fig. 2.

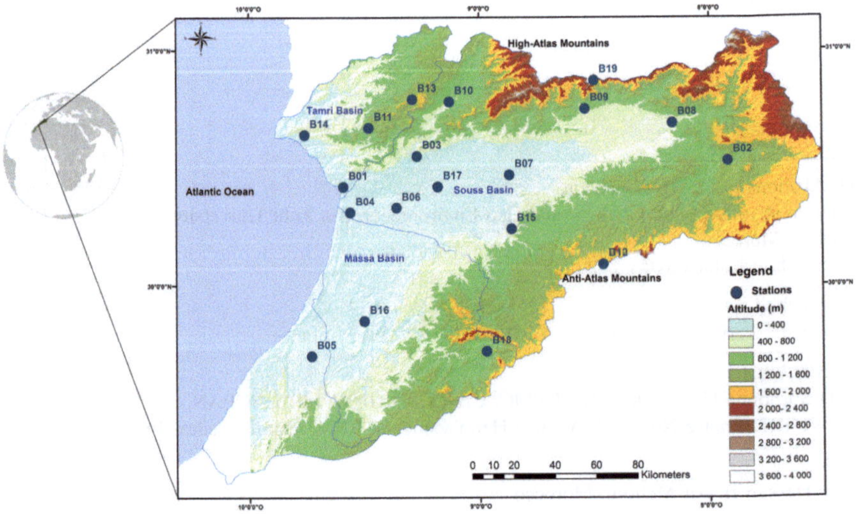

Fig. 1 Distribution of meteorological stations with altitudes in meters

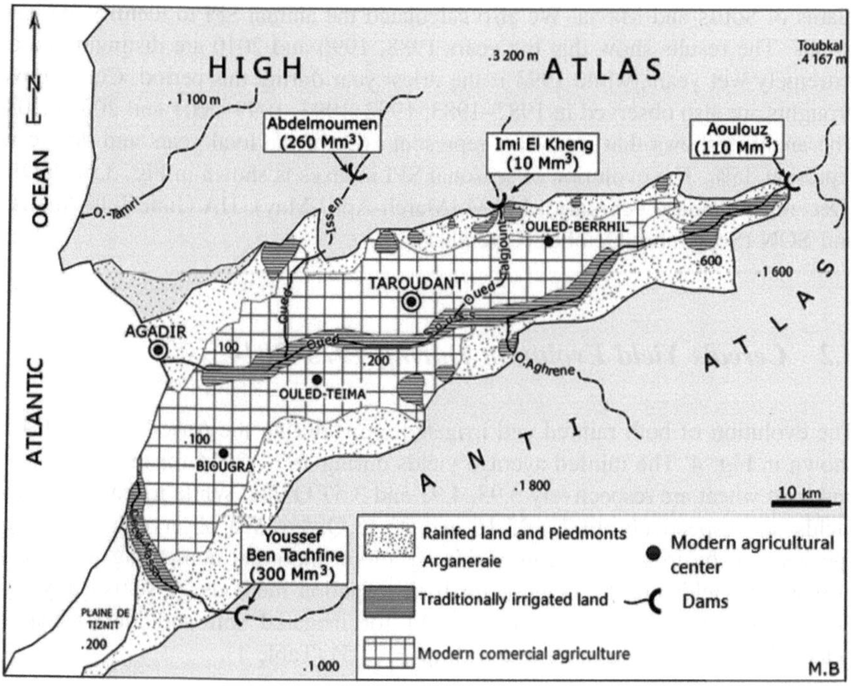

Fig. 2 Location map and agrarian landscapes of the Souss valley [modified from Boujnikh and Humbert (2010)]

2.3 Methods

Standardized Precipitation Index (SPI) was calculated to represent the evolution of interannual precipitation in the Souss-Massa region during the studied period. Mann-Kendall test (Pettitt 1979; Mann 1945) was applied to detect possible trends of the three species yields. While correlation between crop yields and inter-annual precipitations were calculated using the Pearson's coefficient and Bravais-Pearson table to evaluate the significance of computed values. Calculated cereals yields are expressed in quintals per hectare (Qx/ha), where 1 quintal is equal to 100 kg.

3 Results and Discussions

3.1 Rainfall Evolution Since the 1970s

The SPI for meteorological seasons is calculated during the period 1973–2010 from 19 stations covering the region of Souss Massa to be used in correlation analysis. The stations are located in both high and low altitude and the most are located in the plains of Souss and Massa. We also calculated the annual SPI to identify extreme events. The results show that the years 1988, 1996 and 2010 are distinguished as extremely wet years, while 1993 is the driest year during this period. Consecutive droughts are also observed in 1982–1983, 1992–1993, 1999–2001 and 2007–2008. The analysis shows that wet years represent 16% of the total years and dry years represent 35%. The evolution of seasonal SPI indexes is shown in Fig. 3, with DJF (December-January-February), MAM (March-April-May), JJA (June-July-August) and SON (September-October-November).

3.2 *Cereals Yield Evolution During 1973–2014*

The evolution of both rainfed and irrigated crops during the period 1973–2014 is shown in Fig. 4. The rainfed average yields during 1973–2014 for the barley, soft and hard wheat are respectively 3.93, 4.91 and 3.57 Qx/ha. While irrigated average yields are respectively 14.77, 23.16 and 22.15 Qx/ha. We observe for the three species that the irrigated cereals yields are more important comparing to rainfed. Soft wheat yields are the highest, for both irrigation methods. A statistically significant positive trend during 1973–2014 for irrigated soft and hard wheat is detected using Mann-Kendall test with ($p = 0.05$) (Table 1).

3.3 *Relationship Between Crop Yields and Seasonal Standardized Precipitation Index*

Before proceeding to correlations analysis, de-trended time series of soft and hard wheat are calculated. Table 2 is representing the coefficients of correlations between seasonal standardized precipitation indexes and yields of barley, soft and hard wheat for both rainfed and irrigated crops. The most significant correlations are positive and are observed between rainfed crops and DJF SPI with ($p = 0.05$). The results show significant positive correlations between MAM SPI and rainfed crops and also between MAM SPI with irrigated barley. This positive correlation implies a strong dependence of cereal crops of precipitation in the region of Souss-Massa.

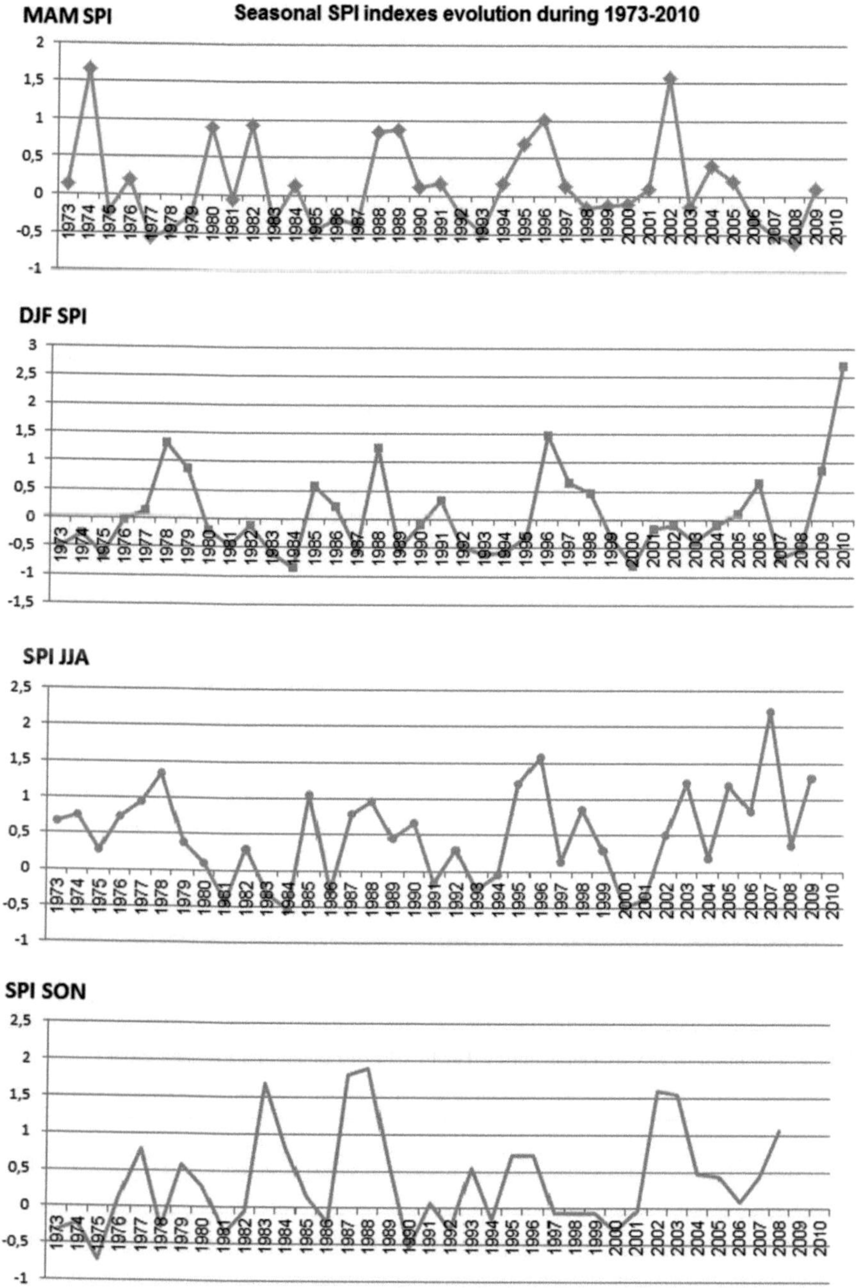

Fig. 3 Evolution of cereals yields during 1973–2014 of rainfed and irrigated crops

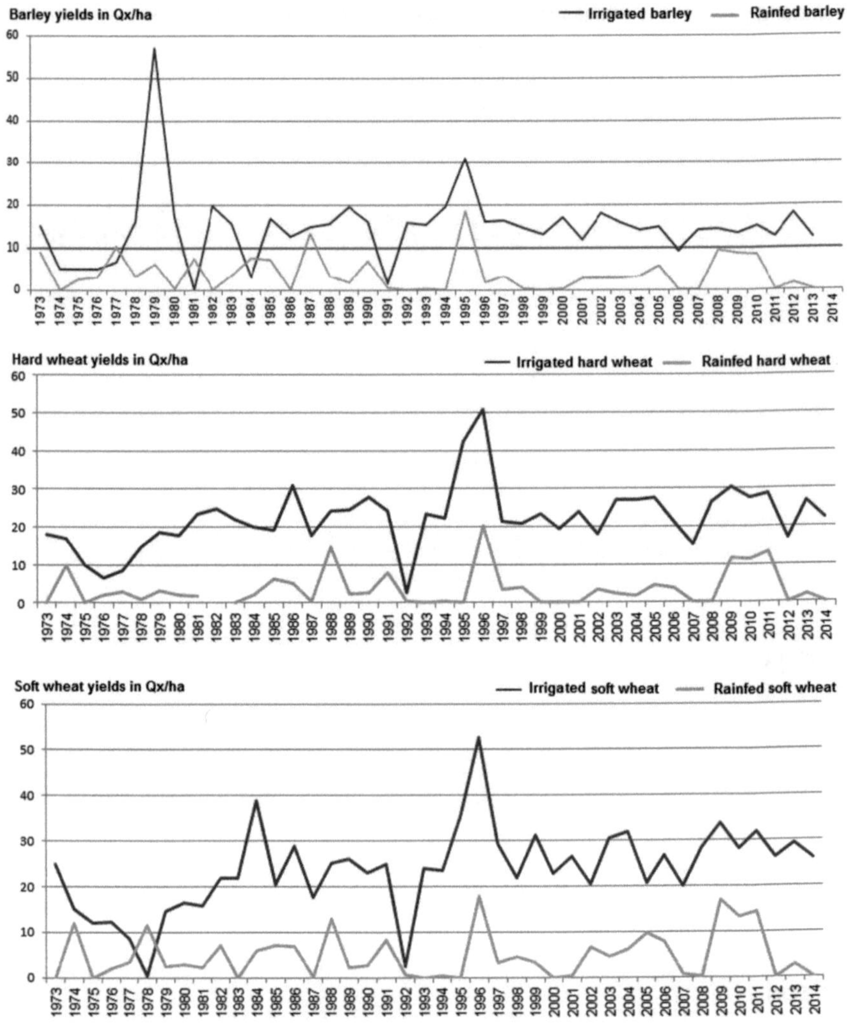

Fig. 4 Evolution of cereals yields during 1973–2014 of rainfed and irrigated crops

Table 1 Computed values of Mann-Kendall test results for crop yields during 1973–2010. Where S is the Mann-Kendall's statistic and Z is the standard test statistic (Gilbert 1987)

	Soft wheat	Hard wheat	Barley
Irrigated			
S	344	272	12
Z	3.7174	2.9371	0.11924
p-value	0.00020125**	0.0033129**	0.90509
Rainfed			
S	51	18	−57
Z	0.54396	0.19173	0.60924
p-value	0.58647	0.84795	0.54237

**Statistically significant with $p = 0.05$

Table 2 Coefficients of correlation between irrigated and rainfed cereals yields and SPI indexes for the seasons DJF, JJA, MAM and SON

	Irrigated barley	Rainfed barley	Irrigated hard wheat	Rainfed hard wheat	Irrigated soft wheat	Rainfed soft wheat
DJF SPI	0.06	0.79**	0.26	0.68**	0.06	0.74**
JJA SPI	−0.15	0.35	0.11	0.31	−0.03	0.37
MAM SPI	0.4**	0.4**	0.29	0.5**	0.26	0.4**
SON SPI	0.12	0.11	0.20	0.18	0.25	0.11

**Statistically significant with $p = 0.05$

4 Conclusions

To analyse impacts of inter-annual precipitations variability on rainfed and irrigated cereals, we first analyse annual and seasonal standardized precipitations indexes. The results show that wet years represent 16% of the total years and dry years represent 35% during the studied period.

We observe for the three species that the irrigated cereals yields are more important comparing to rainfed. Soft wheat yields are the highest, for both irrigation methods. And a statistically significant positive trend is detected during 1973–2014 for irrigated soft and hard wheat using Mann-Kendall test with ($p = 0.05$).

This work highlights the relationship between cereals yields and both winter and spring precipitations in the region of Souss Massa region. A strong positive correlation is revealed between winter precipitations and traditionally irrigated crops, particularly barley yields. A significant correlation is also observed for spring precipitations and rainfed crops, but also with irrigated barley yields.

The obvious significance of positive correlation between winter and spring precipitations and rainfed crops, raises the question of the sustainability of the cereals in the region due to the spatial and temporal irregularity of water resources. While the importance of irrigated crops yields refers to an intensive agriculture drawing principally on groundwater resources. The decrease of water resources in the region sounds the alarm of food security of local population and of the regional and national economy.

Acknowledgements This work was financed by ERANETMED (WASA) Project.

References

Bouchaou et al (2011) Climate change and its impacts on groundwater resources in Morocco: the case of the Souss- Massa basin. In: Holger T, Martin-Bordes JL (eds) International contributions to hydrogeology: climate change effects on groundwater resources: a global synthesis of findings and recommendations. CRC Press Taylor & Francis Group, IAH book series, vol 27, Chap 8, 129–144 pp: 414

Boujnikh M, Humbert A (2010) L'eau dans le bassin du Souss: concurrences et désorganisation des systèmes paysans. Norois 214:113–126

Brahim A et al (2016) Assessment of climate and land use changes: impacts on groundwater resources in the Souss-Massa River basin. In: Choukr-Allah R, Ragab R, Bouchaou L, Barcelo D (eds) The handbook of environmental chemistry. Springer, Berlin, Heidelberg, pp 1–22

Gilbert RO (1987) Statistical methods for environmental pollution monitoring. Van Nostrand Reinhold, New York

Mann HB (1945) Non-parametric test against trend. Econometrica 13:245–259. doi:10.2307/1907187

Pettitt AN (1979) A non-parametric approach to the change-point problem. Appl Stat 28:126–135. doi:10.2307/2346729

Groundwater as an Useful Resource in the Adaptation to the Climate Change: The Case of the Sinclinal de Calasparra Aquifer (Murcia, SE Spain)

I. Alhama, G. García and T. Rodríguez

1 Introduction

1.1 The Sinclinal de Calasparra Aquifer

The Segura River basin covers the provinces of Alicante and Murcia, located in the southeast of the Iberian Peninsula. It is an eminently agricultural region, whose main water demand, around 86% (CHS 2015), is used for this purpose and entails a contribution to the national GDP of over 2000 million euros per year, with 90% of the production destined for foreign trade with Europe (MAPAMA 2017). The semi-warm mediterranean climate predominates in the basin, with average annual temperatures around 18 °C and rainfall ranging between 300 and 400 mm/year.

18 of the 57 aquifers defined in the basin are overexploited (Senent and García Aróstegui 2014). This situation is aggravated taking into account the data of the historical series of the Agencia Estatal de Meteorología; in the last 10 years, average temperatures and annual rainfall above 20 °C and below 300 mm, respectively, have been registered in different parts of the basin.

The Sinclinal de Calasparra aquifer (MAsub 70.022, Fig. 1) is located in the southern sector of the Segura River basin, northwest of the Region of Murcia. It has a total area of 332 km^2 and annual resources of 12 hm^3, sufficient to enable the Confederación Hidrográfica del Segura (CHS, hereinafter), which is the Basin Management Organism, to operate in a controlled and sustainable way. The river

I. Alhama (✉) · G. García
Departamento de Ingeniería Civil, Universidad Politécnica de Cartagena, Paseo Alfonso XIII, 52, 30203 Cartagena, Spain
e-mail: ivan.alhama@upct.es

T. Rodríguez
Departamento de Ingeniería Minera, Geológica y Cartográfica, Universidad Politécnica de Cartagena, Paseo Alfonso XIII, 52, 30203 Cartagena, Spain

M. L. Calvache et al. (eds.), *Groundwater and Global Change in the Western Mediterranean Area*, Environmental Earth Sciences,
https://doi.org/10.1007/978-3-319-69356-9_2

recharges the aquifer in some stretches, while in others the discharge occurs from the aquifer, as is the case of the Gorgotón spring area.

The aquifer is constituted by jurassic and cretaceous limestone and dolomitic rocks of medium-high permeability that reach 300 m of thickness (Fig. 2) and are connected to each other by tectonic accidents (Rodríguez Estrella 1974; IGME 1985). A third of its total extension behaves as a free aquifer, while the rest remains confined under miocene marls up to 500 m of thickness. The piezometric levels oscillate between 170 and 210 m above sea level, depending on the extractive regime and rainfall.

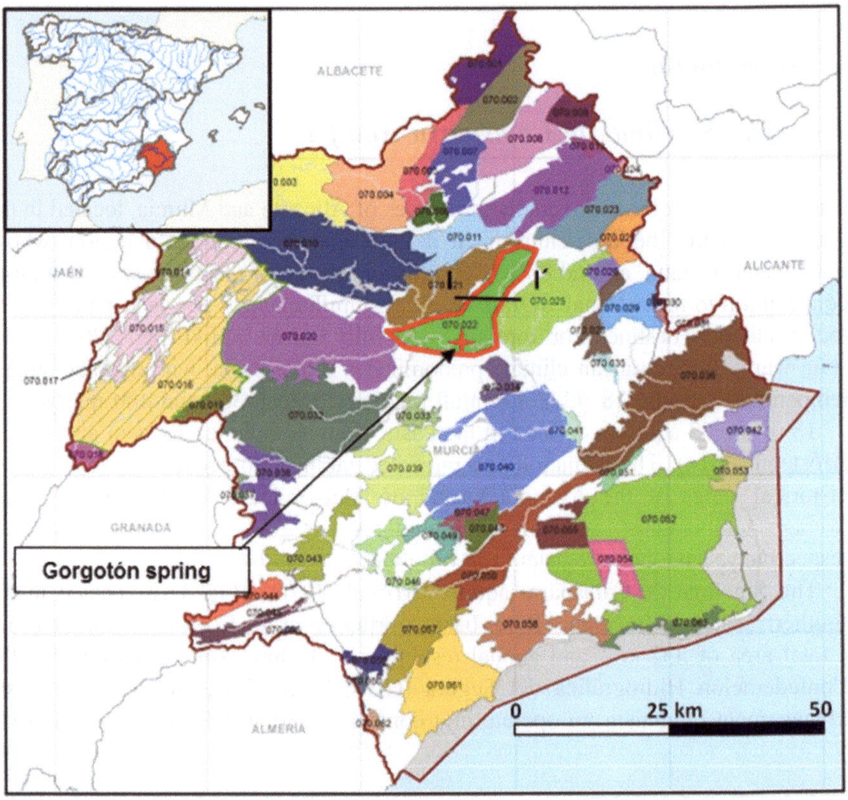

Fig. 1 Groundwater masses in the Segura River basin. The MAsub 70.022 and the Gorgotón spring are marked in *red*. *Source* https://www.chsegura.es and http://hispagua.cedex.es/datos/hidrografia

1.2 Exploitation Regime

The Sinclinal de Calasparra aquifer has been temporarily used in times of drought, taking advantage of the great hydric potential that it presents. The first time the CHS pumped groundwater in the Sinclinal was in the years 1984–85, doing it again from 1992 to the present, although not continuously (Fig. 3). The withdrawn groundwater is spilled to the Segura River so that, downstream, it can be derived from it for agricultural use. This also allows the river to maintain its ecological flow.

The groundwater abstractions have given water supply for irrigation and human consumption for the different watering communities and the populations of Murcia and Alicante. One of the most outstanding actions was carried out by the Mancomunidad de los Canales del Taibilla, dependent on the Ministry of Environment between the years 2004–2006, when a total volume of 110 hm^3 were

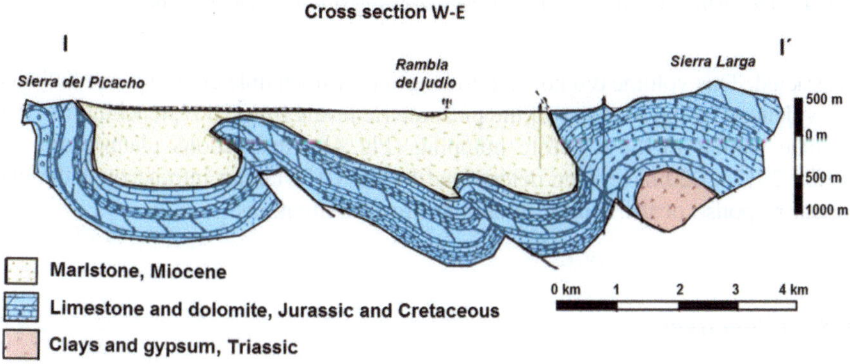

Fig. 2 Schematic geological cross section between the Sierra del Picacho and the Sierra Larga (see Fig. 1). Modification from Rodríguez Estrella and Conradi (2006)

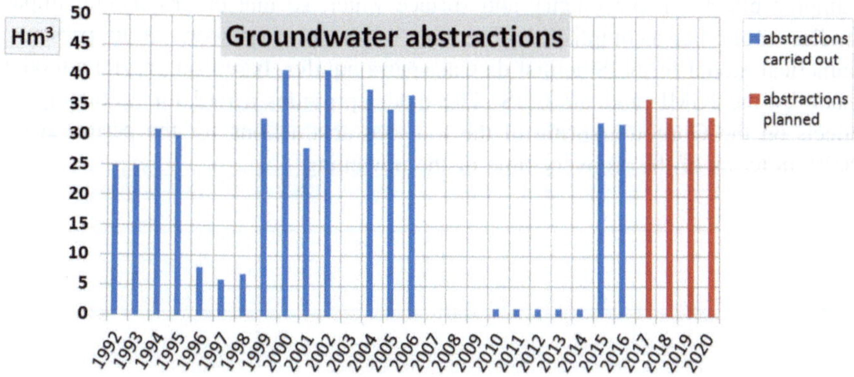

Fig. 3 Abstractions during the period 1992–2016. Also included the forecast for 2017–2020. 2003 and 2007–2009 data are not available

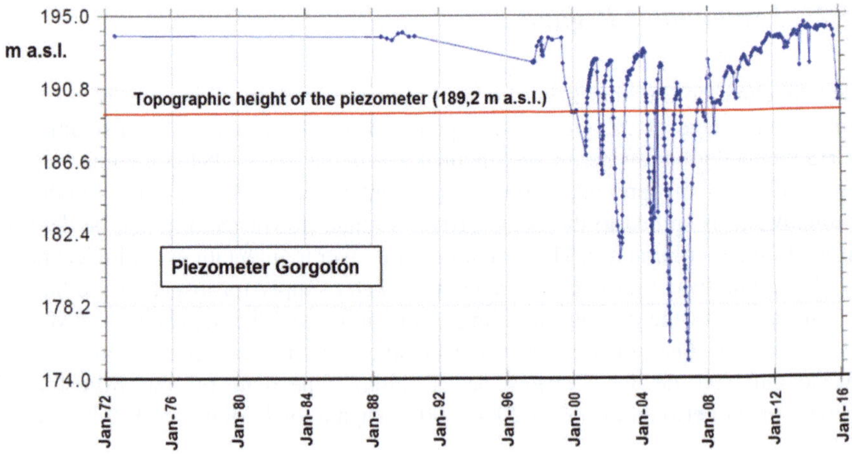

Fig. 4 Piezometric evolution in a piezometer located next to the Gorgotón spring

abstracted. This volume did not lead to long-term irreversible environmental effects, as evidenced by the recovery of the piezometric levels from 2007, reaching in 2010 a value similar to those before pumping (Fig. 4). Abstractions planned for the period 2017–2020 have been determined by the CHS taking into account the historical response of the aquifer in similar drought situations.

1.3 Objectives

The announced drought situation in the territorial scope of the CHS requires the adoption of exceptional measures for water resources management, including the use of groundwater storage. Numerical modelling should be carried out to quantify pumping effects on piezometry and surface water, so that environmental impact studies could be accomplished. The aim of the present work is to present a numerical model of the Sinclinal de Calasparra aquifer, based on previous models made by the IGME and the CHS. The model provides a tool for predicting the effects on the Gorgotón spring of the foreseen abstractions for the period 2017–2020, in terms of the recovery time of the upwelling.

2 Numerical Model of the Sinclinal de Calasparra Aquifer (2010–2030 Period)

It is known that at least two mathematical models of the Sinclinal de Calasparra aquifer have been carried out to date (IGME-DGA 2015). Since 2003, with special attention to the long periods of drought of 2006–2008 and 2011–2015 and with the idea of preserving ecological flows in the Segura River basin, updates and calibrations of the original models have been carried out. The models have simulated satisfactorily changes in piezometry and the behaviour of the spring during and after the studied periods.

The model presented in this work is based on the previous ones, although it has been recalibrated with recent piezometric data and incorporates hydraulic parameters from field tests. It has been made with Modflow (United States Geological Survey), in transient regime and without salt transport. The main characteristics of the model are presented in Table 1.

The studied period covers the years 2010–2030, with actual data of piezometry and rainfall from 2010 to 2016. The temporal distribution of the abstractions planned by CHS (the most important), together with the withdrawals from private wells and watering communities, is shown in Fig. 3. From 2020, only the volume granted to private wells is considered, 1.316 hm^3/year. The aquifer is highly regulated and controlled, so that the current and planned abstractions are supervised by the CHS. Findings and data related to aquifer management are available to the public through its website.

The Gorgotón spring has been simulated by means of a drain whose topographic elevation is similar to that of the ground surface. This tool allows the discharge of water from the aquifer to the river only when the piezometric level is above the ground surface elevation. The model itself calculates the discharge flows from the aquifer to the drain in each period. The model sensitivity mainly depends on the recharge from the rainfall, so that, in order to place the results in an unfavorable context, half of the one indicated in the basin hydrological plan has been

Table 1 Summary of the main characteristics of the model

Surface extension	346 km^2	Hydraulic conductivity *INPUT*	1.2–80 m/d
Deep extension	300 m	Hydraulic conductivity *OUTPUT*	2.4–100 m/d
Number of files	15	Rainfall recharge	10 hm^3/year
Number of columns	18	Recharge from irrigation	2 hm^3/year
Unit cell size	2000 × 2000 m^2	River infiltration	10.5 hm^3/year
No. calibration piezometers	6	Abstractions from pumping wells	0.3–5.9 hm^3/year
No. pumping wells	17	Effective porosity	0.02–0.26
Time period simulated	2010–2030 year		

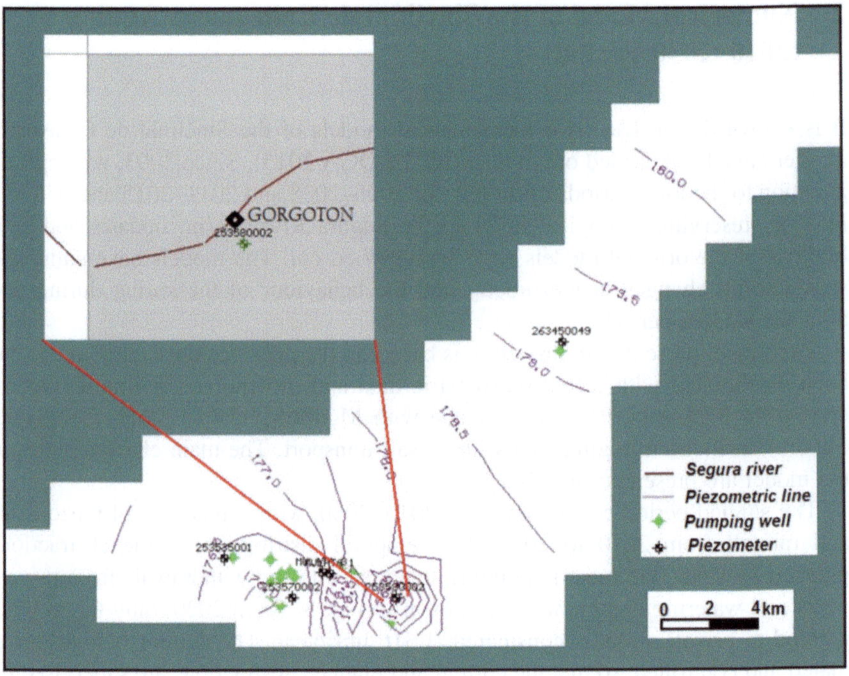

Fig. 5 Location of the calibration piezometers and pumping wells in the model. The piezometric lines correspond to the simulated situation in December 2015

considered. A detail of the extension of the model and the location of wells, piezometers and the spring can be seen in Fig. 5. The figure also shows the simulated piezometry in December 2017.

3 Results of the Simulation and Discussion

The time changes in the Gorgotón upwelling is shown in Fig. 6. The spring remains dry when the level of the piezometer of the Gorgotón, located 300 m from it, is below its elevation (189.2 m above sea level). It is estimated that the reappearance of the upwelling will occur a year and a half after the stop of the main groundwater abstractions. The recovery of the levels will take place five years later. These results are in agreement with the response of the aquifer in the action carried out in 2004–2006, when the recovery of the upwelling and the piezometric levels occurred one year and four years after the cessation of the pumping, respectively.

The average discharge volume of the spring before the abstractions that took place in 2015 (Fig. 7), ranges from 64,000 to 68,000 m^3/d, depending on the pumping regime of the private wells. In the period July–November 2015, there is a

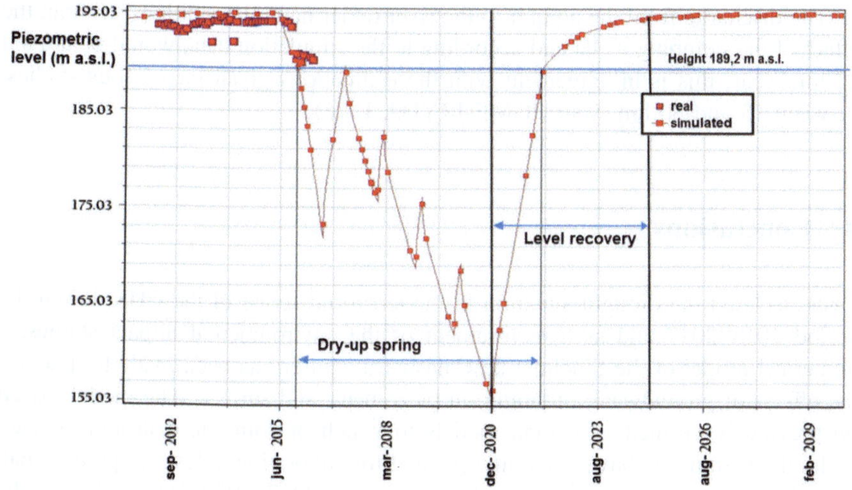

Fig. 6 Actual and simulated piezometry after the calibration process

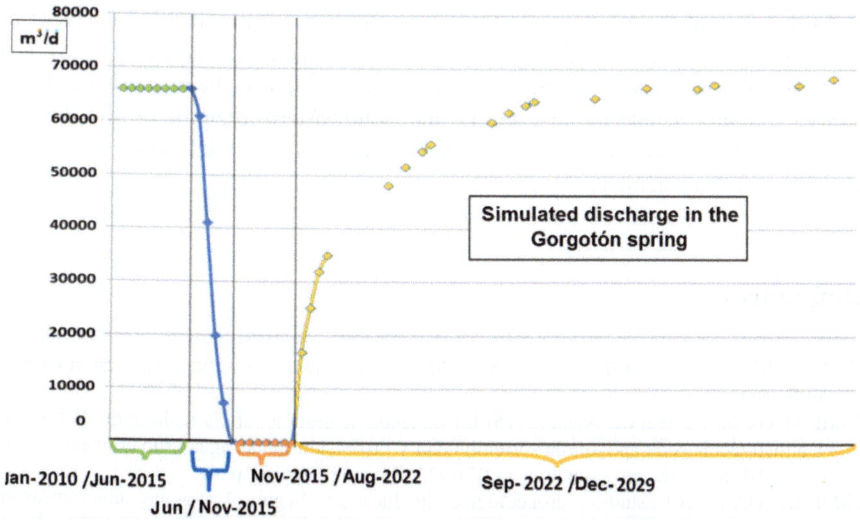

Fig. 7 Evolution of the discharge in the Gorgotón spring according to the simulation

gradual fall of the upwelling until its disappearance. This situation is maintained until September 2021, when a progressive increase of upwelling becomes noticeable until reaching, around year 2024, a steady value between 64,000 and 68,000 m^3/d.

The discharge from the aquifer to the river under natural conditions, without the effect of the pumping wells and according to the simulation data, would be around 22 hm^3/year; this is in agreement with the estimations given in previous studies, between 15 and 30 hm^3/year (IGME-IRYDA 1976).

4 Conclusions

Once declared the drought situation in the territorial scope of the CHS during the period 2015–2017 and as a requirement of the environmental impact studies, a numerical model of the Sinclinal de Calasparra aquifer has been made to quantify the effects of groundwater pumping on piezometry and surface. The model, based on previously verified numerical models to which piezometric data and updated hydraulic parameters have been incorporated for calibration, allows to predict that, according to pumping estimations for the period 2017–2020 (34 hm^3/year), the Gorgotón spring will discharge groundwater again in December 2020, a year and a half after the stop of the pumping. The spring will recover its usual discharge level (around 68,000 m^3/d) four years after the stop of the pumping.

In the case of the Sinclinal de Calasparra aquifer, its sustainable management through the intervention and supervision of the competent authorities (CHS), has made it possible to alleviate the worsening effects of drought in the agricultural sector, a major economic engine in the southeastern regions of the Iberian Peninsula. Numerical models are, once again, an essential tool for proper groundwater management.

References

IGME (1985) Gestión coordinada de recursos hídricos superficiales y subterráneos en la Cuenca del Segura

IGME-Dirección General del Agua (2015) Encomienda de gestión para la realización de trabajos científico-técnicos de apoyo a la sostenibilidad y protección de las aguas subterráneas. T.: 32 Cap.: VIII. Masa de agua subterránea 071.022. Sinclinal de Calasparra

IGME-IRYDA (1976) Estudio hidrogeológico alto Júcar alto Segura. Las aguas subterráneas en Albacete. Ref.: 32621

CHS and Ministerio de Agricultura, Alimentación y Medio Ambiente (2015) Plan hidrológico de la demarcación del Segura; Anejo 3 de usos y demandas

MAPAMA. Ministerio de Agricultura y Pesca, Alimentación y Medio Ambiente (2017) Producción agrícola del sector de frutas y hortalizas, comercio exterior. URL: http://www.mapama.gob.es/es/agricultura/temas/producciones-agricolas/frutas-y-hortalizas/informacion_general.aspx#para0

Rodríguez Estrella T (1974) El sistema hidrogeológico del Sinclinal de Calasparra. V Coloq. De Inv. sobre el Agua 189–203

Rodríguez Estrella T, Conradi C (2006) Uso sostenible y controlado en el acuífero kárstico del sinclinal de Calasparra (Murcia). Capítulo del libro: Karst, cambio climático y aguas subterráneas. (AQUA in MED, Málaga). Publicaciones del IGME. Editores: JJ Durán, B Andreo y F Carrasco. Serie: Hidrogeología y Aguas Subterráneas, nº 18. Madrid. 173–183. ISBN: 84-7840-628-X

Senent Alonso M, García Aróstegui JL (2014) Sobreexplotación de acuíferos en la cuenca del Segura. Evaluación y perspectivas. Instituto Euromediterráneo del Agua. Murcia

An Index-Based Method to Assess Impacts of Global Change on Seawater Intrusion Problems

L. Baena Ruiz, D. Pulido-Velazquez, A. Renau-Pruñonosa, I. Morell, C. Llopis-Albert, A. J. Collados-Lara and J. Senent-Aparicio

1 Introduction

In the future it will be more difficult to supply the demand in coastal systems due to impacts of the global change (GC), which is expected to produce a decrease in recharge and an increase in sea level and crop irrigation requirements (Fujinawa 2011; Unsal et al. 2014). The main objective of this work is to assess hydrological impacts in global change for the Plana de Oropesa-Torreblanca aquifer. It attempts to summarize seawater intrusion (SWI) problems at different spatial scales including maps, 2D conceptual cross sections and global indices that reflect the aquifer global status and vulnerability regarding to SWI. The proposed methodology intend to contribute with the Water Framework directive (WFD) (EU-WFD 2000) to identify groundwater bodies in risk. To this end, the proposed methodology involves the following steps: generation of consistent plausible future scenarios of global change; simulation of hydrological impacts in the aquifer (including quantity and quality) through a sequential coupling of mathematical models: rainfall-recharge models, agronomic water requirements and irrigation returns models, and fluid density-dependent flow models; and a global assessment of aquifer status and vulnerability to SWI at different spatial scales.

L. Baena Ruiz · D. Pulido-Velazquez (✉) · A. J. Collados-Lara
Spanish Geological Survey (IGME), Urb. Alcázar Del Genil, 4-Edif. Zulema Bajo, 18006 Granada, Spain
e-mail: d.pulido@igme.es

C. Llopis-Albert
Universitat Politècnica de València, Valencia, Spain

J. Senent-Aparicio
UCAM, Murcia, Spain

A. Renau-Pruñonosa · I. Morell
Jaume I University, Castellón, Spain

M. L. Calvache et al. (eds.), *Groundwater and Global Change in the Western Mediterranean Area*, Environmental Earth Sciences,
https://doi.org/10.1007/978-3-319-69356-9_3

2 Methodology

Figure 1 shows the flowchart of the proposed methodology.

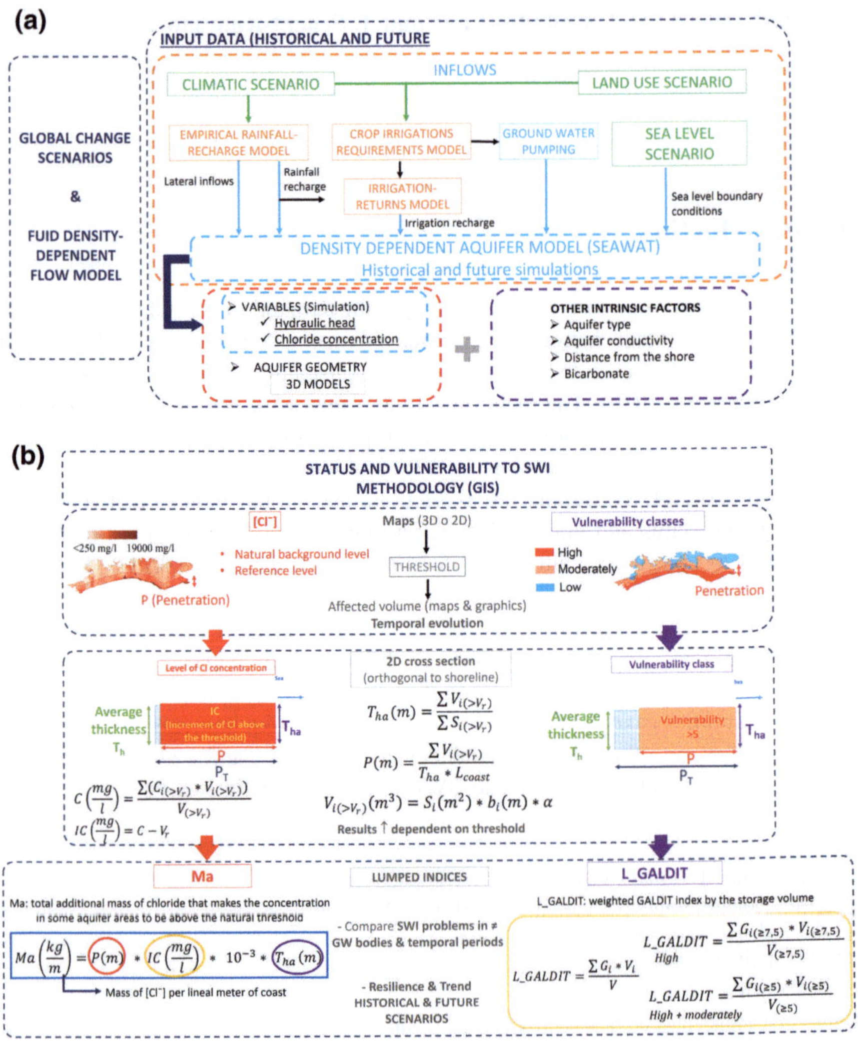

Fig. 1 Flow chart of methodology: **a** Generation of global change scenarios and fluid density-dependent flow model. **b** Assessment of status and vulnerability to SWI

2.1 Generation of Global Change Scenarios

We generate consistent pictures of monthly future climate scenarios by applying different ensemble hypothesis (equifeasible and non-equifeasible) based on a multi-criteria analysis of the approximations obtained with different corrections techniques (including bias and delta change approaches) considering statistics of the historical series and Regional Climatic Models simulations (Pulido-Velazquez et al. (u.r.)). Future Land Use and Land Cover (LULC) scenarios are established according to the General Town Plan for Torreblanca.

2.2 Aquifer Impacts (Quantity and Quality). Fluid Density-Dependent Flow Model Fed with the Outputs of Some Auxiliar Models to Propagate GC Scenarios

A modelling framework is defined to assess hydrological impacts on the coastal aquifer based on a density-dependent simulation whose inputs are defined by sequential coupling of different models: rainfall-recharge models, crop irrigations requirements and irrigation-return models (Pulido-Velazquez et al. (u.r.)).

2.3 Summary and Representation of Aquifer Status and Vulnerability to SWI

From model results we obtain field maps of chloride concentration and saturated thickness for each simulated time. A threshold of chloride concentration is established taking into account the natural background of the aquifer. This threshold indicates the chloride concentration level from which SWI problems exist. From maps we calculate the affected volume by SWI and chloride concentration in the affected volume. We also assess aquifer vulnerability from model results by employing GALDIT method (Chachadi and Lobo Ferreira 2007). GALDIT method considers six weighted factors which represent intrinsic characteristics of the aquifer and control vulnerability to SWI. The GALDIT index is calculated using an empirical formulation and finally the aquifer is classified in three vulnerability classes (high, moderate and low). From GALDIT index we can estimate the historical evolution of aquifer storage for each vulnerability class. To summarize SWI problems (status and vulnerability) at any time, we propose a conceptual average cross section (orthogonal to coast line) which represents average affected geometry (affected thickness and penetration of the intrusion) (see Fig. 1). Finally we define a lumped index to summarize global status (Ma) and vulnerability (L_GALDIT) to SWI. Ma index is defined as the total additional mass of chloride that makes the concentration in some aquifer areas to be above the natural threshold. L_GALDIT is defined as the weighted GALDIT index by the storage volume.

3 Site Description and Data

The Plana de Oropesa-Torreblanca aquifer (Fig. 2) is a detritcal mediterranean aquifer which extends over 75 km^2. In this area there have been important LULC and are expected to continue changing in the future.

The following historical data for the period 1973–2010 were used to define the inputs of the integrated modelling framework:

- Changes in LULC (Fig. 2), obtained from both fieldwork undertaken in the area and from the European CORINE Land Cover database (Feranec et al. 2010).
- Historical rainfall and temperature for the Plana Oropesa-Torreblanca and Maestrazgo aquifers were taken from the Spain02 project dataset (Herrera et al. 2016).
- Historical evolution of total pumping in the Plana Oropesa-Torreblanca aquifer was deduced from historical data.
- An infiltration rate coefficient of 14% for the historical period, which was obtained from previous lysimiter readings from a neighbouring aquifer (Plana de Castellón) (Tuñon 2000).

The LULC he already-approved tourist developments (the public urbanization work (PAI) for the Marina d'Or Golf in Oropesa and Cabanes, and the General Town Plan (PGOU) for Torreblanca) anticipate an increase in population of more than 130,000 inhabitants, as well as the disappearance of most of the agricultural activity in the area. A more detailed description of future land uses and water resources can be found in Pulido-Velazquez et al. (u.r.). The data to generate future climate scenarios for a short-term horizon (2011–2035) come from climate simulations in the framework of the CORDEX project for the emission scenario RCP8.5.

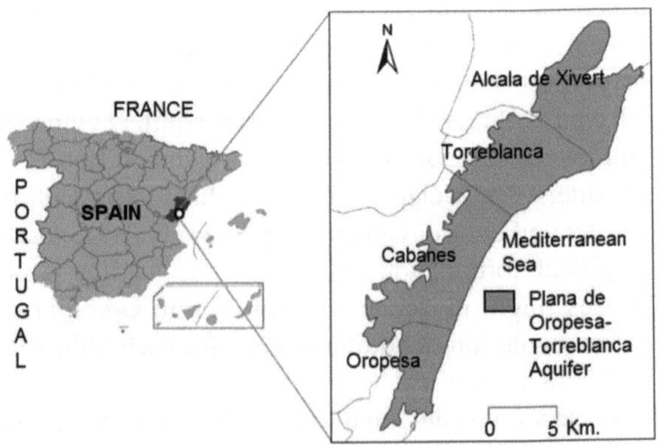

Fig. 2 Situation of the study area

4 Results and Discussion

Figure 3 shows the average characteristics for the future climate scenarios.

Figure 4 summarizes the mean annual values for each of the main component of the hydrological balance in the aquifer with the proposed sequential simulation of the 4 future scenarios.

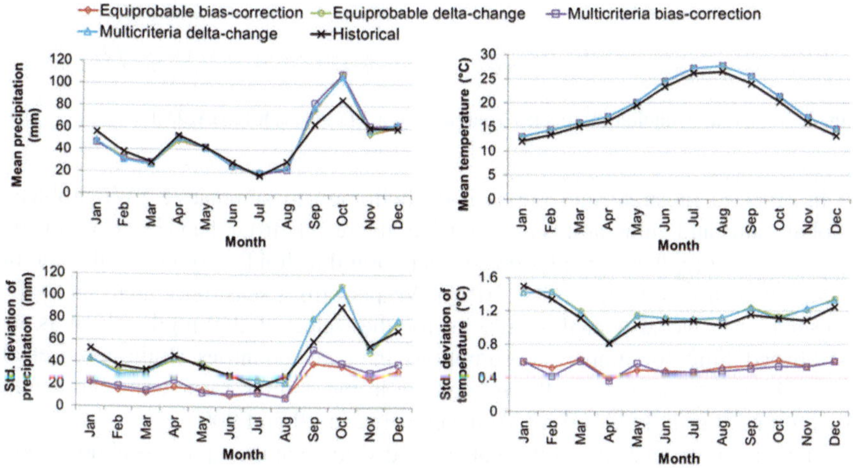

Fig. 3 Monthly mean and standard deviation of the future series (rainfall and temperature) for the four ensemble options (modified from Pulido-Velazquez et al. u.r.)

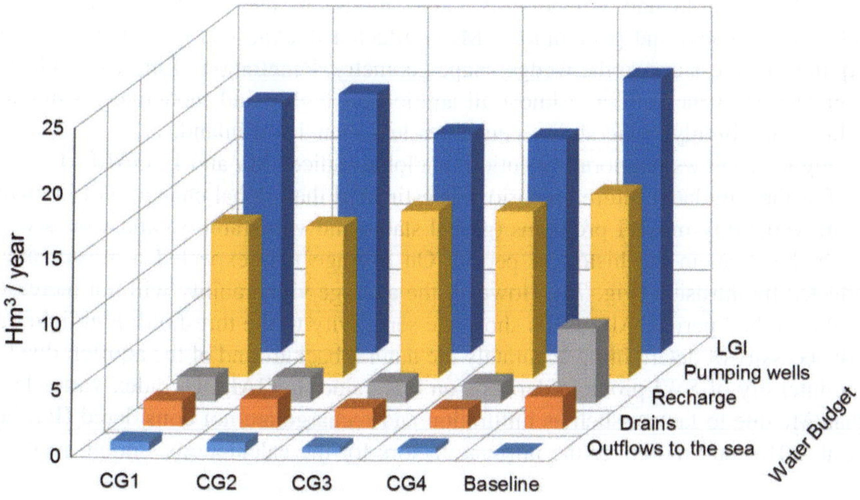

Fig. 4 Mean inflows and outflows for various global scenarios (GC1, GC2, GC3, GC4) (modified from Pulido-Velazquez et al. u.r.)

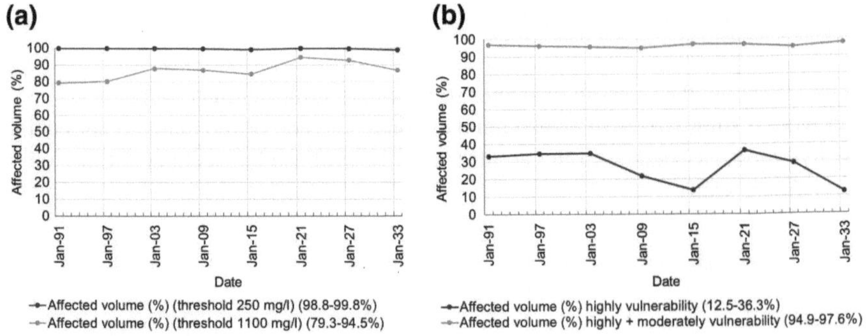

Fig. 5 Temporal evolution of affected volume (%): **a** Ma index, **b** L_GALDIT

In order to illustrate the proposed methodology to summarize SWI problems (global status and vulnerability) we consider the results from the GC1 scenario. We establish 1100 mg/l as the chloride concentration threshold according to the natural background in this aquifer (CHJ 2016). We perform a sensitivity analysis to this threshold and we also analyse results for the value of 250 mg/l, which is the established limit for human consumptions in many countries (Ballesteros et al. 2016). Figure 5 shows that affected volume in the aquifer considering the natural background threshold (1100 mg/l) ranges between 80 and 90%.

The sensitivity analysis shows that affected volume in the aquifer is almost 100% when the threshold 250 mg/l is considered. Regarding vulnerability, the affected volume by highly vulnerability is less than 40% but considering both vulnerability classes (moderately and highly), the affected volume is almost 100%.

Figure 6 shows the conceptual average cross section and affected area for each chloride threshold and vulnerability class (minimum and maximum values for affected thickness and penetration). Mean affected thickness is greater than mean aquifer thickness due to the wedge-shape geometry. Penetration of intrusion affects a great part of the aquifer. Almost all aquifer has highly and moderately vulnerability while highly vulnerability penetrates less than 1 km inland.

Figure 7 shows temporal evolution of global indices Ma and L_GALDIT.

For the simulated future scenario it is estimated that global change will involve more variability in SWI problems (global status and vulnerability) and values will be higher than in the historical period. On average it is expected a greater area affected by intrusion (Fig. 7a). However the average vulnerability will not increase in the studied period. All results show the sensitivity to the threshold, highlighting the necessity of determining accurately the natural background of the aquifer, due to the intensity of SWI problem depends on this value. L_GALDIT index varies less than Ma due to factors such as infiltration and recharge are not considered (Benini et al. 2016) and moreover the intrinsic factors for the calculations smooth results.

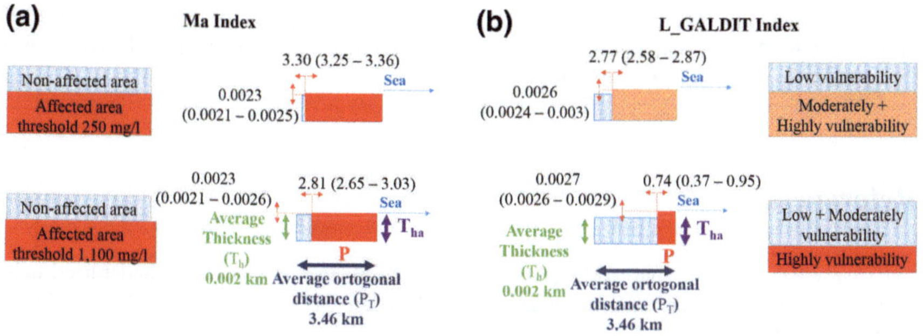

Fig. 6 Average cross sections for different thresholds: **a** Ma index, **b** L_GALDIT. Vertical exaggeration scale: 500

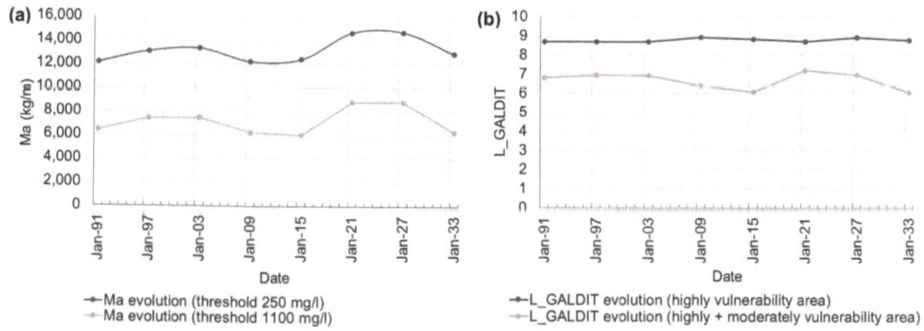

Fig. 7 Temporal evolution of lumped índices: **a** Ma, **b** L_GALDIT

5 Conclusions

In this work we propose a method to assess and summarize impacts on global change on SWI in coastal aquifers at different spatial scales. We generate future scenarios taking into account climate change, LULC and sea level. We simulate the impacts using a fluid density-dependent flow model which is fed with the outflow of auxiliary models which allow the propagation of future climatic scenarios. Results of chloride concentration and head levels are used as input in an index-based method to assess and summarize the global status and vulnerability of the aquifer to SWI. This method includes maps, conceptual 2D cross sections and global indices which synthesize global aquifer status and vulnerability. Results show that the Plana de Oropesa-Torreblanca aquifer has a great affected area and the global status to SWI is bad. Moreover this aquifer has a high vulnerability indicating that aquifer recovery would be difficult due to their hydrological characteristics. For the simulated future scenario (2015–2035) it is expected that global change will involve a great variability in SWI problems and values will be higher than in the historical period. On average, although an increasing in affected area by SWI is estimated, it is not expected to increase vulnerability in this period.

Acknowledgements This work have been partially supported by the CGL2013-48424-C2-2-R project and the Plan de Garantía Juvenil from Mineco, co-financing by BEI and FSE. We thank to the projects Spain02 and CORDEX for the data provided.

References

Ballesteros BJ, Morell I, García-Menéndez O, Renau-Pruñonosa A (2016) A standardized index for assessing seawater intrusion in coastal aquifers: the SITE index. Water Resour Manage 30:4513–4527

Benini L, Antonellini M, Laghi L (2016) Assessment of water resources availability and groundwater salinization in future climate and land use change scenarios: a case study from a coastal drainage basin in Italy. Water Resour Manage 30:731–745

Chachadi AG, Lobo Ferreira JP (2007) Assessing aquifer vulnerability to seawater intrusion using GALDIT method. Part 2-GALDIT indicators description. IAHS-AISH Publication, pp 172–180

CHJ (2016) (Júcar Water Agency) Júcar River Basin Plan. Demarcación hidrográfica del Júcar. Confederación Hidrográfica del Júcar. Ministry of Agriculture, Food and Environment, Spain

Feranec J, Hazeu G, Soukup T, Jaffrain G (2010) Determining changes and flows in European landscapes 1990–2000 using CORINE land cover data. Appl Geogr 30:19–35. doi:10.1016/j.apgeog.2009.07.003

Fujinawa K (2011) Arthroscope of Mediterranean coastal area in the context of hydrogeology: projected impacts of climate change. Sustainable land management 311–332

Herrera S, Fernández J, Gutiérrez JM (2016) Update of the Spain02 gridded observational dataset for Euro-CORDEX evaluation: assessing the effect of the interpolation methodology. Int J Climatol 36:900–908. doi:10.1002/joc.4391

Pulido-Velazquez D, Renau-Pruñonosa A, Llopis-Albert C, Morell I, Collados-Lara AJ, Senent-Aparicio J (ur) Integrated assessment of future potential global change scenarios and their hydrological impacts in coastal aquifers. A new tool to analyse management alternatives under uncertainty in Plana Oropesa-Torreblanca aquifer. Hydrol Earth Syst Sci Discuss. doi:0.5194/hess-2017-262, in review, 2017

Tuñon J (2000) Determinación experimental del balance hídrico del suelo y evaluación de la contaminación asociada a las prácticas agrícolas. PhD thesis, Universitat Jaume I de Castellón, Spain

Unsal B, Yagbasan O, Yazicigil H (2014) Assessing the impacts of climate change on sustainable management of coastal aquifers. Environ Earth Sci 72:2183–2193

Assessing the Impacts of Climate Change on Groundwater Recharge for the Chiba Basin in Tunisia

S. Benabdallah, H. Mairech and F. M. Hummel

1 Introduction

Adaptation to climate change requires a good knowledge of the local impacts related to the induced changes, particularly in vulnerable and semi-arid regions with high climate variability and scarce water resources. Prevailing studies on the potential impact of climate change relates to either global/regional levels or to large watershed specific investigations. The global and regional level studies consist in general of potential future trends for climatic factors or for water resources.

Climate studies on Tunisia have shown that the country is very exposed to climate change and that its economy, population and ecosystems are therefore very vulnerable (MESD 2015). Thus, local studies at the basin level can be of importance to local decision makers. For instance, surface water and groundwater recharge are influenced not only by hydrologic processes, but also by the physical characteristics of the land surface and soil profile (Kumar 2012).

Water scarcity already forces anthropogenic actions in the Chiba basin due to irrigation of agricultural land and overexploitation of Korba groundwater aquifer (Kerrou et al. 2010). This facilitated a drop of groundwater levels near the coast and thus sea water intrusion with deteriorating effects on water quality (Kouzana et al. 2009; Zghibi et al. 2011). In this paper, we are concerned by the assessment of the impacts of climate change under two different scenarios on water resources of the Chiba basin with a particular focus on the groundwater recharge component. A hydrological modeling approach is presented relying on the application of the distributed physically-based model ArcSWAT (Soil and Water Assessment Tool) to

S. Benabdallah (✉) · H. Mairech
Centre de Recherches et des Technologies des Eaux, BP 273, Soliman, Tunisia
e-mail: sihem.benabdallah@certe.rnrt.tn

F. M. Hummel
Ludwig–Maximilians–University, Munich, Germany

M. L. Calvache et al. (eds.), *Groundwater and Global Change in the Western Mediterranean Area*, Environmental Earth Sciences,
https://doi.org/10.1007/978-3-319-69356-9_4

assess the model performance and the effect on the overall water balance and the spatio-temporal recharge component under two climate change scenarios between a reference (1971–2000) and future (2041–2071) periods.

2 Materials and Methods

2.1 Study Site Description

Chiba basin is located in the east of Cap-Bon peninsula in the north east of Tunisia (Fig. 1). It falls within the following coordinates: 36° 20′: 37° 10′N and 10° 30′: 10° 10′E. The basin is bounded by the Jbel Abderahman anticline in the west draining no perennial rivers eastward to the Mediterranean Sea where the land flattens and transitions into a coastal plain. Elevation in the river basin ranges between 2 m and 450 m. In total the watershed extends to an area of approximately 200 km^2. Roughly 64 km^2 of the upstream are still natural and controlled by the Chiba dam that was constructed in 1963 mainly to accumulate surface water for irrigation. The reservoir has an area of 2.72 km^2 with a maximum water volume of 7.86 million m^3 (CRDA 2005).

The aquifer of Korba encompasses an area of 430 km^2 in a band of around 40 km length and 10 km width along the shoreline (Paniconi et al. 2001). Its formations are composed of sands, sandstone, and limestone. The Plio-Quaternary formation constitutes the shallow unconfined aquifer in the lower lying region of

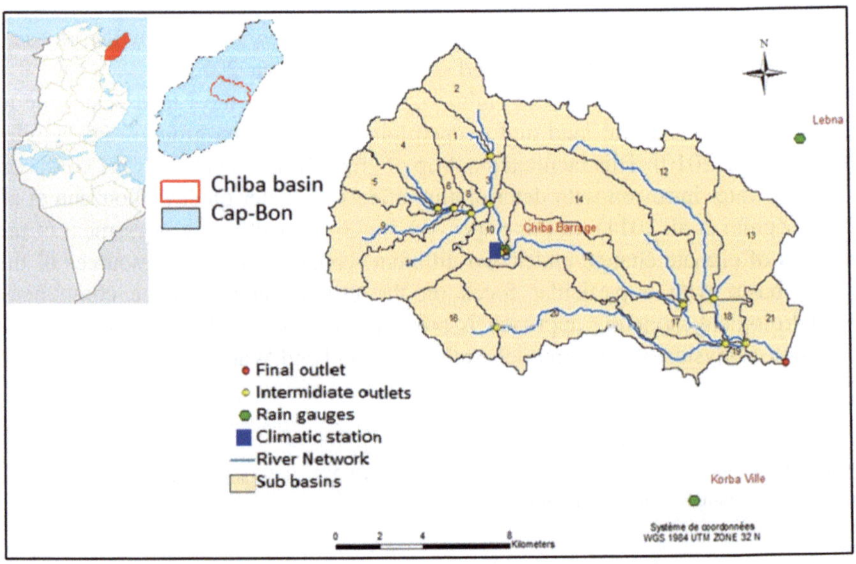

Fig. 1 Site location

the Chiba catchment. A thick clay layer separates the shallow aquifer from the confined aquifers (Chemingui et al. 2013).

Most intensive agricultural activities concentrate on the eastern part of the catchment near the Mediterranean Sea, which coincides with the extent of the Korba groundwater aquifer. The groundwater is heavily exploited with shallow wells in order to irrigate industrial crops, mainly tomatoes, throughout the year. Around 90% of the agricultural land in this area is under drip irrigation.

Model runoff from daily rainfall is estimated based on local land use, soil type, and digital elevation model (DEM) and antecedent moisture condition. DEM with a scale of 1:50,000 was generated using contours lines, created for the purpose of this study, from National Topographic Maps. The initial cell resolution has an interval of 50 m. The soil and the land use layers were produced by the Soil and Water Conservation Agency from soil maps at a scale of 1:50,000 and from Landsat Thematic Mapper images for the year 2000. Soil properties were extracted from several technical studies realized by the National Ministry of Agriculture.

2.2 *Methodology*

The overall approach is summarized in Fig. 2. The model ArcSWAT, is a continuous-time and physically based hydrological model developed to predict the impact of land management practices on water, sediment and agricultural chemical yields in large complex catchments. ArcSWAT operates by dividing the watershed into sub-basins. Each sub-basin is further discretized into a series of hydrologic response units (HRUs), which are unique soil-land use combinations. The water in each HRU is stored in soil profile, shallow aquifer and deep aquifer. Surface runoff from daily rainfall is estimated using a modified SCS curve number method, which estimates the amount of runoff based on local land use, soil type, and antecedent moisture condition (Arnold et al. 1998).

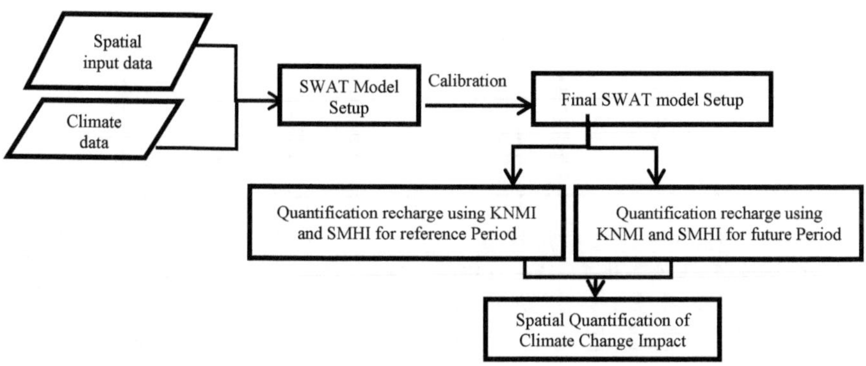

Fig. 2 Methodology

The model was setup and calibrated based on in situ climatic data for the period 1998–2010. Figure 1 showed the subdivision of the model into sub-basins and the location of the climatic station. Climate data of global and regional climate models from the EU (European Union) project Ensembles was applied (Ensembles 2011). Thus, KNMI and SMHI provide climate data for these timeframes on a daily basis. KNMI is driven with the general circulation model ECHAM5-r3 and regional climate model RACMO2. SMHI runs on GCM ECHAM5-r3 as well but employs the RCM Rossby Centre regional atmospheric climate model RCA3. Both models have a spatial resolution of 25 km and use the A1B scenario of the IPCC to project future climate. The global data were then introduced to the model to quantify the recharge component for both climate scenarios without the application of any corrections to the downscaled reference period.

3 Results and Discussions

ArcSWAT includes an automated calibration procedure that has been developed by Van Griensven and Bauwens in 2003. This procedure is based on a Shuffled Complex Evolution Algorithm that was widely used for the calibration of hydrological models. The calibration of the model was made over the period of 2000–2010 based on the observed discharge located upstream of the basin at the Chiba dam (Fig. 3). It led to values of the Nash-Sutcliffe efficiency (NSE), the ratio of the root mean square error to the standard deviation (RSR) and of the percent bias (PBIAS) of 0.75, 0.5 and −12.61% respectively.

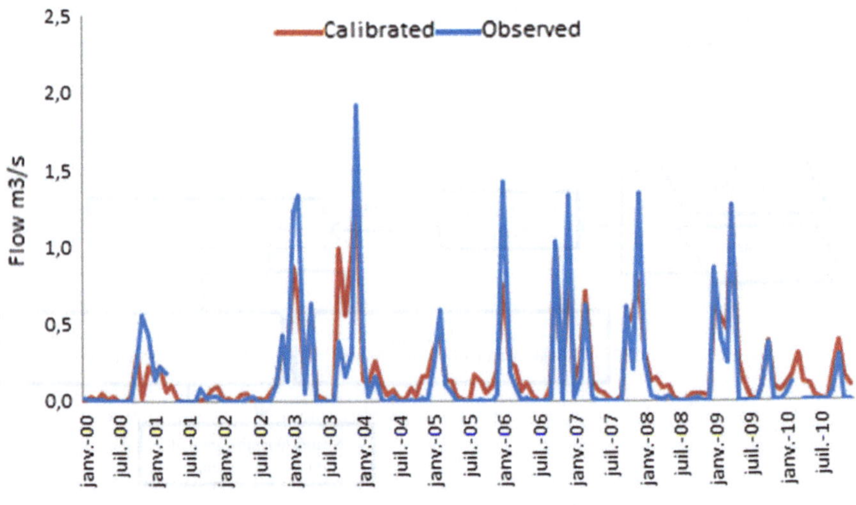

Fig. 3 Observed versus simulated flow

Further, the comparison between the various simulations showed that the increase in temperature and the decrease in precipitation estimated by both climate change scenarios KNMI and SMHI had an impact on the water balance of the study area. Thus, the estimated annual mean values of temperature in Chiba basin have to rise by 1.8 ± 0.1 °C between the reference and scenario period. Future climate trends translate into a reduction in rainfall of −16.59 and −16.75% according to the KNMI and SMHI scenarios respectively.

However, we observe an underestimation in precipitation for both reference scenarios that attended −46% for KNMI and −40% for SMHI compared with in situ rainfall observations. On the other hand, it has to be considered that climate model data are processed to cover the dimensions of Chiba basin. Hence, the comparability of such data sets to point measurements has to be treated with care. Besides that deviations in climate model data are systemic and affect the reference and scenario period in the same way.

The findings presented in Table 1 relative to the main water balance components for the Chiba basin results in a decrease of the surface flow (9.29%/−34.29%) and in the total aquifer recharge (−29.15%/−39.26%). Such possible reductions in aquifer recharge will induce a decline in groundwater storage. At this stage, it is not possible to quantify the reduction in groundwater level or resources without improving our understanding of the mechanisms and feedback loops that are most likely to drive changes in the groundwater conditions. Hence, ongoing research is set up on modeling techniques, aquifer characteristics, recharge rates, seawater intrusion, as well as monitoring of groundwater over-abstractions for the Chiba coastal basin.

The effects in the future climate trends manifested as well at the level of potential evapotranspiration (PET), with an increase of 10.65% in PET according to the KNMI scenario and 6.56% according to SMHI scenario. This increase will take place in response to the expected increase in temperature. Actual evapotranspiration, in turn, will decline to −9.60% according to the KNMI scenario and −10.03% according to SMHI scenario. This decline is explained by the decrease in rainfall.

Table 1 Average annual water balances for different scenarios for reference and future periods

Some water balance components	KNMI		SMHI	
	1971–2000	2041–2070	1971–2000	2041–2070
Precipitation (mm)	330.9	276	349.2	290.7
Surface runoff (mm)	3.23	2.93	3.5	2.3
Lateral soil flow contribution (mm)	8.4	6.55	8.82	6.06
Groundwater contribution to stream flow (mm)	10.85	7.01	12.71	7.73
Total aquifer recharge (mm)	23.6	16.72	27.66	16.8
Percolation out of soil (mm)	23.46	16.52	27.65	16.65
Actual evapotranspiration (mm)	233.3	210.9	236.2	212.5
Potential evapotranspiration (mm)	1461.3	1617	1387.9	1479

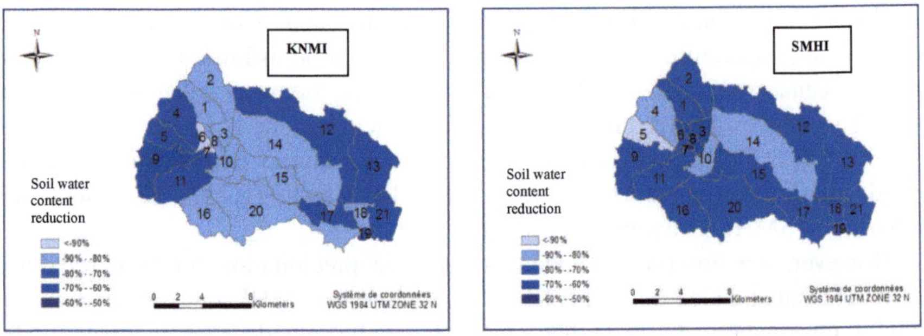

Fig. 4 Soil water content change between reference and future periods

Figure 4 presents the spatial rate of decrease in soil water content according to the sub basins. For both scenarios, this rate may reach as high as 90% with a minimum of 40% spatially variable depending on the change of rainfall, the land use type and the soil type. During the 2010–2070 period, the soil water content show considerable variability between soil profiles indicating that loamy sand areas, located in the center of the basin, are the most affected areas.

The reduction in these parameters indicates the critical future that water resource runs in the basin of Chiba and indicates clearly the need for an adaptation strategy for the water distribution, the agriculture management and the irrigation needs.

4 Conclusions

The proposed approach enabled us to quantify the recharge and the induced climate change based on the application of a physically based hydrology model. Through the application of two scenarios of climate change projections, we highlighted the importance of considering changes in the distribution of daily precipitation and temperature in estimating the impacts of climate change on groundwater recharge, soil moisture content and runoff.

Both climate change scenarios projects a decrease in runoff and in groundwater recharge. The percentage of change differs for both scenarios accounting for the projected changes in rainfall and temperature of both models. Thus, the analysis was based on comparison with the reference periods.

In this paper, we showed that there is a significant decline in rainfall leading to a marked decline in surface water and groundwater recharge. The future climate trends for soil moisture content spatial variability should be considered for future agricultural development.

Acknowledgements This work was performed under the project "Climate induced changes on the hydrology of Mediterranean basins" funded by the FP7 programme of the European Commission.

References

Arnold JG, Srinavasan R, Muttiah RS, Williams JR (1998) Large-area hydrologic modeling and assessment: part I. Model Dev JAWRA 34:73–89

Chemingui A, Horriche F, Paniconi C (2013) Implementation of a hydrological model of groundwater recharge for the Chiba catchment (Cap-Bon, Tunisia). In: 5th international conference on modeling, simulation and applied optimization ICMSAO, Hammamet, Tunisia, 28–30 April 2013

CRDA (2005) Carte des perimetres irrigues de la cote orientale de Cap-Bon CRDA Nabeul

Ensembles (2011) Ensemble-based predictions of climate changes and their impacts. http://www.ensembles-eu.org/

Kerrou J, Renard P, Tarhouni J (2010) Status of the Korba groundwater resources (Tunisia): observations and three-dimensional modeling of seawater intrusion. Hydrogeol J 18: 1173–1190

Kouzana L, Mammou AB, Felfoul MS (2009) Seawater intrusion and associated processes: case of the Korba aquifer (Cap-Bon, Tunisia) Surface geosciences (hydrology-hydrogeology). CR Geosci 341:21–35

Kumar CP (2012) Climate change and its impact on groundwater resources. Int J Eng Sci 1(5): 43–60

MESD (2015) Intended nationally determined contribution of Tunisia—2015. National Report to UNFCCC. Aug 2015, 18p

Paniconi C, Khalifi I, Giacomelli A, Tarhouni J (2001) Modeling and analysis of seawater intrusion in the coastal aquifer of eastern cap-bon Tunisia. Transp Porous Med 43:3–28

Van Griensven A, Bauwens W (2003) Multiobjective autocalibration for semidistributed water quality models. Water Resour Res 39(12):1348

Zghibi A, Zouhri L, Tarhouni J (2011) Groundwater modeling and marine intrusion in the semi-arid systems (Cap-Bon, Tunisia). Hydrol Processes 25:1822–1836

The Importance of the Groundwater Governance in the Global Change Context: A Proposal for a Mediterranean Aquifer (Llanos de la Puebla, Spain)

J. Berbel, A. Expósito and L. Mateos

1 Introduction

Groundwater (GW) governance has acquired relevant economic, social and political importance in the last decades, especially for developing countries where groundwater abstractions have become critical to sustain its growth (Siebert et al. 2010). Nevertheless, the majority of aquifers are still overdraft worldwide (Vaux 2011). GW is a classic common pool resource, defined by substractability and non-excludability (Ostrom 2005). Hardin's tragedy of the commons predicts that individualistic competitive exploitation of a common pool such as an aquifer with results in extracting too much groundwater too soon, leading to overabstraction rates as users ignore the social cost (or user cost) of their own abstractions. Further, climate change is introducing additional costs and risks hard to manage, including increased demand for groundwater and reduced recharge rates, with consequent heightened risk of conflict (Brouyere et al. 2004; Jyrkama and Sykes 2007). This fact is extremely important for countries such as Spain that suffer frequent severe droughts and where groundwater constitutes a strategic resource to maintain water supply during dry periods and to solve conflict among users (Custodio 2010).

Groundwater resources have enabled the development of an intense wealth-creating agricultural economy in many countries around the world. The economic development of many semi-arid areas of Spain has been due to or was started by intensive aquifer exploitation, in agriculture (most of the Mediterranean

J. Berbel (✉)
University of Córdoba, Córdoba, Spain
e-mail: julio.berbel@gmail.com

A. Expósito
University of Sevilla, Sevilla, Spain

L. Mateos
CSIC-IAS, Córdoba, Spain

M. L. Calvache et al. (eds.), *Groundwater and Global Change in the Western Mediterranean Area*, Environmental Earth Sciences,
https://doi.org/10.1007/978-3-319-69356-9_5

areas of southern and eastern Spain), industry (valleys of Catalonia and Valencia), and tourism (mainly in the south and the archipelagos). Llamas and Martínez-Santos (2005) estimate current aquifer storage depletion in the Iberian Peninsula at about 15 km^3, causing water table drop, water quality degradation, land subsidence and other negative ecological impacts.

2 Groundwater Governance (State of the Art)

Groundwater governance is included by many authors as being among the most important challenges for the future of water sustainability in the world (Grabert and Narasimhan 2006). Further, groundwater is inherently more complicated to govern than surface water because: (a) it is easily self-appropriated, with no need for cooperation and infrastructure management, (b) it is difficult to measure and control, (c) impacts of excessive pumping are usually detached in space and time from the actions that caused the problem. These unique characteristics explain that groundwater is weakly governed and underfunded within water policy frameworks. However, the role played by groundwater has increased as it serves as an essential resource to guarantee socio-economic development of some areas, especially those with arid or semi-arid climates. Groundwater governance requires therefore a drastic shift. Governance should be understood as the operation of rules, instruments and institutions that, built within a multi-actor context, can align stakeholders behaviour and actual outcomes with policy objectives in a multilevel framework with the use of multiple instruments. As a result of top-down and bottom-up processes, management decisions may benefit all parts (groundwater users included) and serve the implementation of longer-term integrated water resources management (IWRM) principles.

The governance framework proposed in this paper to illustrate the analysed case studies is based on Wiek and Larson (2012). This framework was developed within the "Water Partnership Program" led by the World Bank (Wijnen et al. 2012). It follows several phases (or levels) that should be accomplished in its practical implementation. Firstly, national and regional policies establish the objectives for groundwater governance within an integrated legal and management framework coordinated with other water, land and environmental related policies (i.e. agricultural, decentralization, etc.). The aim is to guarantee efficiency; equity and sustainability of groundwater uses (Fig. 1).

At a strategic level, an IWRM plan must be developed, including all necessary instruments to align stakeholder behaviour and outcomes with the previously established policy goals. Information constitutes here a critical issue, as it is crucial for the involvement of local actors (i.e. users); while obtaining reliable knowledge about the groundwater behaviour (recharge rates, transmissivity, etc.), withdrawals and uses is difficult since specialize research and modelling is required and accurate information is not guaranteed. Effective management of groundwater resources will therefore depend on the knowledge and attitudes of main actors, as stated by Allan

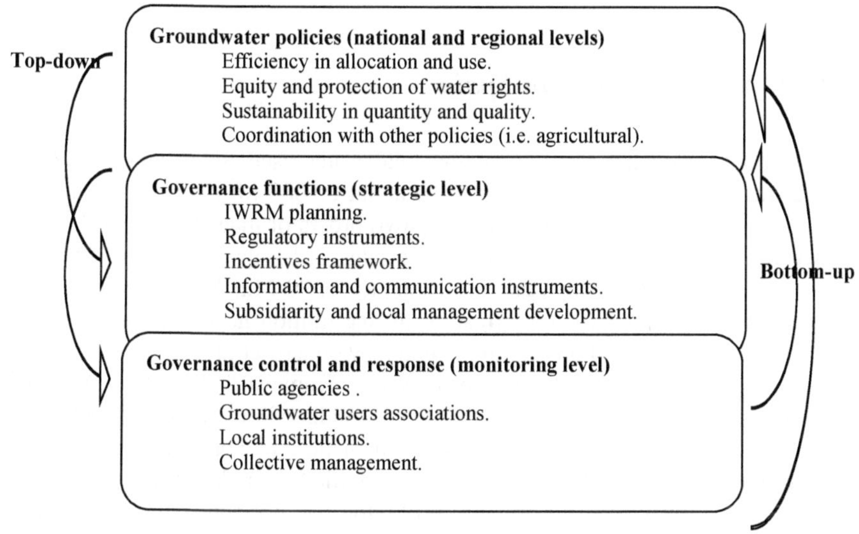

Fig. 1 Framework for assessing groundwater governance (adapted from Wijnen et al. 2012)

(2007), "more important than knowledge of the volumes and rates of use of renewable groundwater… is the knowledge constructed by political classes and by the major users of water in the region". Finally, local level governance involves the institutions that control outcomes locally and respond to the instruments set at the strategic level. Monitoring (and control) in this phase are critical, as it guarantees the access to information about quantity and quality of the resource and permits adjustments backwards at higher levels, leading to bottom-up management initiatives. At this local level, advisory instruments developed through collective participation may be considered as supplementary tools for groundwater management. Thus, once the groundwater governance framework is in place, different management instruments can be implemented and tested.

There are several examples where delegation and collective management have proved to be effective, once top-down initiatives have been developed. Based on analysis of water governance systems in the Netherlands, Australia and South Africa, Huntjens et al. (2012) argue that groundwater water users associations (GWUAs) are able to establish and protect water rights. According to Ostrom (2005), most individuals affected by operational rules should be able to participate in modifying them and contribute to the "command and control measures". In large-scale resource systems (i.e. river basin, major aquifer), it is important to enhance the participation of those involved in making key decisions about the system (Huntjens et al. 2012). Petit (2004) explained that in a situation of water stress, local stakeholders may recognize more easily the common characteristics of groundwater resources and try to find solutions to reduce/avoid over-exploitation.

3 Case Study

3.1 GW Management in Spain

GW in Spain accounts for 20% of the total water used for irrigation. Average aquifer recharge has been estimated to be about 30,000 Mm^3/year while the total amount of stored groundwater is probably two orders of magnitude higher that the yearly renewable resources. In the last decades, groundwater use in Spain has increased from 2000 Mm^3/year in 1960 to 6500 Mm^3/year nowadays and approximately 75% is used for irrigation of one million hectares, which is about 30% of the total irrigated area in the country (Molinero et al. 2011).

The implementation of the European Water Framework Directive (WFD) in 2000 implied the obligation for Member States to identify and classify groundwater bodies as part of the "Initial Characterisation Stage". In Spain, 699 groundwater bodies officially identified 259 (37%) were classified "at risk" of not attaining the environmental objectives set by the WFD for the horizon 2015. The aquifer Huéscar-Puebla analyzed in this work was within this group.

Historically, groundwater abstraction rights in Spain are tied to land ownership (Water Law of 1879). The Water Law of 1985 redefined abstraction rights and declared all aquifers as public domain, being the River Basin Authorities (RBAs) responsible for groundwater abstractions regulations. Further, a Register of Public Water and a Catalogue of Private Water was created as instruments for groundwater management. Registration was compulsory for all well owners. Control of wells and abstractions has not been easy due to limited resources at the RBAs but also to the unclear differentiation of water management competencies between RBAs and regional governments. Nevertheless, the new legal framework put in place in 1985 grants RBAs capacity to enforce pumping restrictions in both the public and private property regimes as well as the creation of groundwater uses associations.

3.2 "Llano de la Puebla" Hydrological System

The hydrological system of Huéscar-Puebla (MAS 05.04) is located under a plateau called "Llano de la Puebla" (or "El Llano"), surrounded by the Béticas mountain range in the north of the province of Granada (Region of Andalusia, southern Spain) (Fig. 2). The hydrogeologic system is composed by two sub-unities or sub-aquifers (Aljibe 2014): the carbonated aquifer of Parpacén (in the west) and the detrital carbonated aquifer of Fuencaliente (in the east). This system belongs to the Guadalquivir Basin. Puebla de Don Fabrique (to the north) and Huéscar (to the west) are the two urban concentrations that give the name to this aquifer system. There are two wellsprings, called Fuencaliente and Parpacén, both located in the surroundings of the town of Huéscar.

Fig. 2 Location map

The plateau "El Llano" has been devoted to rainfed agriculture since centuries, mainly to the cultivation of cereals and fodders. The development of horticulture in the Segura basin (60 km Eastwards) during the 80s and 90s, opened the market for cultivating high value crops such as broccoli, cauliflower, lettuces, etc. in the open air. By producing during the summer, this region covers the season when Murcia and Almeria greenhouse and intensive productions stop. Therefore, the abstraction of GW reserves affecting Fuencaliente spring was driven by the high profitability of cultivating high value vegetable crops. Figure 3 illustrates Fuencaliente's spring flows evolution. Farmers and citizens in Huéscar, who have used the spring since Roman time, made a claim to the Ministry of Agriculture and Environment for controlling the aquifer. This action led to an urgent intervention of the Guadalquivir RBA in order to re-establish the water balance in this groundwater system, as well as to manage the rising conflicts among the two towns (Huéscar and Puebla de Don Fabrique) and farmers. Consequently, the RBA made a complete hydrogeological study that conducted to an agreement between all affected parties in order to limit the abstractions and the cultivated area affecting the Fuencaliente aquifer. These measures did not affect the neighbour (East) Parpacén aquifer.

3.3 "Llano de la Puebla" Management Plan and Proposals

The terms of the agreement with the RBA were (a) abstractions decrease, from the estimated 8.7 hm^3/year (2003) to 5.6 hm^3/year in 2008 and 4.7 hm^3/year in 2013; (b) controlling annual withdrawals with individual well water meters; (c) annual water allocation to each farm (61 farmers have granted rights); (d) limitations of cultivated area from 2855 ha in 2003 to 1219 ha in 2013; (e) maximum of 20 ha

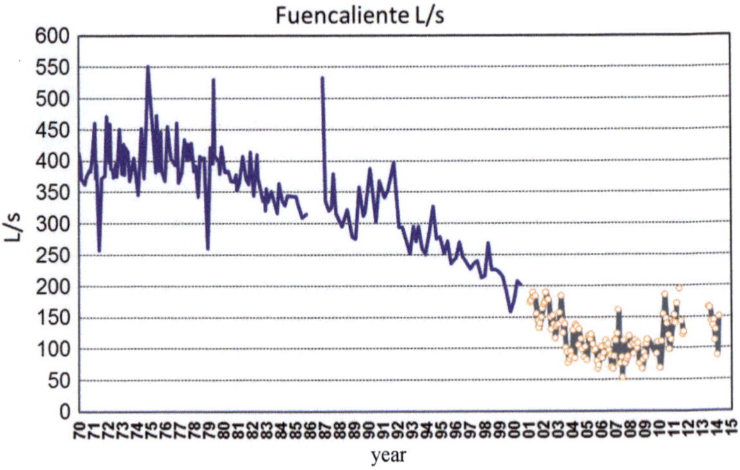

Fig. 3 Fuencaliente spring flow (*source* CHG)

per authorized well and no rotation or changes in the location of irrigated land; (f) creation of an unique interlocutor in the form of a GWUA and finally, (g) payment of a tariff to finance the water supply to former users of the "Fuencaliente" wellspring taken from the near San Clemente reservoir. The plan has allowed the recovery of piezometric level as illustrated in Fig. 4.

This specific payment of a groundwater tariff is exceptional in Spain. The Water Act does not foresee this possibility, thus, a reform of the Law would be required to introduce it. The financial resources for the RBAs come from the government budget and from the tariffs paid by users of regulated surface water. In the case of "Los Llanos" the payment of a water tariff is justified by the complementary use of surface water (from the San Clemente reservoir), but it is exceptional in Spain.

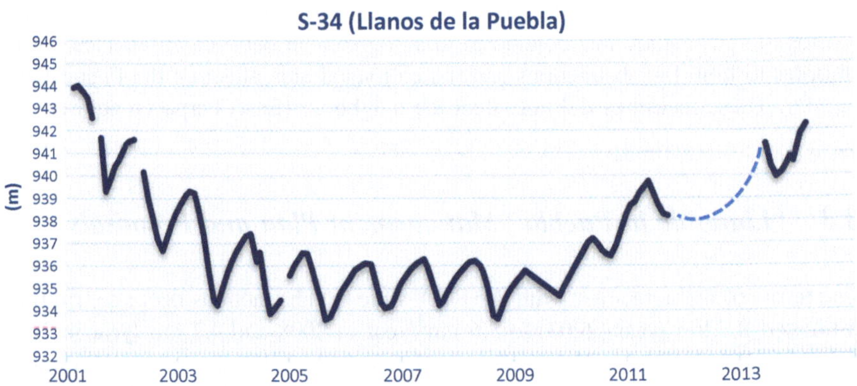

Fig. 4 Piezometic level 'Llanos de la Puebla' (*source* CHG)

4 Discussion and Concluding Remarks

The lack of an effective groundwater governance has usually resulted in the depletion of groundwater levels due to unsustainable agricultural development. The case described in this work serves as one example of effective governance and shed some light on the possible paths of sustainable governance of groundwater resources. The circumstances enforced sustainability in abstractions but further changes to improve flexibility and self-governance are required to guarantee an optimal management of the GW resources.

The aquifer "Llanos La Puebla" has reached a level of sustainability thanks to the governance model applied, including the use of innovative cost recovery instruments (tariff) that have not precedent in GW management in Spain. There are proposals to improve current management plan that should be attended (e.g. water rights trade, land rotation). Another avenue of research is the analysis of the different treatments given to this specific case and the nearby "Parpacén" (West) and "La Zarza" (East) aquifers. Important lessons about effective governance and the sustainability of groundwater resources use may be concluded from this research.

One easily implemented lesson in developed countries is the use of modern technology to monitor and map groundwater abstraction (e.g. meters, remote sensing), what needs to be implemented and placed on a cooperative ground between private and public agents. Thus, the implementation of governance frameworks with bottom-up initiatives, together with command and control measures (top-down direction) seems promising.

Acknowledgements This research has been financed by the Sec. Gral. Ciencia, Tecnología e Innovación (MINECO AGL2014-53417-R).

References

Aljibe (2014) Servicios para la reordenación, actualización de la información y elaboración del modelo matemático de funcionamiento hidrogeológico de la masa de agua subterránea 05.04 Huéscar-Puebla (Granada), Confederación Hidrográfica del Guadalquivir, Sevilla

Allan JT (2007) Rural economic transitions: groundwater uses in the Middle East and its environmental consequences. In: Giordano M, Villholth, KG (eds) The agricultural groundwater revolution: opportunities and threats to development. CABI Press, pp 63–78

Brouyère S, Carabin G, Dassargues A (2004) Climate change impacts on groundwater resources: modelled deficits in a chalky aquifer, Geer basin, Belgium. Hydrogeol J 12(2):123–134

Custodio E (2010) Intensive groundwater development: a water cycle transformation, a social revolution, a management challenge. In: Martínez-Cortina L, Garrido A, López-Gunn E (eds) Rethinking water and food security. CRC Press, pp 259–277

Grabert VK, Narasimhan TN (2006) California's evolution toward integrated regional water management: a long-term view. Hydrogeol J 14(3):407–423. doi:10.1007/s10040-005-0005-O

Huntjens P, Lebel L, Pahl-Wostl C, Camkin J, Schulze R, Kranz N (2012) Institutional design propositions for the governance of adaptation to climate change in the water sector. Global Environ Chang 22(1):67–81

Jyrkama MI, Sykes JF (2007) The impact of climate change on spatially varying groundwater recharge in the grand river watershed (Ontario). J Hydrol 338(3):237–250

Llamas MR, Martínez-Santos P (2005) Intensive groundwater use: silent revolution and potential source of social conflicts. Water Sci Technol 51(8):167–174

Molinero J, Custodio E, Sahuquillo A, Llamas MR (2011) Groundwater in Spain: legal framework and management issues. In: Findikakis AN, Sato K (eds) Groundwater management practices. IAHR Monograph, CRC Press, pp 123–137

Ostrom E (2005) Understanding institutional diversity, vol 241. Princeton University Press, Princeton

Petit O (2004) La nouvelle économie des ressources et les marchés de l'eau: une perspective idéologique? VertigO-la revue électronique en sciences de l'environnement 5(2)

Siebert S, Burke J, Faures JM, Frenken K, Hoogeveen J, Döll P, Portmann FT (2010) Groundwater use for irrigation—a global inventory. Hydrol Earth Syst Sci 14(10):1863–1880. doi:10.5194/hess-14-1863-2010

Vaux H (2011) Groundwater under stress: the importance of management. Environ Earth Sci 62 (1):19–23. doi:10.1007/s12665-010-0490-x

Wiek A, Larson KL (2012) Water, people, and sustainability—a systems framework for analyzing and assessing water governance regimes. Water Resour Manag 26(11):3153–3171. doi:10.1007/s11269-012-0065-6

Wijnen M, Augeard B, Hiller B, Ward C, Huntjens P (2012) Managing the invisible. Understanding and improving groundwater governance. Water Papers, World Bank, Washington

Effects of Climate Variability on Groundwater Resources in Coastal Aquifers (Case of Mitidja Plain in the North Algeria)

A. Bouderbala

1 Introduction

Water is indispensable for life, but its availability at a continual quality and quantity is vulnerable by many factors, of which climate plays a leading role (Kumar 2012). In many regions in Algeria, groundwater is considered as a main resource for fresh water supply. The Intergovernmental Panel on Climate Change (IPCC) indicates that global earth surface temperature has increased and it will increase of 2–4 °C over the next 100 years, which has a direct effect on the hydrologic cycle by increasing evaporation and indirectly impact on groundwater. In addition, there are also other associated impacts, such as seawater intrusion and water quality deterioration.

The relationship between the climate variability and groundwater quality and quantity is more complicated and poorly understood, but the greater variability in rainfall for a longer period has as a consequence a decrease in the potentiometric levels, and saline intrusion in coastal aquifers. Quantifying the impact of climate change on groundwater resources requires not only reliable forecasting of changes in the major climatic variables, but also needs an estimation of groundwater recharge (Kumar 2012). Understanding climate variability and change is vital for society and ecosystems, particularly with regard to complex changes affecting the availability and sustainability of groundwater resources (Bouderbala 2015).

This article presents the impact of climate change on groundwater resources in the alluvial aquifer of Mitidja.

A. Bouderbala (✉)
Department of Earth Sciences, University of Khemis Miliana,
44000 Khemis Miliana, Algeria
e-mail: bouderbala.aek@gmail.com

M. L. Calvache et al. (eds.), *Groundwater and Global Change in the Western Mediterranean Area*, Environmental Earth Sciences,
https://doi.org/10.1007/978-3-319-69356-9_6

2 Study Area

2.1 Presentation

The alluvial Mitidja plain, which has an average altitude of 100 m a.s.l., is an elongated depression extends from the east to west for a length of 100 km and a width between 3 and 18 km. It covers an area of 1450 km^2. It is limited in the north by the Mediterranean Sea in its eastern part, and by the Sahel mountains (260 m a.s.l.) in its western part. It is bordered in the south by the Blida Atlas (1630 m a.s.l.), and in the west by the mountains of the Dahra (1560 m a.s.l.). It lies between latitudes 36° 25′N and 36° 48′N, and between longitudes 2° 32′E and 3° 20′E.

The plain of Mitidja hosts a vivid agricultural economy, which is strengthened by the existing water resources and topography features, endowed with vast fertile and gently-sloping lands. It is occupied by cereals, vegetables, fruit trees and other crops (Khouli and Djabri 2011). The study area has a Mediterranean climate type, characterized by hot and dry summers and rainy winters, with an average rainfall about 600 mm, the average air temperature is 18.5 °C and the annual rate of evapotranspiration is 1200 mm. The hydrology of the basin is characterized by a set of watercourses that drain alluvium outcroppings; they are drained mainly by the four great rivers that flow into the plain to reach the Mediterranean sea are: wadi Nador, wadi Mazafran, wadi El-Harrach and wadi Hamiz.

2.2 Geology and Hydrogeology

The lithological succession in Mitidja plain appears from the bottom to top as follows (Fig. 1): The Pliocene: it is divided into upper and lower Pliocene. The lower Pliocene is formed of gray or blue-gray marls, including a high layer of blue marls, sometimes sandy, attributed to the Piacenzian. The upper Pliocene is composed of yellow marls, sandy, limestone and sandstone limestone, as well as the mollassus attributed to the Astian (Khouli and Djabri 2011).

The quaternary: it is formed by consolidated sedimentary formation, fluvial siliceous gravel and sandstone gravel with red clay of Cretaceous origin, alluvial deposits with silty lens alternating with pebbles; gravel and sand.

The geophysical study conducted in 1973 revealed the existence of two superimposed aquifers in the plain of Mitidja (Fig. 2):

(i) The Pliocene aquifer (Upper Pliocene) is a confined aquifer formed by the sandstone and sandy limestone. Its substratum is composed of the blue marls and its top is composed of semi permeable yellow marls named marls of El Harrach. This aquifer is very deep, generally located between 250 and 300 m in the major part of the plain.

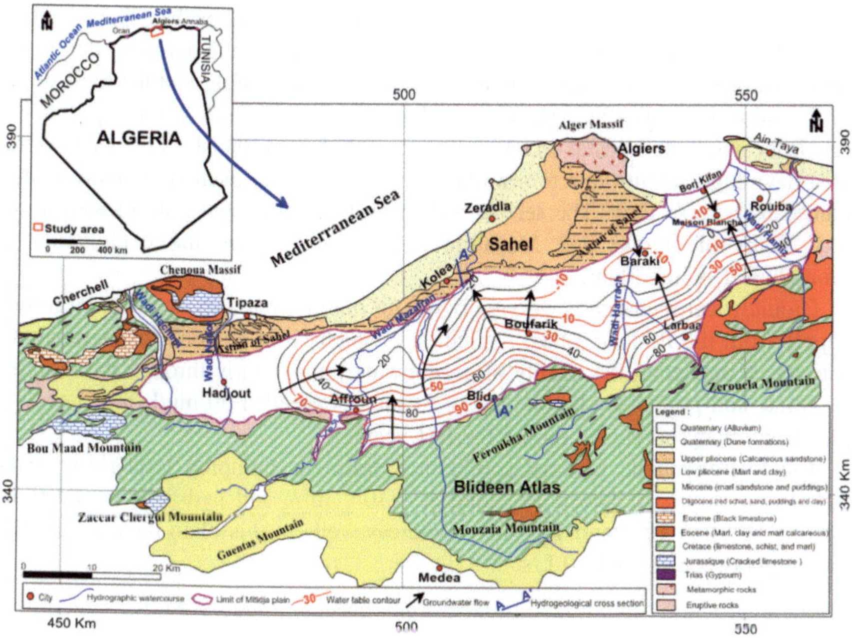

Fig. 1 Map showing the geology of the study area and the water table map of dry season 2013

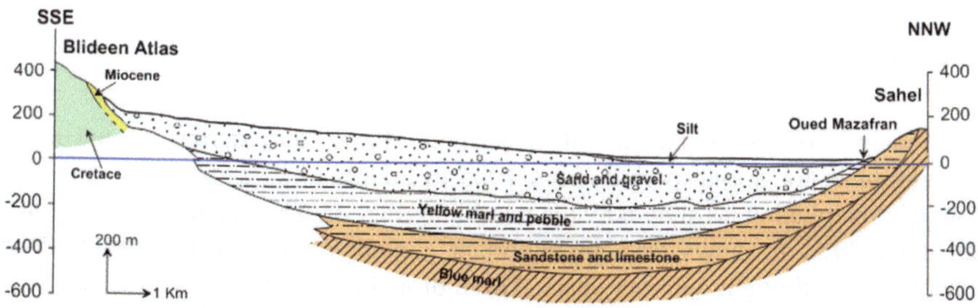

Fig. 2 Hydrogeological cross section A-A′ in the study area

(ii) The quaternary alluvial aquifer is overlain almost the entire basin. It is mainly composed of sand and gravel alternating with silts and clays. Apart from the zone of Mazafran, this aquifer is entirely unconfined and based on the marls of El Harrach wich constitute the substratum of the alluvial aquifer. Its thickness varies from 100 to 200 m. Its eastern and western limit is ensured by the rise of the blue marls of the Pliocene. The depth of the water table ranges between 4 and 30 m. For irrigation and the drinking water supply, the aquifer is exploited by more than 3000 wells.

The water table map (Fig. 1) reveals a general groundwater flow from south to north. It shows also tight isolines in the south part of the alluvial plain due to the influence of the recharge area, the low thickness of the aquifer and the high slopes of the substratum. The map shows also three depression cones in this aquifer. The first is located in the coastal sector, in the city of "Maison Blanche", the wells in this area have high salinity due to seawater intrusion after an overexploitation accompanied by reverse flow. The second and third depression cones are located in the well fields of Baraki and Mazafran respectively, where the exploitation of groundwater is important.

Groundwater recharge is essentially due to rainfall, but also by an underground supply from the Blidean Atlas. Agriculture, which is the region's main economic activity, is impacting the quality of groundwater, particularly through the leaching of nitrate and pesticides. Nitrate (NO_3^-) content exceeded 50 mg/L in water samples from many boreholes during the last years.

The hydrodynamic parameters obtained from the pumping tests show transmissivities varying between 10^{-3} and 1×10^{-2} m^2/s and hydraulic conductivity varying between 10^{-4} and 10^{-2} m/s. Whereas wells in this plain have a flow rate ranging between 10 and 60 L/s.

3 Impact of Climate Variability on Groundwater Resources

3.1 Climate Variability

The analysis of the annual rainfall data about Hamiz dam station located in the Mitidja plain for a long period (from 1905 to 2006) shows a marked decrease of annual precipitation, as shown by the trend line (Fig. 3). This reduction is estimated at about 20%. It shows also an important annual irregularity in time, with an alternation of drought and wet years. These drought years had as a consequence a low recharge of the alluvial aquifer and an the overexploitation of groundwater, which deteriorate the groundwater quality and advancement of the seawater intrusion in the coastal area of Mitidja plain. While the wet and rainy years contributed to the recharge and dilution of the aquifer, and when they are very heavy rainfall, they causes flooding (e.g. flood of Algiers in 2001). We may say that the climate variability observed during the last years is characterized by a rainfall return but with greater intensity.

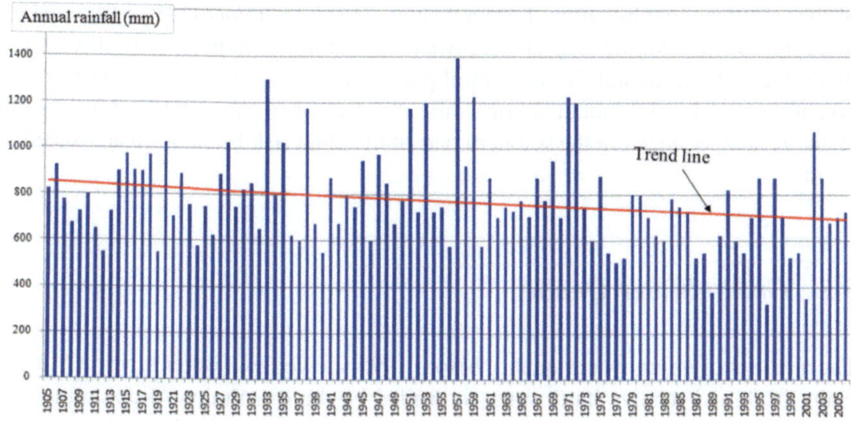

Fig. 3 Annual rainfall variability for Hamiz dam rainfall station (1905–2006)

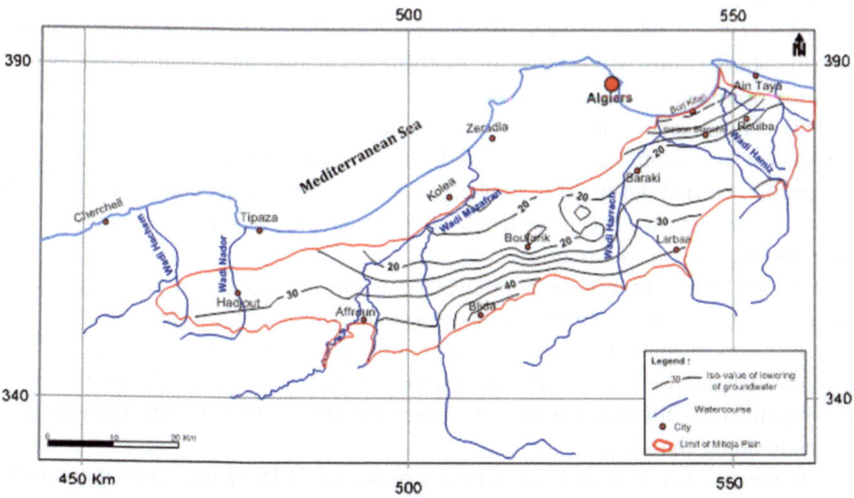

Fig. 4 Drawdown of groundwater levels from 1973 to 2013

3.2 *Potentiometric Study*

The proliferation of wells and the persistent drought (large rainfall deficit) in Mitidja plain for the last two decades had a negative impact on the quality and quantity of groundwater on the alluvial aquifer. The principal groundwater direction, follows the main depression oriented from West to East, towards the Mediterranean Sea.

The map of groundwater level decline in this aquifer from the year 1973–2013 (Fig. 4) shows a marked decrease in the potentiometric level. This is a consequence of the significant exploitation of groundwater. The potentiometric level declined by more than 10 m near the coastal area, where the seawater intrusion was observed at more than 1.5 km inland. In the central part of the plain, the recorded decrease of potentiometric level is about 20 m, while the decline of groundwater level in the recharge area localized in the southern part of the plain (Blida and Larbaa) exceeded 35 m.

3.3 Groundwater Quality

The analysis of groundwater quality before 1995 showed two water types in Mitidja plain: fresh and relative hard water controlled by the dissolution of dolomitic and calcitic sediments of the Atlas Mountains, mainly observed in the recharge areas and in the most of plain. The second water type observed is between brackish and saline water, with important influence of the sea and containing a considerable fraction of chlorides.

The process of salinization in this coastal sector is due to the seawater intrusion into the aquifer, this is the result of its overexploitation and large rainfall deficit, with the dominance of sodium-chloride water type.

This coastal sector shows also high values of electrical conductivity with the dominance of sodium-chloride type facies of groundwater. The geoelectrical prospection carried out in 2001 showed a low resistivity in the coastal sector of the plain, which confirms the persistence of seawater intrusion phenomenon, existed since the eighties in this region.

After the salinization of the boreholes located near to the coast, they were closed and only a few wells at small depths remained. These wells were monitored seasonally by the National Agency of Hydraulic Resources (ANRH-Blida), and the groundwater samples are usually taken from the surface of water table and not after a pumping time, for this reason we cannot use these groundwater analyses as a means to study the seawater intrusion.

The physic-chemical analyses for the dry period 2015 (Table 1) of the aquifer, show pH ranges from 7.6 to 8.3, indicating that the groundwater in this area is slightly alkaline. It shows more than 80% of samples have TH > 40 °F which indicates a very hard groundwater quality in this aquifer, while the electrical conductivity (EC) shows a measured values ranging from 900 to 2600 μS/cm, with an average of 1709 μS/cm. So, the majority of water samples exceed the standard recommend by WHO (2008), i.e. 1500 μS/cm, thus undrinkable. The Cl^- concentration ranges from 84 to 395 mg/L, SO_4^{2-} from 46 to 391 mg/L, HCO_3^- from 61 to 458 mg/L and NO_3^- from 22 to 106 mg/L. The Na^+ from 19 to 144 mg/L, Mg^{2+} from 1 to 94 mg/L; and Ca^{2+} from 115 to 258 mg/L.

The high bicarbonate concentrations in this area can be due to the occurrence of oxidation of organic matter of the soil layers at emerging land (Bouderbala 2015). The contents of chlorides in the study area indicate more than 33% exceed the standard limit 250 mg/L. The very high concentrations of chlorides are probably

Table 1 Statistics of the physic-chemical analyses made in 32 wells in the aquifer of Mitidja (dry season 2015)

	Ca^{2+} (mg/L)	Mg^{2+} (mg/L)	Na^{+} (mg/L)	K^{+} (mg/L)	Cl^{-} (mg/L)	SO_4^{2-} (mg/L)	HCO_3^{-} (mg/L)	NO_3^{-} (mg/L)	pH	EC (µ/Cm)	TH (°F)
Min	115	1	19	2	84	46	61	22	7.6	900	32
Max	258	74	144	9	395	391	458	106	8.3	2600	95
Moy	188.3	28.3	73.8	3.6	222.4	190.6	260.3	57.7	8.14	1709	59
Norm OMS	100	75	150	12	250	200	–	50	6.5–8	1500	40
% > norm	100	0	0	0	40	30	–	60	0	70	80

due to marine pollution. The high values of nitrates exceeding the limit recommended for drinking water (50 mg/L), indicate the degradation of groundwater quality, and they can be mainly attributed to the excess of fertilizers and pesticides used in agricultural activities in this area (Saida et al. 2017). The most common source of the high values of calcium in the study area is due to the lithological alluvial deposits, and to the dissolution of the consolidated detrital sedimentary rocks. The high concentrations of sulphates are due probably to the use of some fertilizers in agriculture or the discharge of untreated wastewater.

3.4 *Vertical Conductivity Profiles*

The electrical conductivity profiles (at 25 °C) were carried out by National Hydraulic Resources Agency in 2009 at the piezometers Pz58 and Pz61 located approximately at a distance of 1.7 km from the coast. These conductivity profiles have made in the goal to confirm the existence of marine intrusion in this coastal aquifer (Fig. 5).

The vertical conductivity profile carried out in the piezometer Pz 58 shows a significant increase in the conductivity, from 3 mS/cm in the top to 15 mS/cm at the bottom of the well. It is explained by the significant flow of water transiting the permeable formations of Quaternary (sand and gravel). Moreover, it seems clear that the piezometer Pz 58 is in an area heavily contaminated by marine intrusion.

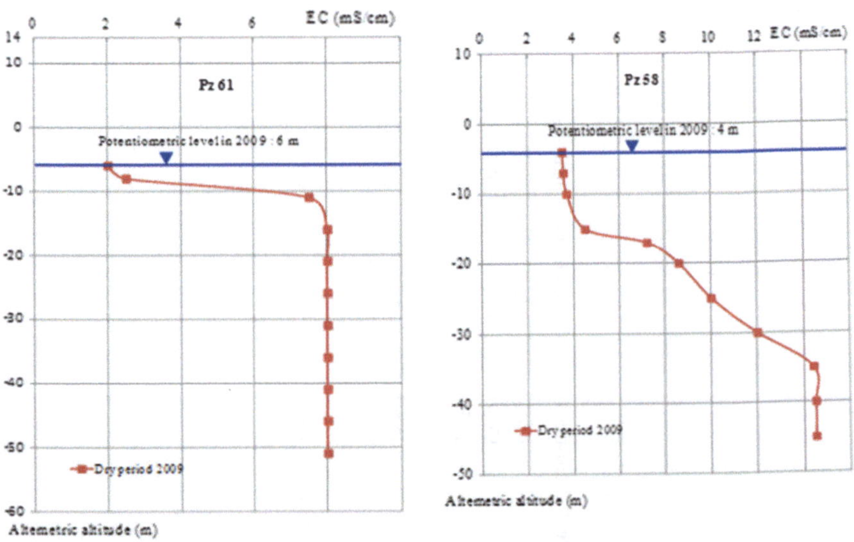

Fig. 5 Conductivity profiles in the piezometers PZ 61 and PZ58

In the case of piezometer Pz 61, the vertical conductivity profile shows a significant increase of the conductivity in this well from the top to the bottom. This conductivity profile has the characteristic of having a linear profile over the entire water column except for the first 4 m where the conductivity is lower, it could be related to the incidence of vertical recharge by less mineralized rainwater. But it confirms that the aquifer is contaminated by the marine intrusion.

4 Conclusion

Mitidja aquifer constitutes an important reservoir of groundwater that used for more than a century for the domestic, agricultural and domestic purposes. However, the anarchical exploitation accompanied with the drought periods in this last decade, caused a decrease of potentiometric levels and deterioration of groundwater quality in this plain.

The analysis of the annual rainfall data of pluviometric station located in this area for a long period shows a marked decrease of annual precipitation, with a reduction about 20%. This caused a decrease of potentiometric level for more than 20 m and seawater intrusion in the coastal sector of the plain near to 2 km inside the aquifer.

References

Bouderbala A (2015) Groundwater salinization in semi-arid zones: an example from Nador plain (Tipaza, Algeria). Environ Earth Sci 73(9):5479–5496

Khouli MR, Djabri L (2011) Impact of use of agricultural inputs on the quality of groundwater case of Mitidja plain (Algeria). Geographia Technica 11(2):35–44

Kumar CP (2012) Climate change and its impact on groundwater resources. Int J Eng Sci 1(5):43–60

Saida S, Tarik H, Abdellah A, Farid H, Hakim B (2017) Assessment of groundwater vulnerability to nitrate based on the optimised DRASTIC Models in the GIS Environment (Case of Sidi Rached Basin, Algeria). Geosciences 7(2):1–23

WHO World Health Organization (2008) Guidelines for Drinking-Water Quality, 2nd edn. Geneva

Development of a GIS Based Procedure (BIGBANG 1.0) for Evaluating Groundwater Balances at National Scale and Comparison with Groundwater Resources Evaluation at Local Scale

G. Braca and D. Ducci

1 Introduction

In the last years water scarcity and drought problems in Italy have become of growing concern (Ducci and Tranfaglia 2008; Fiorillo and Guadagno 2012). These problems seem to be worsened in the near future, both for the increase of water demand, often exceeding the available sustainable water resources, and for the temperature increase and the variation of the quantity and the distribution of precipitation even if in different ways from north to south of Italy (Toreti et al. 2009).

Moreover, the Water Framework Directive (WFD) (2000/60/EC) has introduced a legal framework for sustainable management of water resources across Europe, even though it does not explicitly mention the evaluation of changes in temperature and precipitation and its effects on groundwater resources.

In this context, the Italian National Institute for Environmental Protection and Research (ISPRA) has developed an automatic GIS-based procedure named BIGBANG (acronym of the Italian sentence "Bilancio Idrologico GIS BAsed a scala Nazionale su Griglia regolare" which means "Nationwide GIS-based hydrological budget on a regular grid") version 1.0, in order to evaluate the water budget components at monthly temporal scale for the whole National territory.

This paper, after a description of the structure of the spatially distributed water balance models, presents and discusses some results obtained in the whole Campania region (southern Italy) during the period 1996–2015. The results have been discretized for each groundwater bodies (GWBs).

G. Braca
ISPRA—Italian National Institute for Environmental Protection and Research, Rome, Italy

D. Ducci (✉)
Department of Civil, Architectural and Environmental Engineering, University of Naples Federico II, Naples, Italy
e-mail: daniela@unina.it

M. L. Calvache et al. (eds.), *Groundwater and Global Change in the Western Mediterranean Area*, Environmental Earth Sciences,
https://doi.org/10.1007/978-3-319-69356-9_7

These results have been compared with groundwater balances carried out at local scale, in a carbonate hydrogeological system, for the period 2000–2015.

2 Materials and Methods

2.1 *The BIGBANG 1.0 Procedure*

The *Italian National Institute for Environmental Protection and Research* (ISPRA) has developed a GIS based procedure (BIGBANG 1.0—"*Nationwide GIS-Based hydrological budget on a regular grid" version 1.0*) for evaluating all the factors of the monthly water balance at National scale in a spatially distributed way, using Python programming language, at moment implemented in a proprietary GIS platform (ESRI ArcGIS 10.1, ESRI 2012)

The procedure is based on the spatial environmental information at very high resolution, also available on the WEB, in formats understandable by the main Geographic Information Systems. This model allows the estimate of the spatially distributed hydrological factors, such as total precipitation, potential and actual evapotranspiration, surface runoff and groundwater recharge. The hydrological factors of total precipitation, actual evapotranspiration, surface runoff and groundwater recharge are evaluated for a 1 km resolution grid, in the ETRS89 Datum, using a LAEA (Lambert Azimuthal Equal Area) projection (Fig. 1). The grid is based on the recommendation at the 1st European Workshop on Reference Grids in 2003 and later INSPIRE geographical grid systems (EEA, European Environmental Agency) and each cell is univocally identified by an ID. The 1 km

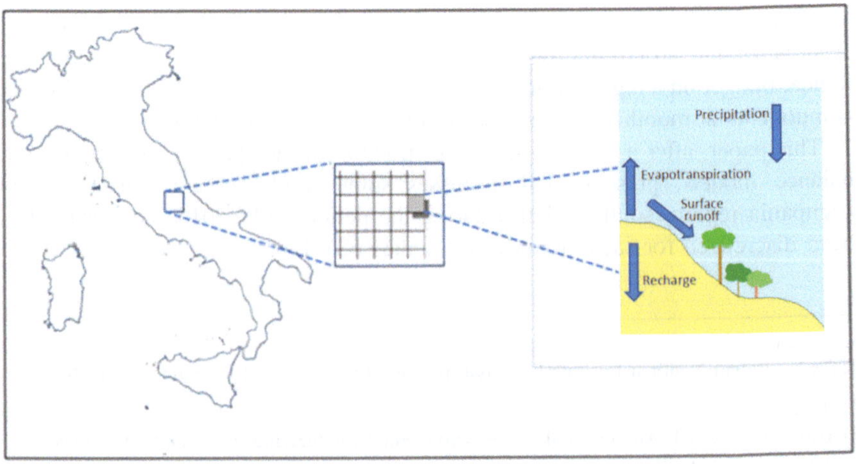

Fig. 1 BIGBANG calculation scheme. The hydrological factors are evaluated for a 1 km resolution grid covering the whole National territory

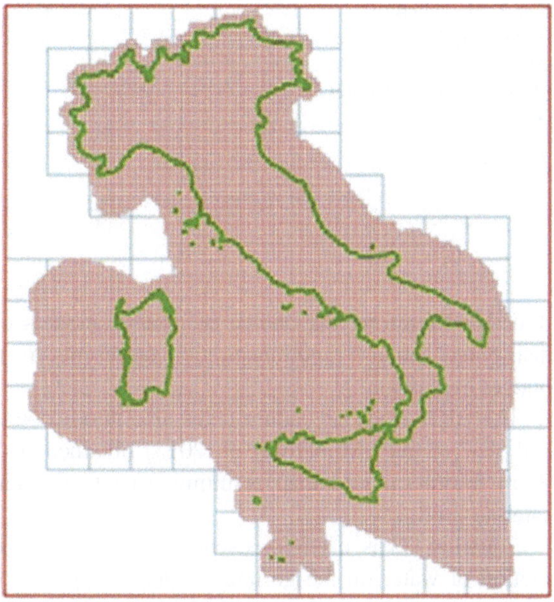

Fig. 2 European environmental agency reference grids for Italy (100 km in gray, and 10 km, in red, resolution)

resolution grid is not shown because too much small, but it is nested in the 10 km resolution one (Fig. 2).

The water budget model used in BIGBANG follows the approach suggested by Thornthwaite and Mather (1955) and it simulates on each grid cells: soil moisture variations, actual evapotranspiration, groundwater recharge and surface runoff, using a set of climatic data, as precipitation and temperature, soil and land-use data, hydraulic and geological properties, etc.

The governing equation is based on mass balance:

$$P - E = R + G + \Delta V \tag{1}$$

where P is the total precipitation, E is the actual evapotranspiration, R is the surface runoff, G is the groundwater recharge and ΔV is the change in soil moisture storage volume.

All the factors have been evaluated in millimeters per month.

The factor (P − E) is also defined as "internal flow" and it represents the total volume of river runoff (R) and groundwater (G) in a territory (Eurostat/OECD 2014). The quantity (R + G) is also defined as "surplus" as the water which does not evaporate or remain in soil storage and it is available to generate surface and subsurface runoff and groundwater storage (Westenbroek et al. 2010). The difference between potential and actual evapotranspiration is also indicated as the water "deficit" that represents the amount of water that should be supplied to the vegetation as irrigation (Westenbroek et al. 2010). Actual evapotranspiration is one of the main water balance components, and its value is very difficult to measure

Fig. 3 Hypothetical relationship between the actual evapotranspiration (AET) and the water storage (WS)

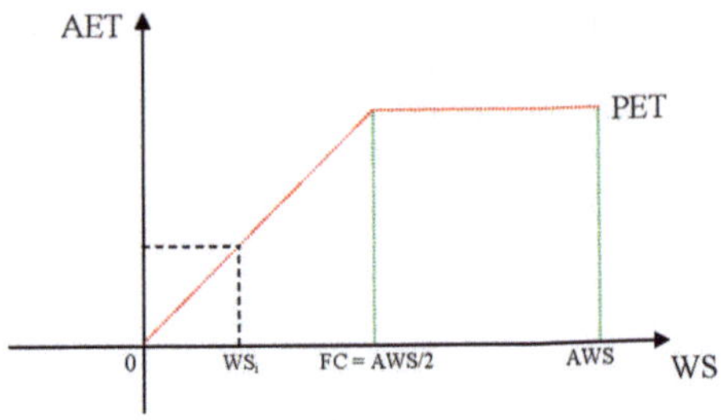

directly. Therefore, the choice of reliable models capable of predicting spatially distributed actual evapotranspiration represents a critical aspect for groundwater budget evaluation.

The BIGBANG procedure takes also into account in each cell the effect of the soil sealing rate (Munafò et al. 2013) obtained by lumping together the ISPRA 20 m resolution grid map resulting from Copernicus earth observation program satellite products.

Equation (1) is used for each 1 km grid cell without consider the horizontal motion of water on the ground-surface, or in the soil. The BIGBANG procedure schematizes as a reservoir a volume of soil of 1 km grid for 1 m deep, whose maximum capacity is given by the available water storage (AWS) depending on soil texture. The variable representing the soil moisture at the end of the month is the water storage (WS).

In the soil model, rainfall is assumed to infiltrate into the soil from which moisture is depleted by the actual evapotranspiration (AET). When the soil storage is full, it is assumed to be saturated and the exceeding rainfall becomes surface runoff and recharge, according to the recharge scheme. Evapotranspiration is assumed to continue at its potential rate (PET) until the soil water storage reaches the value of field capacity FC that we assume as half of AWS (Kandel et al. 2005). Afterwards, evapotranspiration (AET) decreases linearly to zero until the storage is empty, reaching the water quantity known as wilting point (Fig. 3).

In the present version, the BIGBANG procedure uses the 1 km grid of AWS from the LUCAS_TOPSOIL data grid (Toth et al. 2013) of the Joint Research Center of UE (Fig. 4).

2.2 *The BIGBANG Procedure Steps*

The automatic GIS procedure consists in the following steps:

(1) Spatial interpolation of monthly time series of hydrological variables which control water budget components: total precipitation, mean temperature, minimum temperature, maximum temperature, solar radiation, and others. Spatial interpolation is performed on the reference grid covering Italy by using tools

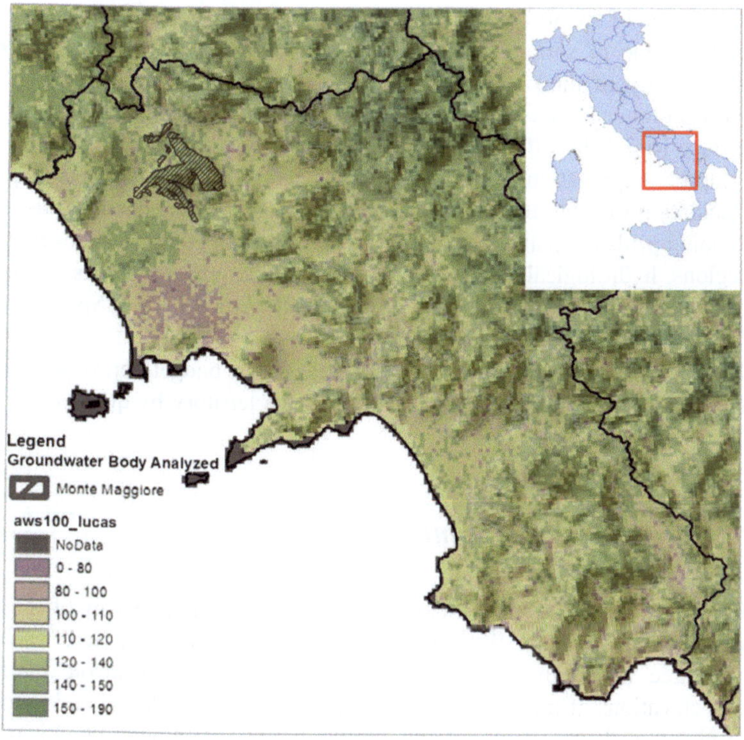

Fig. 4 AWS in mm for 1 km resolution grid derived from JRC LUCAS_TOPSOIL data grid and location of the Monte Maggiore groundwater body

available in ESRI Spatial Analyst Package (kriging, IDW, etc.). Nevertheless, BIGBANG can use grids available from reliable web sources. In BIGBANG procedure, for example, monthly mean temperature grids are derived from ISPRA SCIA System (Fioravanti et al. 2010), available from the web site, and transformed in the ETRS89-LAEA coordinate system of the grid reference.

(2) Estimation of the grid of monthly snow precipitation, snow accumulation as snow water equivalent, using a simple model based only on precipitation, mean temperature and elevation derived from a DEM (McCabe and Markstrom 2007). Snow melt is estimate using a simple degree day model (DeWalle and Rango 2008).

(3) Calculation of the grid of monthly potential evapotranspiration by selection of different formulations: simplified Turc (1961) formula, Thornthwaite (1948) formula and Hargreaves and Samani (1982) formula. The choice of the appropriate formula depends mainly on the hydrological data availability.

(4) Calculation of the grid of monthly actual evapotranspiration on the basis of potential evapotranspiration grid through a soil water balance.

(5) Calculation of the grid of groundwater recharge estimated as a percentage of the total volume of surface runoff (R) and groundwater (G) in function of the permeability of the outcropping hydrogeological units. The shapefile of the hydrogeological units is downloadable at the ISPRA website. Calculation of the Surface Runoff by difference.
(6) Temporal aggregation of balance components grids (seasonal, yearly, etc.).
(7) Calculation of long term annual average of balance components grids.
(8) Clipping grids of spatial distribution of hydrological components over territory (regions, hydrological basins and sub-basins, groundwater bodies, etc.).
(9) Calculation of spatial statistics using ESRI ArcGIS 10.1 Spatial Analyst Package tools.
    Definition of tables summarizing all the water budget terms for different intervals of time step and different parts of the territory by queries.

2.3 *Hydrogeological Features*

The Campania region (13,500 km^2) is located in the southern part of the Italian peninsula and it shows at W a coastline along the Tyrrhenian Sea. The region presents three main landscapes: the Apennine carbonate Mesozoic mountains, reaching elevations of more than 2000 m (about 32%), the alluvial and pyroclastic coastal plains of the rivers Garigliano, Volturno and Sele (about 18%) and finally hills and valleys constituted by prevalently impervious sediments, while a small part is represented by the Roccamonfina and Vesuvius volcanoes and pyroclastic hills of the Phlegrean Fields. The main aquifers with copious springs are the carbonate mountains, which have a very high permeability due to a well-developed karstic network.

The Campania Region is partitioned in 80 significant groundwater bodies: 10% volcanic GWBs, 25% alluvial GWBs, 30% karstic GWBs and 35% mixed, at low-moderate permeability. The carbonate GWB of Monte Maggiore (Fig. 3) (180 km^2) is composed prevalently by Cretacic limestone, and only in little part by older dolomite. The GWB feeds two springs located at the foot Triflisco Springs and Pila Springs (total mean discharge about 4.6 m^3/s) toward south, while the remaining part of the flow is directed toward another GWB.

3 Results

The yearly and monthly evaluations have been compared with groundwater balances carried out at local scale, in the carbonate groundwater body of Monte Maggiore. The yearly comparison is shown in Table 1. The most evident difference is in the area, due to the dimension of the cells at national scale that creates an error.

Table 1 Yearly water budget for Monte Maggiore groundwater body

Monte Maggiore—period 2000–2015									
Method	Area	Mean elevation (h)	Temperature (T)	Precipitation (P)	Actual evapotranspiration (E)	Surface runoff (R)	CIP	Groundwater recharge (G)	
	km^2	m a.s.l.	°C	mm	mm	mm	%	mm	10^6 m^3/year
Local scale	180	338.9	14.8	1271	706 (1)	564	90	508	91.4
				1271	620 (2)	650	90	585	105.4
National scale	186			1116	556 (3)	560	90	504	93.8

Actual evapotranspiration evaluated by (1): Turc; (2) Hargreaves and Samani; (3) Thornthwaite and Mather. CIP = percentage of (P-E) that becomes groundwater recharge (G) in function of the permeability of the outcropping hydrogeological units

At local scale T has been evaluated on the basis of the Digital Elevation Model considering the equation: T = 16.8–0.0059 h. The difference in precipitation is due to the different interpolation model used. At the end the differences among methods and scale are low and all less than 15%. The monthly comparison also shows a good accordance between the different scales, individuating the same periods for deficit and exceedance.

4 Conclusions

The comparison between a groundwater balance carried out at National scale and at local scale shows that the lithological and hydrogeological features, much more detailed at local scale, can influence some factors of the balance, but the final amount of the recharge is comparable

Future developments of the proposed procedure BIGBANG (that is only at version 1.0) will regard improvements especially about spatial interpolation method of precipitation that should be taken into account the auxiliary variables as topographic elevation, coastal proximity, facet orientation, and others like PRISM method (Parameter-elevation Relationships on Independent Slopes Model, Daly et al. 1997), hydraulic soil properties, evapotranspiration model and the parameters calibration. Moreover the water balance performed by BIGBANG should be improved with the introduction of the water surface volumes stored in the natural lakes and artificial reservoirs.

As historic time series of monthly precipitation and temperature data are retrieved from hydrological year books, another application of BIGBANG procedure will be to reconstruct historic monthly water balance in Italy in order to analyze the long time series (more than fifty years) of water budget terms and their variability throughout the time.

Finally in the future it will be very helpful that BIGBANG produced dataset, maps and tables, could be available to users through a dedicated web site.

References

Daly C, Taylor G, Gibson W (1997) The PRISM approach to mapping precipitation and temperature. In: 10th AMS Conference on Applied Climatology, Reno, NV, pp 10–12

DeWalle DR, Rango A (2008) Principles of snow hydrology. Cambridge University Press

Ducci D, Tranfaglia G (2008) Effects of climate change on groundwater resources in Campania (Southern Italy). Geol Soc Sp Publ 288:25–38

ESRI Environmental Systems Research Institute Inc. (2012) ArcGIS Desktop: Release 10.1 Redlands, CA

Eurostat/OECD (2014) Data collection manual for the OECD/Eurostat Joint Questionnaire on Inland Waters Version 3.0—Sept 2014

Fioravanti G, Toreti A, Fraschetti P, Perconti W, Desiato F (2010) Gridded monthly temperatures over Italy. EMS Annual Meeting Abstracts, 7, EMS2010-306, ECAC Conference, Zurich, 13–17 Sept 2010

Fiorillo F, Guadagno F (2012) Long karst spring discharge time series and droughts occurrence in Southern Italy. Environ Earth Sci 65(8):2273–2283

Hargreaves GH, Samani ZA (1982) Estimating potential evapotranspiration. J Irrig Drain Eng 108:223–230

Kandel D, Chiew F, Grayson R (2005) A tool for mapping and forecasting soil moisture deficit over Australia. Technical Report 05/2 Mar 2005

McCabe GJ, Markstrom SL (2007) A monthly water-balance model driven by a graphical user interface. U.S. geological survey open-file report 2007-1088, 6 p

Munafò M, Salvati L, Zitti M (2013) Estimating soil sealing rate at national level—Italy as a case study. Ecol Ind 26:137–140

Thornthwaite CW (1948) An approach towards a rational classification of climate. Geogr Rev 38:55–89

Thornthwaite CW, Mather JR (1955) The water balance. Laboratory of climatology, 8, Centerton NJ

Toreti A, Fioravanti G, Percontia W, Desiato F (2009) Annual, seasonal precipitation over Italy from 1961 to 2006. Int J Climatol 29:1976–1987

Turc L (1961) Estimation of irrigation water requirements, potential evapotranspiration: a simple climatic formula evolved up to date. J Ann Agron 12:13–14

Toth G, Jones A, Montanarella L (eds) (2013) LUCAS topsoil survey. Methodology, data, results. JRC Technical Reports. Luxembourg. Publications office of the European Union, EUR 26102—scientific, technical research series ISSN 1831-9424 (online), ISBN 978-92-79-32542-7; doi:10.2788/97922

Westenbroek SM, Kelson VA, Dripps WR, Hunt RJ, Bradbury KR (2010) SWB—a modified Thornthwaite-Mather soil-water-balance code for estimating groundwater recharge. U.S. geological survey techniques, Methods 6–A31, 60 p

Webgraphy

Copernicus earth observation program satellite products, http://www.copernicus.eu/

EEA, European Environmental Agency, http://www.eea.europa.eu/data-and-maps/data/eea-reference-grids-2

ISPRA website, http://www.isprambiente.it

ISPRA website hydrogeology, http://www.sinanet.isprambiente.it/it/sia-ispra/download-mais/complessi-idrogeologici/view

ISPRA SCIA System, http://www.scia.isprambiente.it/home_new.asp

Groundwater Salinity and Environmental Change Over the Last 20,000 Years: Isotopic Evidences in the Lower Sado Aquifer Recharge, Portugal

P. M. Carreira and J. M. Marques

1 Introduction

Steady increase in the salinity of most of the major aquifers being used for water supply in coastal regions, in particular in areas under arid and semi-arid conditions, provide evidences of water quality deterioration (Jalali 2007; Bouchaou et al. 2008). This increase of mineralization in groundwater resources is often due to inflow of saline dense water during heavy withdrawals of fresh water from coastal aquifers and/or mobilization of saline formation waters by over-exploitation of inland aquifer systems. Not only seawater mixing is responsible for water resources degradation. Because of the different income sources for groundwater quality deterioration, it is necessary to identify and characterize the specific processes involved. Among the different approaches, isotope techniques are particularly effective for identifying the source of salinity and renewability of groundwater (Carreira et al. 2014; Re and Sacchi 2017).

The Lower Tagus-Lower Sado basin is located in Setúbal-Lisbon region and represents an important water resource for a vast region. The highly populated urban and industrialized areas of Setúbal and Lisbon are supplied by this system, which has been extensively exploited over the last decades. In order to find out the source of salinization in these groundwater systems chemical (Cl^-, HCO_3^-, SO_4^{2-},

P. M. Carreira (✉)
Centro de Ciências e Tecnologias Nucleares (C2TN), Instituto Superior Técnico, Universidade de Lisboa, Estrada Nacional 10, km 139, 7, 2695-066 Bobadela, LRS, Portugal
e-mail: carreira@ctn.tecnico.ulisboa.pt

J. M. Marques
Centro de Recursos Naturais e Ambiente (CERENA), Instituto Superior Técnico, Universidade de Lisboa, Av. Rovisco Pais, 1049-001 Lisbon, Portugal

M. L. Calvache et al. (eds.), *Groundwater and Global Change in the Western Mediterranean Area*, Environmental Earth Sciences,
https://doi.org/10.1007/978-3-319-69356-9_8

Ca^{2+}, Mg^{2+}, Na^{+} and K^{+}) and isotopic ($\delta^{2}H$, $\delta^{13}C$, $\delta^{18}O$, ^{3}H and ^{14}C) analyses were performed on groundwater samples collected in 43 boreholes. In the region under investigation there is a growing concern that these groundwater systems maybe threatened by further exploitation (not controlled) due to mixing with the shallow aquifers (highly polluted), seawater intrusion processes in the coastal areas, or either by brine dissolution detected at depth by geophysical studies (Astier 1979).

Isotope methodology applied (stable and radioactive environmental isotopes) were particularly effective for identifying the source of salinity and renewability of groundwater. Since most of the recharge is derived from direct infiltration of precipitation, the groundwater will reflect the isotopic composition of that precipitation. However, if most of the recharge is derived from surface water (rivers or lakes) instead of precipitation, the groundwater will reflect the mean isotopic composition of the contributing river or lake and possible mixture between different aquifer units. The difference in isotopic composition allows for differentiation of precipitation sources, and hence of recharge mechanisms and in some case the presence of paleowaters (Fig. 1).

The isotope techniques were used to distinguish the importance of the following processes which may lead to the salinization of groundwater: (i) leaching of salts by percolating water; (ii) seawater intrusion (present or past) of salt water bodies such as seawater, brackish surface water or brines; and (iii) concentration of dissolved salts through evaporation. Environmental isotope studies were applied in order to assess the origin of salinization, being complement by geochemical investigations, which, in some cases, were not able to undoubtedly solve the questions. The case study that will be presented is situated south of Lisbon in a very industrialized region, the Lower Tagus-Sado sedimentary basin (central Portugal), where groundwater salinization occurs, and reaching values of several grams of Total Dissolved Solids (TDS)/L. The source of this high mineralization could be: (a) actual seawater intrusion; (b) dissolution of a brine structure at depth or mixing with ancient seawater trapped during the basin formation.

Stable isotopes provide an effective label for seawater and freshwater to enable tracing of seawater intrusion, as well as identifying other processes that may be responsible for groundwater salinization. Evolution of stable isotope concentration of water during different processes related to water salinization is presented in Fig. 1. In studies dealing with seawater intrusion and identification of groundwater salinization process, it is often common to consider both isotopic and hydrochemical evolution. Such an approach will enable a clear distinction to be made of the salinization process (or processes), for cases where freshwater salinity may be caused by direct seawater intrusion, leaching of salt formations, mineral dissolution, or salt accumulation due to evaporation, as often encountered in irrigated areas (Edmunds and Droubi 1998). During the processes of salt formations or mineral dissolution leaching, the stable isotope content of the water is not affected while the salinity of water increases (Fig. 1). This is a unique feature which will enable identification of such processes based on isotopic and geochemical data.

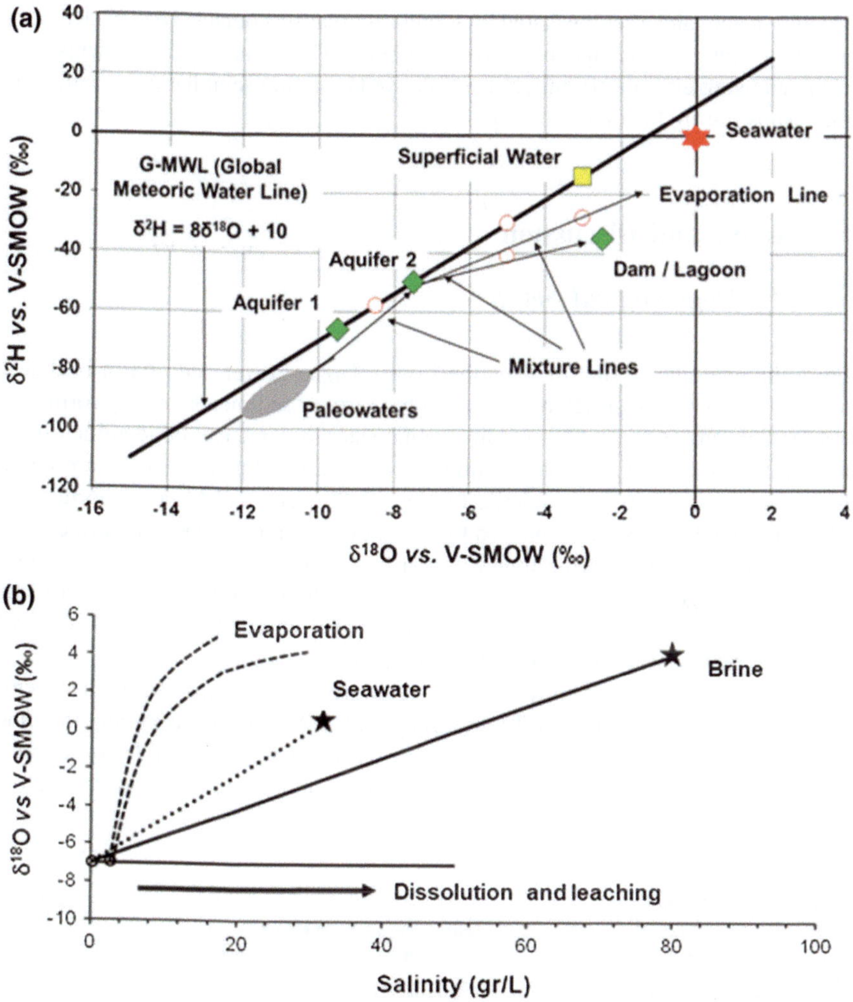

Fig. 1 **a** δ^2H versus δ^{18}O: change in isotopic composition of the groundwater associated with different processes. The symbols Aquifer 1, Aquifer 2, Dam/Lagoon and superficial water stand for hypothetic mean isotopic composition of different water systems. **b** δ^{18}O versus salinity: change in isotopic composition of water, ascribed to different salinization processes

2 Analytical Procedures

At Lower Tagus-lower Sado basin a total of 45 water samples were collected. Stable isotopes (δ^2H, δ^{13}C and δ^{18}O) were measured using a SIRA 10 mass spectrometer, the results are in delta notation (δ), the ^{18}O and ^{2}H are reported to V-SMOW, while ^{13}C reported to V-PDB. Carbon-14 determinations were carried out in the precipitate barium carbonate of the total dissolved inorganic carbon

(TDIC) in situ, and after measured by liquid scintillation counter. The ^{14}C activity is given in pmC (percentage of modern carbon). The tritium measurements were obtained using electrolytic enrichment followed by liquid scintillation counting and the results are in TU (Tritium Units).

3 Results and Discussion

3.1 Hydrogeological Setup

From the geological point of view the Lower Tagus-Lower Sado basin is characterized by a synclinal structure composed by Tertiary sediments, mainly formed by marine deposits (Fig. 2). Three main groundwater systems can be identified in the region: (i) a shallow Quaternary aquifer constituted by alluvial deposits presenting high transmissivity values, underlie by the (ii) Pliocene and Miocene formations. The Miocene deposits are composed by sandstones and limestones of marine origin related with different marine transgression and regression events. These deposits show an average thickness around 200–300 m, although in the central part of the basin these values increase up to 800 m. Finally, (iii) fluvial terraces made of sands and clays represent the Pliocene layers (Fig. 3).

Geophysical studies performed in the region, reveal two important fault systems. The first located in the Lower Tagus valley with a N30ºE direction, and the other the so-called Setúbal-Pinhal Novo fault runs N-S and are responsible for a graben structure which allows the rising of a brine formation or ancient seawater trapped in the sediments (Fig. 3).

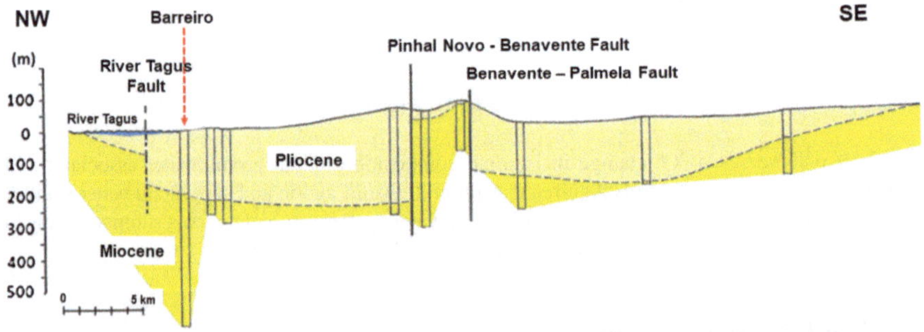

Fig. 2 Schematic cross section of the Lower Tagus-Lower Sado basin (adapted from Simões 1998)

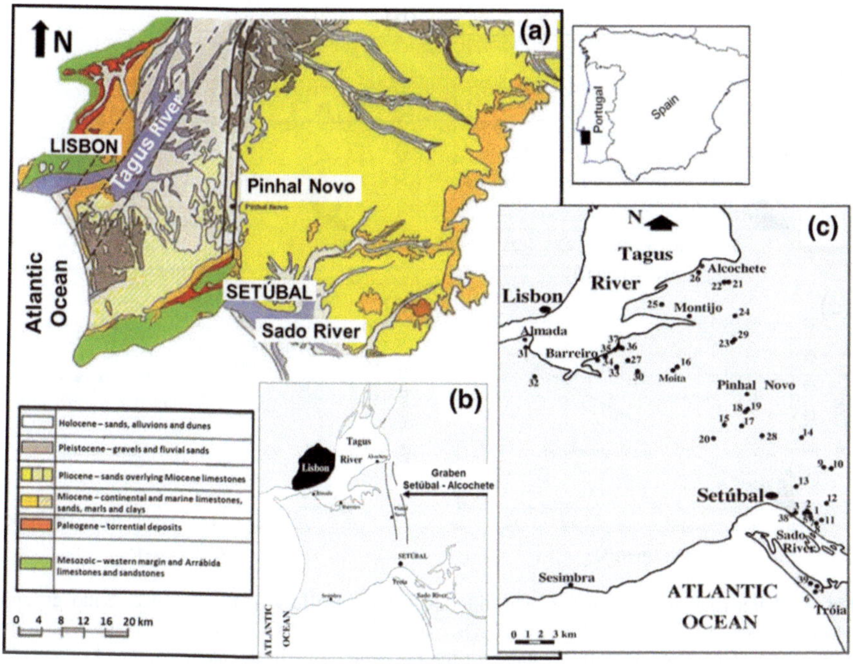

Fig. 3 Sketch of the Lower Tagus-Lower Sado basin. **a** Setubal-Alcochete graben; **b** schematic map of the region; **c** location of the sampling sites (adapted from Simões 1998)

3.2 Hydrogeochemistry

The hydrochemical evolution, of groundwaters ascribed to the different aquifer systems, is characterized by a progressive increase in the TDS varying from 80 mg/L up to 2565 mg/L in the Pliocene aquifer while in the Miocene the mineralization ranges between 200 mg/L and 7800 mg/L.

In the central part of the basin, groundwaters are used for human and agriculture supply, and an increase of the water salinization was identified. Two different evolution trends were recognized in the ratio Ca^{2+}/Na^{+}, both apart from the seawater—fresh water mixing line, reflecting flow paths with different Ca^{2+} content (carbonate minerals). Besides the trend observed can be due also to different ionic exchange proportions as the ion exchange mechanism is able to influence the ion concentration of the groundwater, and since the adsorption by the aquifer matrix of Na^{+} with release of Ca^{2+}, is a process which is triggered by intruded seawater (Fig. 4).

The magnitude of ion exchange processes in groundwater chemical composition can be assessed by the equation presented by Pennisi et al. (2006). Those authors combine the Ca^{2+} and Na^{+} with the conservative ion Cl^{-} content of the groundwater samples. Through this approach, the difference between the analytical data and the

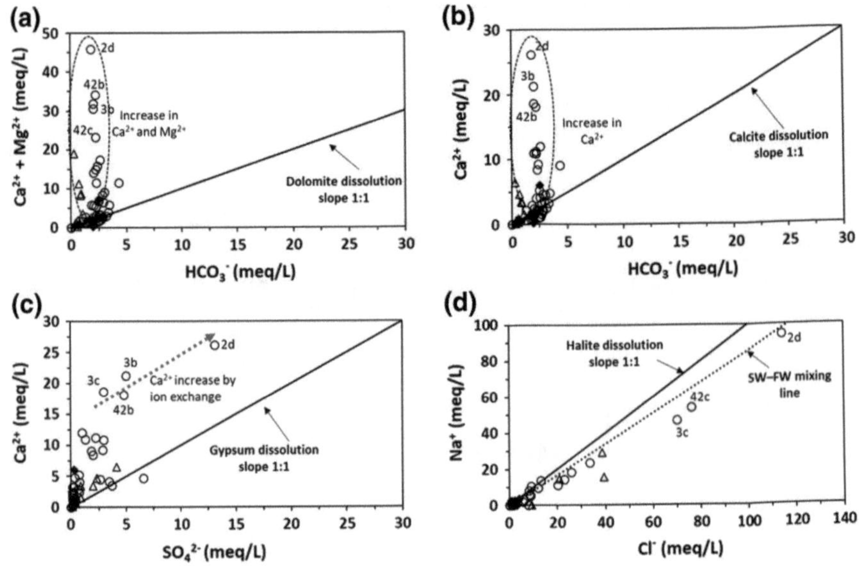

Fig. 4 Water-rock interaction between different solutes for Lower Tagus-Lower Sado groundwater samples: **a** $Ca^{2+} + Mg^{2+}$ versus HCO_3^-; **b** Ca^{2+} versus HCO_3^-; **c** Ca^{2+} versus SO_4^{2-} and **d** Na^+ versus Cl^-. The (Δ) symbols stand for the Pliocene groundwater samples, the (♦) stands for the Mio-Pliocene groundwater samples and (O) for the Miocene groundwater samples from Lower Tagus-Lower Sado basin

expected ion content in the groundwater is calculated if a seawater intrusion mechanism is assumed to be present in the groundwater system (Eq. 1).

$$\begin{aligned}\Delta nX &= nX(m) - nX(c) \\ &= nX - [nX(0) + (nCl(m) - nCl(0))(nX(sea)/nCl(sea)]\end{aligned} \qquad (1)$$

The $n_{(m)}$ stands for the measured concentration and $n_{(c)}$ for the expected content by mixing with seawater for ion X, based on the conservative ion Cl^-, where $n_{X(0)}$ and $n_{Cl(0)}$ stand for the concentrations of ions X and Cl^- in groundwater not affected by salinization. In the Sado basin, these values represent the mean content of the groundwater samples with an electrical conductivity above 300 μS/cm. The research area is part of a sedimentary basin where cation exchange process can easily occur with the uptake of sodium dissolved in the groundwater and release of calcium, by the aquifer matrix. Most of groundwater samples present Na^+ deficit ($\Delta n_{Na} < 0$) and an excess of Ca^{2+} ($\Delta n_{Ca} > 0$), supporting the hypothesis that the missing in sodium and increase of calcium is ascribed to cation exchange mechanisms.

3.3 Isotopic Signatures

For most of the groundwater samples representing the Miocene aquifer no tritium was found, only in three water samples an average content around 1.2 ± 0.5 TU was determined. For all the remaining samples (26) the tritium content was zero TU. Moreover, in the shallow aquifer (Pliocene aquifer) the tritium content varies between 1 and 3 TU.

Radiocarbon was determined in 13 boreholes on the Total Dissolved Inorganic Carbon (TDIC). The ^{14}C content varies from 71.9 ± 0.7 pmC to 88 ± 0.8 pmC in the TDIC of the Pliocene groundwater samples, and from 2.9 ± 0.3 pmC up to 45.6 ± 0.9 pmC in the Miocene groundwater samples.

A common and easy way is to estimate the initial ^{14}C activity, relating the $\delta^{13}C$ content of the TDIC in the groundwater to the mixed carbon from carbonate rocks, with carbon from soil CO_2 and the fractionation factor between the different carbonate phases function of temperature (Salem et al. 1980; Gonfiantini and Zuppi 2003). This simple model was chosen since, in all analysed water samples the Saturation Index (SI) for calcite is lower than −1.69, and in parallel, the $\delta^{13}C$ measured in the TDIC of the groundwater samples is around −10‰.

The estimated apparent carbon-14 ages varies between modern (borehole 5 and 27) in the Pliocene aquifer system, up to 27.9 ± 3.4 ka Before Present (BP) in the Miocene aquifer system (borehole 3). Looking to the apparent groundwater ages, estimated for the deepest aquifer, all the values (with the exception of borehole 31) present ages higher than 15.9 ± 3.2 ka BP, with an average age around 20 ka BP, indicating the presence of paleowaters.

A dispersion groundwater samples with the increase of salinization and apparent groundwater age can be observed pointing out to brine dissolution mechanism in the southern part of the basin (boreholes AC1, JK1 and 38), as the mechanism responsible for the deterioration of the water quality. However, in the northern part of the basin modern seawater intrusion seems to play the major role in the groundwater mineralization in the shallow aquifer (Fig. 5), since the increase of salinization is not followed by a groundwater ageing increase.

In order to corroborate the origin of the groundwater mineralization in the Lower Sado-Lower Tagus basin, the isotopic composition ($\delta^{18}O$) was plotted as a function of the Cl^- content (Fig. 6a). Borehole 2 is plotted on the seawater—fresh water mixing line, followed by the JK1 from the depth aquifer and from Pliocene samples 5 and 27, only the sample AC1 is displaced from this trend. Looking to this distribution pattern the oxygen-18 and deuterium content and chemistry are not sufficient to clarify the groundwater mineralization origin.

Nevertheless, combining all the available information (isotopes and chemistry), is possible to establish a different approach to explain the salts origin: mixing with modern seawater should lead to a ^{14}C activity increase. However, in Sado-Tagus basin this relationship (salinization and age) is not observed. The increase of salts in the groundwater system should be ascribed to ancient seawater trapped in the sediments and coeval of the sedimentary basin formation, allowing distinguishing

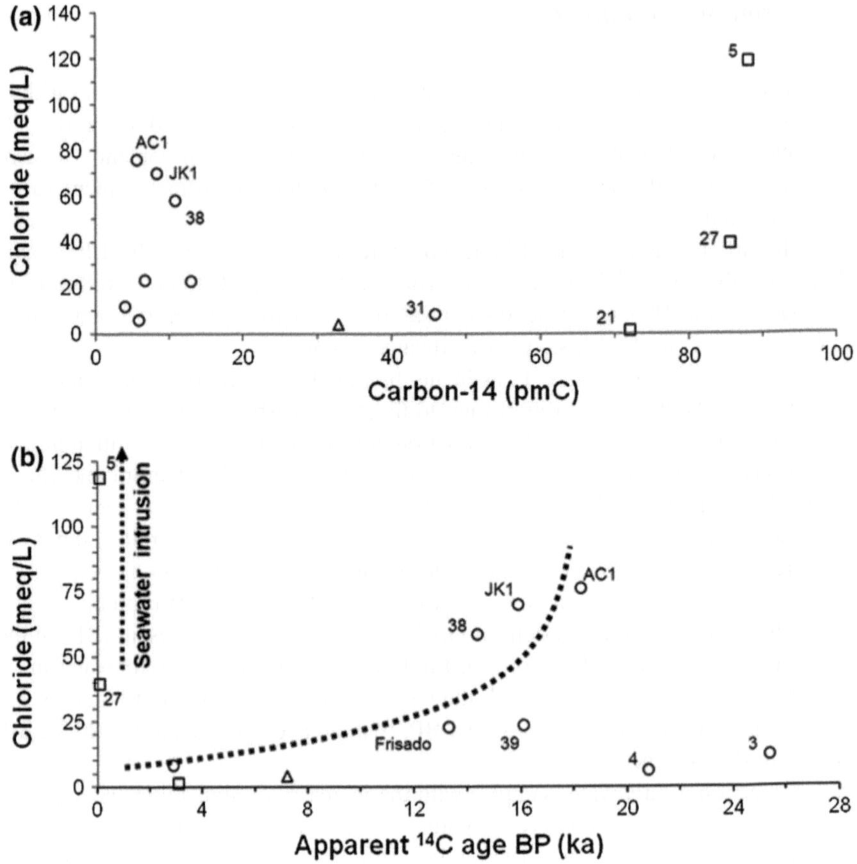

Fig. 5 Results from Lower Tagus-Lower Sado basin: **a** Cl^- versus carbon-14 content; **b** Cl^- versus apparent carbon-14 groundwater age. The (□) symbols stand for the Pliocene groundwater samples, the (Δ) stands for the Mio-Pliocene groundwater samples and (O) for the Miocene groundwater samples from Lower Tagus-Lower Sado basin

paleo-marine and modern seawater intrusion by combining radiocarbon and hydrogeochemical data.

A small enrichment in the heavy isotopes is observed within the groundwater samples, of about 2‰ in deuterium and 0.2‰ in oxygen-18 in Miocene groundwater samples, indicating that the recharge have occurred under different climatic conditions (Fig. 6b).

Similar isotopic enrichment has been also observed in another coastal aquifer in Portugal (Aveiro Cretaceous Aquifer (Carreira et al. 1996) and Morocco (Edmunds 2005). The authors indicate a possible explanation to the heavy isotope enrichment related to the isotopic enrichment of the global ocean during glacial time, due to the preferential storage of isotopically depleted water in the polar ice caps. Aveiro being a coastal aquifer, like the Lower Tagus-Lower Sado, is recharged mainly by

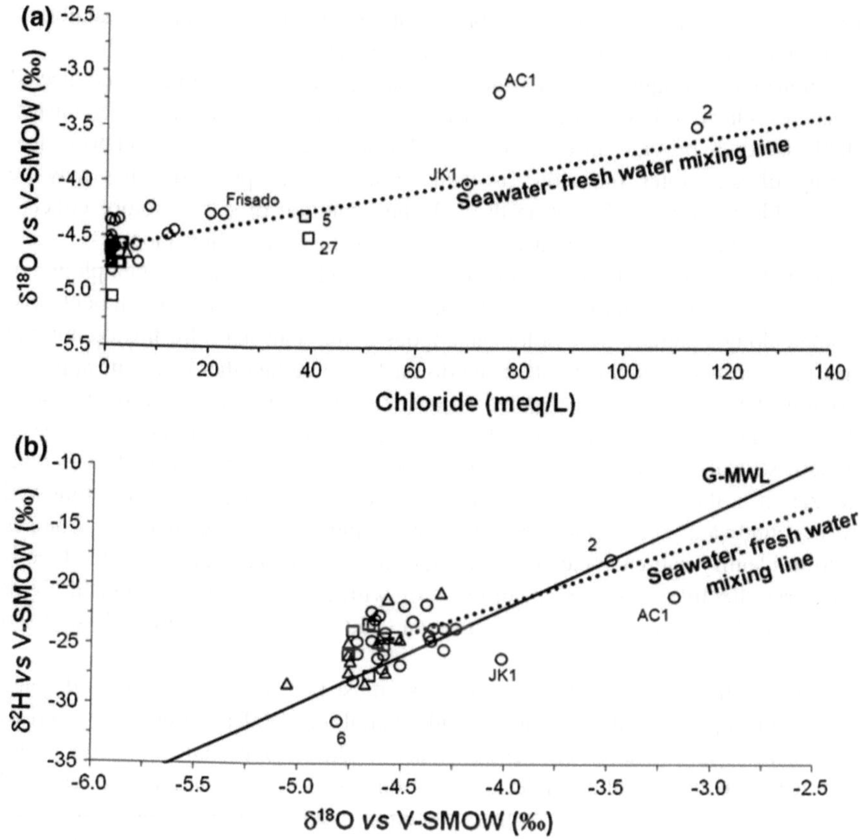

Fig. 6 Results from Lower Tagus-Lower Sado basin: **a** Cl versus $\delta^{18}O$ content; **b** $\delta^{18}O$ versus $\delta^{2}H$, seawater—freshwater mixing line is plotted, G-MWL stands for the Global Meteoric Water Line. Symbols as in Fig. 5

precipitation originating from the first step condensate, this newly formed groundwater may reflect, in the first instance the fluctuations of the isotopic composition of the ocean in the source regions.

4 Final Remarks

Groundwater resources mineralization can be ascribed to different sources such as: agriculture activities; seawater intrusion (active or ancient); dissolution of brines at depth; dissolution of evaporitic minerals dispersed in the geological matrix of the aquifer or even concentrated in saline domes, for example. Under a climate change perspective, groundwater resources of good quality must be considered as strategic resources. A different climatic scenario, under higher mean annual temperatures, associated with a decrease in the precipitation amount, will lead to additional environmental arid conditions in the Mediterranean region.

Salinization of aquifers in coastal areas can be the result of different connected processes related to both seawater intrusion and water-rock interaction mechanisms. Stable isotopes provide an effective label for sea water and freshwater to enable tracing of sea water intrusion, as well as identifying processes that may be responsible for water salinization through the evolution of stable isotope concentration of water during different processes related to water salinization.

In the Lower Tagus-Lower Sado basin the groundwater samples when plotted in a Cl^- versus Ca^{2+} logarithmic diagram, are displaying a Ca^{2+} enrichment, distributed along a pattern more or less analogous to the seawater—fresh water mixing line, most probably ascribed with brine dissolution. On the other hand, the apparent groundwater ages estimated for the Miocene aquifer present an average around 20 ka BP indicating the presence of paleowaters. In the northern part of the basin modern seawater intrusion seems to play the major role in the groundwater mineralization in the shallow aquifer, since the increase of salinization is not followed by a groundwater ageing increase. The use of single tracers (chemical or isotopes data) to identify and characterize salts origin in groundwater systems used alone is much less effective than, used in combination with each other. The combined use of geochemical and isotopic tracers (stable and radioactive environmental isotopes) has proved to be highly effective.

Climate changes, on which seawater rise must be included, add an additional demanding topic in water resources, with spatial-temporal uncertainty as well as vulnerability assessment. The assessment of the evolutional trend on water resources near the coastline under different climate scenarios and spatial-temporal uncertainties is essential for decision making, such as preservation and good management. Demand for good water quality will by consequence, rapidly increase, leading to salinization processes.

Acknowledgements The C^2TN/IST author gratefully acknowledge the FCT support through the UID/Multi/04349/2013 Project, and CERENA/IST author acknowledges the FCT support through the UID/ECI/04028/2013 Project.

References

Astier JL (1979) Etude des resources em eaux souterraines de la péninsule de Setúbal—Portugal. Geophysique. Nouveaux résultats de la prospection électrique. UNDP—Unesco, Dir. Geral Serv. Hidraulicos

Bouchaou L, Michelot JL, Vengosh A, Hsissou Y, Qurtobi M, Gaye CB, Bullen TD, Zuppi GM (2008) Application of multiple isotopic and geochemical tracers for investigation of recharge, salinization, and residence time of water in the Souss-Massa aquifer, southwest of Morocco. J Hydrol 352:267–287

Carreira PM, Soares AMM, Silva MAM, Araguás-Araguás L, Rozanski K (1996) Application of environmental isotope methods in assessing groundwater dynamics of an intensively exploited

coastal aquifer in Portugal. In: Isotopes in water resources management, IAEA, Vienna, vol 2, pp 45–58
Carreira PM, Marques JM, Nunes D (2014) Source of groundwater salinity in coastline aquifers based on environmental isotopes (Portugal): natural vs. human interference. A review and reinterpretation. Appl Geoch 41:163–175
Edmunds WM (2005) Groundwater as an archive of climatic and environmental change. In: Aggarwal PK, Gat JR, Froehlich KFO (eds) Isotopes in the water cycle. Past, present and future of a developing science. Springer, The Netherlands, pp 341–352
Edmunds WM, Droubi A (1998) Groundwater salinity and environmental change. Isotope techniques in the study of environmental change. IAEA, Vienna, pp 503–518
Gonfiantini R, Zuppi GM (2003) Carbon isotopic exchange rate of DIC in karst groundwater. Chem Geol 197:319–336
Jalali M (2007) Salinization of groundwater in arid and semi-arid zones: an example from Tajarak, Western Iran. Environ Geol 52:1133–1149
Pennisi M, Bianchini G, Muti A, Kloppmann W, Gonfiantini R (2006) Behaviour of boron and strontium isotopes in groundwater–aquifer interactions in Cornia Plain (Tuscany, Italy). Appl Geochem 21:1169–1183
Re V, Sacchi E (2017) Tackling the salinity-pollution nexus in coastal aquifers from arid regions using nitrate and boron isotopes. Environ Sci Pollut Res 24:1–15
Salem O, Visser JM, Deay M, Gonfiantini R (1980) Groundwater flow patterns in western Lybian Arab Jamahitiya evaluated from isotope data. In: Arid zone hydrology: investigations with isotope techniques. IAEA. Vienna, pp 165–179
Simões MMM (1998) Contribuição para o conhecimento hidrogeológico do cenozoico na Bacia do Baixo Tejo (Contribution to the hydrogeological knowledge of the Cenozoic in the Baixo Tejo Basin). Ph.D. dissertation, Faculdade de Ciências e Tecnologia (FCT), Universidade Nova de Lisboa (in portuguese)

Sowing Water in Monchique Mountain: A Multidisciplinary MAR Project for Climate Change Adaptation

R. C. Carvalho, T. Carvalho, F. R. Sousa and S. Gil

1 Introduction

Portugal is among the southern European countries that are most vulnerable to the impacts of climate change (IPCC 2014a, b, ENAAC 2015). The SIAM II Project (Santos and Miranda 2006) predicts, regarding water resources, a greater seasonal asymmetry of precipitation, with a marked decrease in summer and serious consequences in the water balance, namely recharge and surface runoff. This will hinder water management efforts, and it has become necessary to opt for a more holistic view, incorporating knowledge of the dynamics of economic, social and meteorological systems (LaMoreaux et al. 2009).

Induced (or artificial) recharging of aquifers (MAR—Managed Aquifer Recharge) is a water management tool increasingly studied and applied (e.g. Espin Pinar et al. 2010; Dillon et al. 2009; Escalante et al. 2015). In Portugal, although there were financial resources already allocated to these solutions (e.g. GABARDINE—see Lobo Ferreira et al. 2006; MARSOL 2014) and good scientific knowledge, its implementation remains in the scientific/research sphere.

At Monchique, a small town of 6000 inhabitants in southern Portugal, the population gets its water supply from groundwater, mostly by horizontal wells, galleries and springs. This system is very common in the Iberian mountain regions as it's very cost-effective due to low energy needs and residual costs for water

R. C. Carvalho (✉) · T. Carvalho · F. R. Sousa
TARH Lda, Rua Forte do Monte Cintra nº1 2°C, 2685-140 Sacavém, Portugal
e-mail: rita.carvalho@tarh.pt

S. Gil
Câmara Municipal de Monchique, Travessa da Portela n.º2, 8550-470 Monchique, Portugal

M. L. Calvache et al. (eds.), *Groundwater and Global Change in the Western Mediterranean Area*, Environmental Earth Sciences,
https://doi.org/10.1007/978-3-319-69356-9_9

treatment although it is extremely vulnerable to meteorological variability. The Municipality of Monchique (CMM) is faced with surpluses during periods of high rainfall and lack of water during the dry season (between June and October), due to higher demand and lower flow availability. It is anticipated that this situation will be further aggravated by climate change. SOWAMO's main aim was to verify the benefits that the implementation of MAR structures contributes to water resources (namely the security of supply to the town's water needs), biodiversity and forests, as well as its economic viability for replication in other localities using similar water supply systems. It is, therefore, a multidisciplinary approach. The impacts on biology and ecosystems can be positive if some management is carried out.

2 Study Area

The project was implemented at Serra de Monchique (Algarve—Southern Portugal) on a hillside area (slope of 1:5), part of National Ecological Reserve and Natura 2000 zones. Figure 1 is an overview of the study area's location, and includes one of the town's water extraction structures for public supply (named Penedo do Buraco), MAR structures and piezometers built under SOWAMO, the numerical model boundary conditions and the lithology of the area.

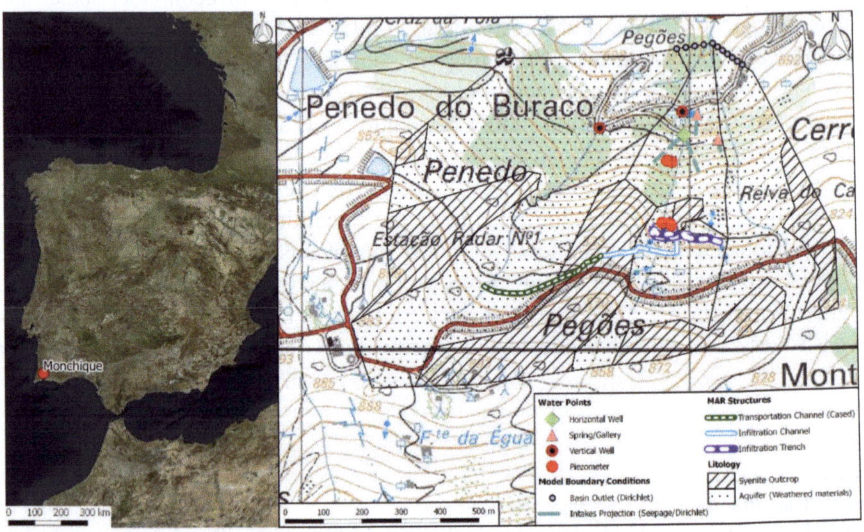

Fig. 1 Location of project SOWAMO and overview of its water management structures, over a cartographic layer of IGEO (1979)

2.1 *Geology*

The area is part of Monchique intrusive, Alkaline Igneous Massif, located in the southwestern extreme of the South Portuguese Zone, paleogeographical unit of the Variscan substrate of Mainland Portugal. This massif has an elliptical shape occupying an extension of about 80 km^2. The enclosing structure is composed of turbidite sequences of the Baixo Alentejo Flysh.

The Monchique Massif consists of two dominant units, the nuclear nepheline syenite and the heterogeneous border syenite. The MAR structures are located on the heterogeneous border syenite. This unit of syenitic nature presents variations both in the percentage of mafic and nepheline minerals and in the texture including variations in grain size and occurrences of foliar syenite. This external unit also has several associations of igneous bodies such as aplito-pegmatitic, dykes of various compositions and enclaves of hornefels (González-Clavijo and Valadares 2003).

2.2 *Hydrogeology*

2.2.1 Climate

Because of its orographic nature (902 m in its highest point), Monchique Mountain presents a milder and more humid climate regime than the rest of Southern Portugal. The annual mean precipitation is 857 mm, concentrated between October and February, December being the wettest month (190 mm) (SNIRH 2017).

2.2.2 Water Balance

The hydrographic network in Monchique is quite broad but most of its small streams are torrential. Other surface water manifestations are almost absent.

Esteves Costa et al. (1985) calculated an infiltration rate of 5%. Considering the mean annual precipitation of 857 mm, the mean annual infiltration in 2014 was 43 mm.

The monitoring data of the horizontal wells operated by the Municipality of Monchique, showed flows of 46,400 m^3/year (2014), for contribution areas of 22 ha, which implies, locally, much higher rates of infiltration than those defined by Esteves Costa et al. (1985). A new water balance for the local (and temporal) scale of the SOWAMO project was determined during the project (Table 1).

Table 1 Water balance for Monchique during SOWAMO's implementation years

Water balance	(mm)	Total (%)
Mean annual precipitation	857	100
Mean evapotranspiration	476	55
Surface runoff	227	27
Mean annual infiltration	154	18

2.2.3 Hydrogeological Units

The mountain of Monchique is constituted by syenitic rocks with predominant fissural permeability with high hydrogeological heterogeneity. The aquifers existing in the Monchique Igneous Massif occur essentially in dependence on the weathered and fissured zones of the massif or rocks with very low permeability (eruptive breccias) (Esteves Costa et al. 1985). There are no regional aquifers but rather dispersed, discontinuous, sometimes weathered and fractured hydrogeological units able to supply local needs through extraction structures (TARH 2014).

2.2.4 Existing Groundwater Extraction Structures

At Penedo do Buraco (see Fig. 1) there are two horizontal wells of about 100 m long, one vertical well with 100 m depth and two excavated galleries). Here, the main aquifer is installed in decomposed and weathered syenite. The contribution area of the existing wells and galleries is 22 ha.

These water extraction structures drain yearly an average of 1.5 L/s, with a minimum of 0.5 L/s (October) and a maximum of 3.5 L/s (April and May). The use of the vertical well depends on the precipitation regime.

2.2.5 MAR Structures

The main guidelines for the developed MAR system involved the maintenance of a fully gravity dependent operation and the easy upkeep of the structures. The built recharge structures included:

- A 330 m transport channel with a 2% gradient, that transfers water from another hydrographic basin to the recharge area;
- A 250 m infiltration channel, with three energy dissipation basins, to insure minimal clogging;
- 320 m of interconnected infiltration trenches located in the terraces, for improved infiltration rates.

3 Methods

After a geological and hydrogeotechnical survey (visualization of slopes, outcrops and excavations), eight surface infiltration tests with single-ring infiltrometers were performed in order to estimate infiltration potential. The drilling of five piezometers (three shallow, to monitor the upper geological system and two deep, for monitorization on the lower level subsystem) took place, and those structures were monitored for conductivity, temperature and water level. Falling head tests under

ISO 22282-2:2012 specifications were performed to access aquifer properties. These results were used for the development of a hydrogeological conceptual model, that was updated with the monitoring information from the piezometers and the municipal groundwater intakes.

In a later phase, this conceptual model allowed a numerical analysis using the commercial software FEFLOW 7.0 (Diersch 2014) and consisted of a finite element mesh for groundwater flow simulation, with exploration with steady and transitory state conditions. The modeling domain corresponded only to the upper subaquifer constituted by soil and decomposed syenites mainly with interstitial circulation.

Calibration was carried out using the monitoring results before induced recharge conditions, but model was explored by assessing the differences between several precipitation scenarios (some corresponding to monitoring periods, other to predictions based on climate change scenarios (see Santos and Miranda 2006)) with and without MAR. The water availability for MAR is, naturally, dependent on precipitation and calculated for each scenario using the water balance.

4 Results and Discussion

4.1 Surface Infiltration Tests with Single-Ring Infiltrometer

Eight infiltration tests were performed, six over soil located in terraces and two near outcrops (weathered and sound rock). The results show that the infiltration rates in the soil (median 4.32 m/day, values ranging between 2.88 and 15.84 m/day) are significantly different from the results obtained near the outcrops (median 1.08 m/day, values ranging between 0.72 and 1.3 m/day).

4.2 Falling Head Tests

For the piezometers drilled in the upper subsystem, a median permeability of 0.28 m/day (values between 0.24 and 0.44 m/day) was calculated, in contrast with the permeability obtained for the piezometers drilled in the lower subsystem (a median of 0.024 m/day, values between 0.022 and 0.027 m/day). The results show that the permeability in the upper subsystem is two orders of magnitude higher than that of the lower subsystem. During drilling, it was observed that water circulates in the transition zone between upper and lower subsystems, demonstrating that lower subsystem works effectively as a barrier or as an aquitard in most of the study area.

4.3 Hydrogeological Monitoring

There is a remarkable correlation between the levels of the SOW1 piezometer, drilled and screened in the shallow subsystem, very close to the horizontal municipal wells, with their total intake flow: a flow delay of one to three months was observed in response to the precipitation (Fig. 2).

This is visible at peak flows between February and May 2016 and February and April 2017. Nonetheless, the increments observed in December 2015 and 2016 only occur after two months of high precipitation. This is because most of the recharge takes place in areas away from abstractions. These results were encouraging for the viability of the induced recharge project.

4.4 Hydrogeological Conceptual Model

The aquifer system is characterized by two overlapping and interconnected subsystems, characteristic of crystalline rocks in temperate and Mediterranean environments (Fig. 3; e.g. Lloyd 1999; Carvalho 2006).

Till depths of about 12–18 m, an unconfined (phreatic) upper subsystem is installed in soil and decomposed syenite with high permeability (≈0.2 m/day). This permeability and the hydraulic gradient of around 0.38 determines low residence times.

Bellow, there is a hard rock subsystem with very low, fissured type permeability, decreasing in depth (≈0004 m/day) till the investigated depth of 85 m. In this lower subsystem, the residence time is presumably higher; the recharge is mostly indirect and comes, by leakage, from the upper subsystem, though it can be direct in the outcrops, with hydraulic gradients in the order of 0.1.

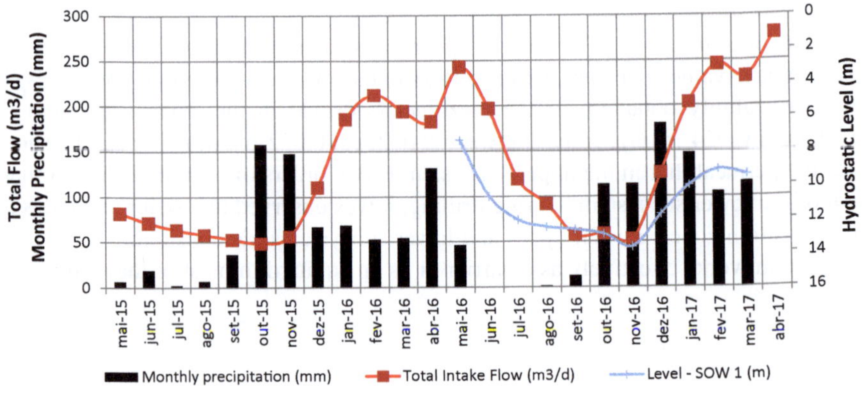

Fig. 2 Relation between monthly precipitation, intake flow and water level at Penedo do Buraco

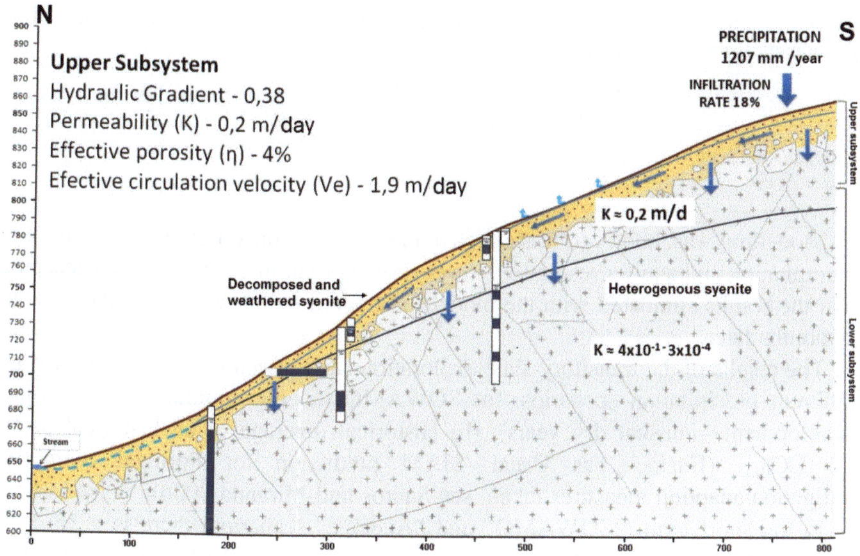

Fig. 3 Simplified conceptual hydrogeological model of the SOWAMO site

4.5 *Numerical Model*

The modeled area includes the recharge basin where the groundwater abstractions of CMM are located, as well as the adjacent two catchments (total area of 81.6 ha), on suspicion that the area of contribution of the gravity dependent systems exceeds the limit of the river basin where they are located.

In terms of geometry, the model was divided into three distinct areas with different characteristics of recharge and transmissivity:

(1) Outcrops (30% of the area), areas with lower recharge rates and low transmissivity. They act as barriers that conduct the groundwater flow preferably through the river basins. Recharge: 2%
(2) The upstream part of the aquifer, which presents higher than normal transmissivity, and higher recharge rates: 16%
(3) The downstream part of the aquifer where the municipal groundwater abstractions are located, with a transmissivity lower than that observed upstream. Recharge: 9%

The water exits were defined by the imposition of Dirichlet boundary conditions in the lowest zone of the aquifer. The water withdrawal in the horizontal abstractions was represented using a "seepage" level boundary condition, which forces the aquifer to reach a piezometric level equivalent to the capture level at the location of the water through the drainage of the system. The water exits in the galleries and horizontal wells was considered the main factor for the model calibration.

Table 2 Variation of extracted water volumes, in different precipitation scenarios simulated by the numerical model, in the dry season

	Sowamo		Historical		CC_{BaU}	
	Δ m^3	%	Δ m^3	%	Δ m^3	%
Jun/Jul/Aug/Sept	4755	35	3831	23	5046	31

Modelling was, however, less effective in simulating the observed delay between precipitation and aquifer response. However, in a preliminary stage, it was accepted that the results simulated with the model for a given month were correspondent to 2 months later.

The results of the transitory-state exploration can be seen in Table 2, divided by different precipitation scenarios: (a) SOWAMO (average rainfall during project development—unusual dry years); (b) Historical (average of all historical data); (c) CC_{BaU} (Business as Usual—SIAM prediction for Portugal, with no mitigation/adaption measures taken, see Santos and Miranda 2006).

In the short term ("SOWAMO" and "Average" scenarios) there is a 23–35% increase in the dryness flows, translating to an increase of 0.3 L/s. For a scenario of Climate Change with no further mitigation or adaptation strategies (BaU), there is an increase of 31% in the exploited water volume.

5 Conclusions

Preliminary results showed that the developed MAR structures are more effective during dry years, although water availability for MAR is lower. However, the effective recharge is more likely to be recovered during these years.

During the months with water scarcity (June–September) there will be an effective increase on the water availability in the horizontal wells and galleries, of up to 35%. Due to the nature of the recharge structures, the positive impacts are extremely dependent on the precipitation distribution. However, the monitoring network (piezometers and CMM water wells) is still active and data will be collected for further recalibration and validation of the numerical model.

These results suggest that similar simple MAR solutions, especially when complemented with other water management methods (e.g. water efficiency and conservation), can be an important tool to adapt groundwater supply systems to Climate Change in the Mediterranean area.

Acknowledgements The research leading to these results has received funding from the EEAGrants and Fundo Português do Carbono, through AdaPT program. The authors thank Agência Portuguesa do Ambiente (APA) and all institutions and individuals that supported, participated and contributed to this project.

References

Carvalho JM (2006) Prospecção e pesquisa de recursos hídricos subterrâneos no Maciço Antigo Português: linhas metodológicas. Doctoral thesis University of Aveiro

Diersch HJG (2014) FEFLOW: finite element modeling of flow. Mass Heat Trans Porous Fract Media. doi:10.1007/978-3-642-38739-5

Dillon P, Pavelic P, Page D, Beringen H, Ward J (2009) Managed aquifer recharge: an introduction. Waterlines Report 13

ENAAC (2015) Estratégia Nacional de Adaptação às Alterações Climáticas (ENAAC 2020)

Escalante E, Gil R, Herrarte F, Sauto J, Campos E, Carvalho T (2015) MAR technical solutions: review and bata base. MARSOL demonstrating managed aquifer recharge as a solution to water scarcity and drought. Deliverable 13.1

Espin Pinar R, Ortiz Moreno E, Guzmán Álvarez JR (2010) Manual del acequiero: Parques Nacional y Natural de Sierra Nevada. Sevilla, Agencia Andaluza del Agua

Esteves Costa F, Amaral Brites J, Pedrosa MY, Vieira da Silva A (1985) Carta Hidrogeológica da Orla Algarvia, Escala 1/100000, Noticia Explicativa. Serviços Geologicos de Portugal, Lisboa

González-Clavijo EJ, Valadares V (2003) A estrutura do complexo e Monchique. Paper presented at VI Congresso Nacional de Geologia, Nova University of Lisbon, Monte de Caparica, 4–6 June 2003

IGEO (1979) Carta Militar de Portugal—Folha n.º 585 (1:25 000), Lisboa

IPCC (2014a) Climate change 2014: synthesis report. In: Core Writing Team, Pachauri RK, Meyer LA (eds) Contribution of working Groups I, II and III to the fifth assessment report of the intergovernmental panel on climate change. IPCC, Geneva, Switzerland

IPCC (2014b) Climate change 2014: impacts, adaptation, and vulnerability. Part A: global and sectoral aspects. In: Field CB, Barros VR, Dokken DJ, Mach KJ, Mastrandrea MD, Bilir TE, Chatterjee M, Ebi KL, Estrada YO, Genova RC, Girma B, Kissel ES, Levy AN, MacCracken S, Mastrandrea PR, White LL (eds) Contribution of working Group II to the fifth assessment report of the intergovernmental panel on climate change. Cambridge University Press, Cambridge, United Kingdom and New York, USA

ISO/IEC (2012) ISO 22282-2:2012(E) Geotechnical investigation and testing. Geohydraulic testing. Part 2: water permeability tests in a borehole using open system

LaMoreaux P, Soliman M, Memon B, LaMoreaux J, Assaad F (2009) Environmental hydrogeology, 2nd edn. IWA Publishing

Lloyd JW (1999) Water resources of hard-rock aquifers in arid and semi-arid zones. Studies Reports Hydrol Unesco Paris 58:1–284

Lobo Ferreira JP, Diamantino C, Moinante M, Oliveira M, Leitão T, Henriques MJ, Medeiros A (2006) GABARDINE: groundwater artificial recharge based on alternative sources of water: advanced integrated technologies and management. DELIVERABLE 51—test sites and their characteristics

MARSOL (2014) Managed aquifer recharge solutions. http://www.marsol.eu. Assessed 30 Jun 2017

Santos FD, Miranda P (2006) Alterações Climáticas em Portugal – Cenários, Impactos e Medidas de Adaptação – Projecto SIAM II. Gravida, Lisboa, p 505

SNIRH (2017) Sistema Nacional de Informação de Recursos Hídricos. http://snirh.apambiente.pt Assessed 22 Jun 2014

TARH (2014) Estudo Hidrogeológico para delimitação dos perímetros de protecção das captações de abastecimento público do Município de Monchique

Concepts on Groundwater Resources

E. Custodio

1 Introduction

Scientific and technical papers and reports on groundwater resources use terms that often are poorly defined. This makes difficult comparing results and knowing what the meaning of the evaluations and calculations is. This leads to erroneous balances and uncertainty in decision making under water stressed conditions. Comparisons are still more difficult when considering surface water resources and groundwater resources and also when other water sources are involved.

Common terms to be considered are those referring to water use, groundwater resources, pathways for making water available, exploitation, quality and origin of water. Also climate related concepts of water resources may be affected by language uncertainties.

In what follows, the terms used for concepts are briefly considered after grouping them. This is done in the framework of water plans resulting from the current Spanish Water Act (TRLA 2001, 2003) and the derived legal norms: the Water Planning Regulations (RPH 2007) and the subsequent Instructions for Water Planning (IPH 2008). The incorporation into the Spanish water legislation on the European Water Framework Directive (DMA 2000) and the Daughter Groundwater Directive (2006) has improved the definition of some concepts but others remain poorly bounded.

It is important to distinguish between extensive and intensive magnitudes in water planning and measure them in the correct units. Extensive magnitudes and variables are commonly given in hm^3/year, and sometimes in m^3/day and L/s. They are for a given area, generally a large one, and may refer to a downstream point or a

E. Custodio (✉)
Royal Academy of Sciences of Spain. Emeritus Prof. at the Civil and Environmental Engineering Department, Technical University of Catalonia (UPC), C/. Jordi Girona 1–3, 08034 Barcelona, Spain
e-mail: Emilio.custodio@upc.edu

M. L. Calvache et al. (eds.), *Groundwater and Global Change in the Western Mediterranean Area*, Environmental Earth Sciences,
https://doi.org/10.1007/978-3-319-69356-9_10

down-flow section for a given river basin or aquifer or along some specified boundary. Intensive magnitudes and variables are the extensive ones relative to a given unit of other magnitude, such as per inhabitant or per unit surface area. In this last case, the resulting common units are m/year or mm/year, although in agriculture the m^3/ha/year is used, which is equivalent to 0.1 mm/year. These magnitudes are rates, as they refer to the unit of time; this is often not explicitly indicated and thus may produce some confusion. Reserve related terms correspond to volumes and are commonly given in hm^3 (10^6 m^3) and in the case of large figures in km^3 (10^9 m^3).

This paper contain mostly personal reflections directed to water plans and related documents in Spain, considering what is common language under the existing hydrological circumstances. So, personal references dominate. Different glossaries deal with hydrological terms, such as those produced by Aquastat (FAO), the US Department of Agriculture, the Dirección del Medio Ambiente (Spain) and the Water Observatory of the Botín Foundation, but often do not address the specificities of groundwater resources, and thus do not provide the required precision. A comparative discussion exceeds the purpose and possibilities of this paper.

The first time a term is presented it is highlighted in **bold**, followed by the translation into ***Castilian*** (Spanish) and *Catalan*, in italics, the first in bold and the second one in normal lettering. This is done as the paper is mostly directed to Spanish and Ibero-American readers.

2 Groundwater Use

Water use related terms refer to the quantification of water utilization. Water **use** (***uso***; *ús*) refers to the water that is devoted to a given activity or process. It has to be distinguished between (a) **total** or **gross** water **use** (***uso total*** o ***bruto***; *ús total* o *brut*), which is the water taken from a source, and includes water that is lost in the way (leakages, evaporation, derivations), and (b) **net** water **use** (***uso neto***; *net ús*), which is the water directly applied to the action or process; this is also called **applied** water (***aplicada***; *aplicada*).

Water **consumption** (***consumo***; *consum*) or **consumptive use** of water (***uso consuntivo;*** *ús consumptiu*) refers to the used water that does not return to the system, irrespective of quality and regime changes. This refers mostly to water that is evaporated, discharged into the sea or to a river below the uptake points, recharged in parts of the aquifer without later use and non-usable environmentally or for environmental purposes. This consumption may be the consequence of the action or process or of artificial disposal that avoids or precludes further use due to non-acceptable or non-treatable quality. This commonly refers to applied water, but may also to gross use; this should be specified. The difference between use and consumption is water that can be used again, although quality or salinity correction may be needed. Urban and industrial consumption is much less than use, but the two values approaches for efficient agricultural irrigation.

Water **demand** (***demanda***; *demanda*) is what a given user wants according to costs and circumstances. It equals or is greater than water use. It should be defined if it is at the source (gross) or at the application area or activity (net).

In Article 3 of the Spanish Water Planning Regulations (RPH 2007), water use is defined as the different classes of resource utilization, as well as any other activity that has a significant impact on water status (*las distintas clases de utilización del recurso, así como cualquier otra actividad que tenga repercusiones significativas en el estado de las aguas*) and demand as the water volume, in quantity and quality, that users are willing to acquire to satisfy a given production or consumption objective; this volume is a function of factors such as service prices, the income level, the kind of activity, technology, and others (*el volumen de agua, en cantidad y calidad, que los usuarios están dispuestos a adquirir para satisfacer un determinado objetivo de producción o consumo; este volumen será función de factores como el precio de los servicios, el nivel de renta, el tipo de actividad, la tecnología u otros).*

The values per person or unit surface, and per unit of time, are often called **dose** (***dotación***, *dotació*). It should be specified whether doses refer to gross or net water use.

Return flow (**agua de rechazo**; *aigua de rebuig*), also called **irrigation excess** (***retorno de riego****; retorn de rec*) in irrigated agriculture, refer to the non-consumed applied water which may be used again, part of which adds to aquifer recharge.

3 Groundwater Resources and Reserves

The concepts of reserve and resource are quite well defined in mining, for non-renewable minerals, including oil and gas. This is more difficult for groundwater as it is renewable in common situations, although slowly in many cases. Thus, reserve and resource concepts have a different meaning, which also do not fully coincide with surface water concepts.

A first concept to be clarified is **recharge** (***recarga****; recàrrega*). Recharge can be defined as the flow of water penetrating into an aquifer (saturated zone), coming from the surface. The penetration is produced through the water table at the unconfined part of aquifers. This water comes from diffuse infiltration of net atmospheric precipitation (rainfall and snowmelt) or by permanent, occasional or sporadic surface water infiltration, in rivers, wetlands, lakes, gullies and ravines. This water may enter directly or through the unsaturated zone. Diffuse recharge is mostly through the unsaturated zone. **Infiltration** (***infiltración***; *infiltració*) is a different concept as it refers to water available at the land surface that penetrates into the ground. Recharge is less than infiltration, often much less, as evaporation and transpiration by plants have to be discounted. Calling infiltration to recharge is an important source of confusion. The exchange of water among aquifers, both over imposed and side ones, should not be properly considered recharge. Also, sea water penetration into coastal aquifers is generally not considered recharge.

Groundwater **discharge** (***descarga***; *descàrrega*) refers to aquifer outflow to rivers as base flow, in springs, to wetlands and to the sea shore. Groundwater withdrawal through drains and tunnels (water galleries) and pumped from wells and boreholes is artificial discharge. Evapotranspiration does not affect the aquifer but soil water, except for shallow water tables and lagoons, in which case this is also considered a discharge from the aquifer.

Groundwater **total resources** (***recursos totales***, *recursos totals*) refers to total inflow into the aquifer or aquifer system, that is to say, recharge and water transferred from over imposed or side aquifers and aquitards. This needs a careful definition, as water inflow from surface water and from other aquifers is outflow from them and thus double accounting of water flow can be done if this is not dully taken into account. The decrease of aquifer water by evapotranspiration due to water table lowering adds to resources, as well as the decrease of reserves when this is part of the aquifer system management plan. **Available resources** (***recursos disponibles;*** *recursos disponibles*) are total resources decreased by flows to be preserved to maintain river and spring flow, lakes, wetlands, shallow water tables, and outflow to the sea to limit seawater intrusion and to keep salinity conditions in particular littoral environments.

Recharge, discharge and resources are not aquifer or aquifer system properties as they depend on exploitation conditions, so their definition should be accompanied by information on current aquifer status.

Groundwater **reserves** (***reservas***; *reserves*), also called **storage** (***almacenamiento***; *emmagazematge*), is the total quantity of water in the aquifer or aquifer system, including that in the aquitards. In thick aquifers, the maximum considered depth has to be specified. When saline water is present it should by specified if reserves refer to total water irrespective of the quality or only fresh water is taken into account. In coastal aquifers, the saline water body is generally included but it should be specified if the offshore part, if it exists, is included or not. Reserves depend on formation volume and total porosity. Water in the unsaturated zone is generally excluded, but can be singled out when this is relevant. Most of groundwater reserve cannot be abstracted as it is held by capillary forces as the water table decreases or is too slowly drained from aquitards or pores in blocks and small fissures in the case of hard rock formations. Groundwater that could be mobilized is that in easily drainable saturated formations, that is, in large connected pores and fissures. **Abstractable reserves** (***reservas captables;*** *reserves extraibles*) or **drainable reserves** (***reservas drenables***; *reserves drenables*) are the part of total reserves than could be abstracted down to a given depth and with a maximum salinity. They depend on specific yield or drainable porosity (also depending on drainage time) and also on exploitation conditions and water quality. This last also depends on the technical and economic possibility of treating abstracted groundwater when its quality impairs, even applying reverse osmosis or reversible electro-dialysis to brackish water.

Renewable groundwater (***agua subterránea renovable;*** *aigua subterrània renovable*) is the water in an aquifer system that has a residence time that is short according to a given time scale, which should be specified. Often it coincides with resources. Groundwater average **renewal time** or **turnover time** (***tiempo de revovación;*** *temps de renovació*) is the ratio of storage to recharge. It is closely related with average **residence time** (***tiempo de residencia;*** *temps de residència*) and **transit time** (**tiempo de tránsito**; *temps de trànsit*).

Groundwater that can be abstracted is water that can be taken for human uses or that sustains ecological functions and the associated services when it outflows or the water table is shallow. This can be called **mobile water** (***agua móvil***; *aigua mòbil*), which is also called **blue water** (**agua azul**; *aigua blava*). Water in the unsaturated soil that can be used by plants or **pedologic** or **soil water** (***agua edáfica***; *aigua edàfica*), which is also called **green water** (***agua verde;*** *aigua verda*), is not properly groundwater. The use of the terms blue water and green water, even if its use is widespread in non-specialized groups and for easier diffusion, do not really improve hydrological terms and in some cases they are less accurate than classical ones.

4 Groundwater Exploitation

Groundwater **exploitation**, also **abstraction**, **development** (***extracción***; *extracció*) is the fact of capturing groundwater and bringing it to the surface as drainage or to be used. It is called intensive **exploitation** (***explotación intensiva***; *explotació intensiva*) when the rate with respect recharge under current exploitation conditions produces significant changes in the aquifer system functioning, including the relationships with surface and seawater, and the associated piezometric and water table drawdown (Custodio 2012). The water table drawdown implies a reduction of drainable reserves.

Strict **groundwater mining** (***minería del agua subterranea;*** *mineria de l'aigua subterrània*) occurs when exploitation exceeds actual recharge and consequently there is a progressive decrease of reserves. The rate of **reserve depletion** (***consumo de reservas;*** *consum de reserves*) includes both the extraction associated drawdown during the long transient stage of large aquifers and groundwater mining (MASE 2015; Foster 1993; Foster and Loucks 2011; Konikow and Leake 2014). A practical approach considers the time, which is relevant for aquifer system management purposes. So, it is considered that there is **groundwater mining** when after ceasing abstraction the recovery time needs a long time to close to initial conditions, which should be specified (MASE 2015; Custodio et al. 2016a, b, 2017). Often this is one or two human generations. In coastal aquifers, the substitution of fresh water by saline water due to abstraction can be considered fresh water mining.

Seawater intrusion (***intrusion marina;*** *intrusion marina*) has two meanings (SASMIE 2017). One is the natural penetration of seawater in continental formations due the greater density of seawater. The other refers to the increased penetration derived from coastal aquifer exploitation and other artificial actions. This should be clarified.

5 Terms that Should Be Used Carefully or Discarded

A term used frequently in literature and reports is groundwater **overexploitation**, and also **overuse** and **overdevelopment** (***sobreexplotación;*** *sobreexplotació*). It is used to point out that there are negative results from groundwater exploitation, so finally any abstraction can be qualified as overexploitation. Positive results are not explicitly taken into account. So, it is currently a colloquial term without precise meaning and negatively tainted. Thus, its use should be discouraged and substituted by exploitation and intensive exploitation accompanied by the explicit presentation of the effects (Custodio 2002). In Spain "sobreexplotación" is a legal term used in the Water Act and the derived dispositions and norms (Molinero et al 2011; Custodio and Dolz 2016), but its meaning is poorly bounded and defined through the appreciated results of exploitation.

When referring to aquifer reserve depletion, the decrease is sometimes called use of **non-renewable water** (***agua no renovable;*** *aigua no renovable*). In most cases, these depleted reserves can be recovered after some time, as recharge exists, even if relatively small, although the recovery time may be often of decades. Only in current very arid environments or in well confined closed aquifers non-renewability is a fact.

Non-renewable groundwater is sometimes also called **fossil water** (**agua fósil**; *aigua fóssil*) as it is considered the result of recharge in geologically past times and since them isolated from the water cycle. Although this is a possible situation, detailed studies are needed to apply appropriately the term. In most cases these studies are no available. It is prudent to avoid this designation.

It is not rare to qualify saline groundwater giving to it different qualifications as **intruded** or **encroached seawater** (agua marina de intrusión; aigua marina d'intrussió), **old seawater** (***agua marina antigua;*** *aigua marina antiga*), **connate marine water** (***agua marina congénita;*** *aigua marina congènita*), **evaporite dissolution water** (***agua de disolución de evaporitas;*** *aigua de dissolució d'evaporites*). All these terms imply a genesis that should be demonstrated. As this is not often the case, the qualification should be avoided.

6 Terms Related to Groundwater and Climate

Groundwater resources and reserves depend on climate (IAH 2012), which determines recharge as one of the major factors. The generally large reserve of aquifers relative to resources is a key property for future **mitigation** (***mitigación;*** *mitigació*) of water resources modification due mostly to the possible but uncertain future changes. This is a key aspect to be carefully managed and a main subject of water resources governance. This is especially important for preserving aquifer water storage for long-term action, avoiding groundwater mining as an unplanned result of trying to mitigate changes by substituting dwindling fast renewal water resources just by increasing groundwater withdrawal.

To look forward, looking backward is important. Medium-to large-size aquifers have been subjected in the past to climate variations, especially in arid and semiarid areas, where future changes are more likely. How have them reacted and the possible remnant situations can be studied, and in fact they are in many cases. This knowledge of past behavior is an optimum proxy of circumstances that likely may happen in the future, if the components can be isolated from each other in order to accommodate the fact that factors and the evolution rate will be different.

When carrying out studies and quantifications, it should be clearly differentiated among different terms, reflecting non-coincident circumstances. **Climate variability** (***variabilidad climática;*** *variabilitat climàtica*) refers to long term changes produced in the past, whose effect may still persist and can be studied through different proxies. **Climate fluctuation** (***fluctuación climática;*** *fluctuació climàtica*) refers to current short term changes at year and decade scale, which conform what we have experienced or reconstructed from climate records and land use conditions; this include cycles of 10 to 50 years that repeat since long time ago and determine that observed groundwater behavior departs significantly from average values; so, changes noticed may be erroneously attributed to climate change or valued not a the due scale. **Climate change** (***cambio climático;*** *canvi climatic*) refers to predicted climate changes in the coming decades affected largely by human impact on the atmosphere, which add to normal fluctuation and the largely unknown long-term derive. **Global change** (***cambio global;*** *canvi global*) is the result of human activities on environmental behavior, which may greatly influence water availability and groundwater aquifer recharge through a series of factors which include land management, land-use actions, forest status, human population, living standards, and trend to circular economy, among others. Often, global change has and will have a greater influence than climate change and thus is should be the focus of research and prognosis.

In flat coastal areas, sea **level change** (***cambio del nivel del mar;*** *canvi del nivell del mar*) may be important for groundwater resources, among other consequences. This happened in the past, is a current fact, and will continue in the future. It is a delayed result of climate variations. There is a major impact on flat coastal areas, mainly through complex coastal area and sediment pattern modification. Sea level rise will affect groundwater resources, not linearly but through complex changes and transient evolution of reserves. The optimum use of coastal aquifer groundwater reserves needs careful management to avoid salinization and water management should be integrated.

References

Custodio E (2002) Aquifer overexploitation, what does it mean? Hydrogeol J 10(2):254–277

Custodio E (2012). Intensive groundwater development: A water cycle transformation, a social revolution, a management challenge. In: Martínez-Cortina L, Garrido A, López-Gunn E (eds) Rethinking water and food security, CRC Press, pp 259–298

Custodio E, Dolz J (2016) Ciencia, técnica y derecho de aguas con referencia a las aguas subterráneas. In: Embid Irujo A (ed) Treinta años de la Ley de Aguas de 1985, Universidad de Zaragoza, Thomson Reuters Proview, pp 209–242
Custodio E, Andreu-Rodes JM, Aragón R, Estrela T, Ferrer J, García-Aróstegui JL, Manzano M, Rodríguez L, Sahuquillo A, del Villar A (2016a) Groundwater intensive use and mining in south–eastern peninsular Spain: hydrogeological, economic and social aspects. Sci Total Environ 559:302–316
Custodio E, Cabrera MC, Poncela R, Puga LO, Skupien E, del Villar A (2016b) Groundwater intensive exploitation and mining in Gran Canaria and Tenerife, Canary Islands, Spain: hydrogeological, environmental, economic and social aspects. Sci Total Environ 557–558:425–437
Custodio E, Albiac J, Cermerón M, Hernández M, Llamas MR, Sahuquillo A (2017) Groundwater mining: benefits, problems and consequences in Spain. Sustainable water resources management. Pon line 04 April 2017.
DOI 10.1007/s40899-017-0099-2
DAS (2006) Directiva Derivada sobre el Agua Subterránea, 2006/118/CE del Parlamento Europeo y del Consejo, de 12 de diciembre de 2006
DMA (2000) Directiva 2000/60/CE del Parlamento Europeo y del Consejo, de 23 de octubre de 2000
Foster SSD (1993) Unsustainable development and irrational exploitation of groundwater resources in developing nations: an overview. Aquifer overexploitation. International association of hydrogeologists, selected papers 3, Heise, 385–402
Foster SSD, Loucks DP (eds) (2011) Non–renewable groundwater resources: a guide book on socially–sustainable management for water–policy makers. IHP–VI Series in groundwater 10, UNESCO/IAH/GW–State Word Bank, pp 1–103
IAH (2012) Climate change effects on groundwater resources. A global synthesis of findings and recommendations. In: Treidel H, Martin-Bordes JL, Gurdak J (eds) Wallingford. IAH Book Series, UK, pp 1–398
IPH (2008). Orden ARM/2656/2008, de 10 de septiembre, por la que se aprueba la instrucción de planificación hidrológica. Boletín Oficial del Estado 229, de 22/9/2008
Konikow LF, Leake SA (2014) Depletion and capture: revisiting the source of water derived from wells. Ground Water 52(1):100–111
MASE (2015) Aspectos hidrológicos, ambientales, económicos, sociales y éticos del consumo de reservas de agua subterránea en España: Minería del Agua Subterránea en España. Preparado por E. Custodio para UPC y AQUALOGY–Cetaqua, Barcelona, pp 1–730. http://h2ogeo.upc.edu/es/proyectos-empresa-ghs#d%C3%A9cada-2010
Molinero J, Custodio E, Sahuquillo A, Llamas MR (2011) Groundwater in Spain: legal framework and management issues. In: Findikakis and Sato (eds) Groundwater Management Practices. CRC Press/Balkema, pp 123–137
RPH (2007) Real Decreto 907/2007, de 6 de Julio, por el que se aprueba el Reglamento de la Planificación Hidrológica. Boletín Oficial del Estado 162, de 7/7/2007
SASMIE (2017) Salinización de las aguas subterráneas en los acuíferos costeros mediterráneos e insulares españoles. Preparado por E. Custodio para UPC y Suez Solutions–Cetaqua, Barcelona, pp 1–800 (in e–press)
TRLA (2001) Real Decreto Legislativo 1/2001, de 20 de julio, por el que se aprueba el texto refundido de la Ley de Aguas. BOE 176, de 24 de julio de 200, 26791–26817. http://www.boe.es/buscar/doc.php?id=BOE-A-2001-14276
TRLA (2003) Artículo 129. Modificación del texto refundido de la Ley de Aguas, aprobado por Real Decreto Legislativo 1/2001, de 20 de julio, por la que se incorpora al derecho español, la Directiva 2000/60/CE, por la que se establece un marco comunitario de actuación en el ámbito de la política de aguas. Ley 62/2003, de 30 de diciembre, de medidas fiscales, administrativas y del orden social. BOE 313, de 31 de diciembre de 2003, 46874–46992

Consequences of Seawater Intrusion in Mediterranean Spain. Project SASMIE

E. Custodio

1 Coastal Aquifers

Coastal aquifers are those in direct or indirect contact with the sea. Consequently, they are susceptible of containing saline water of marine origin. The denser seawater penetrates into the continent or island, forming a wedge whose toe may be close to the shore or penetrate several km inland, depending on local hydrogeological circumstances. This was established early in the 20th century and well-developed after the 1950s, when most of the principles, concepts, calculation methods, and hydrogeochemical and environmental isotopes processes were set (Custodio 1976; Custodio and Bruggeman 1987; Werner et al 2013). Also the management principles were set (Bear 2007). Several detailed examples are available (Post and Abarca 2010), but not many. Some strange and surprising costal circumstances, mostly in coastal karstified carbonate formations, especially in the Mediterranean Region, were known since early times, but were often misinterpreted and to some extent hindered the scientific development of hydrogeology.

Seawater intrusion into coastal aquifers is a natural phenomenon depending on the flow of continental groundwater to the sea. The transition between freshwater and seawater may be sharp, but generally there is a mixing zone whose width varies over a wide range, according to local hydrogeological circumstances. The discharge of continental or island groundwater along the coast and littoral offshore part is important for the existence of coastal wetlands and to create some particular conditions of salinity and nutrient contribution, which is are of ecological and economic relevance. Outflow to the sea is often greater than continental water outflow, as what is discharged is a mixing with seawater, enhanced by tidal effect.

E. Custodio (✉)
Civil and Environmental Engineering Department, Royal Academy of Sciences of Spain, Technical University of Catalonia (UPC), C/. Jordi Girona 1–3, 08034 Barcelona, Spain
e-mail: emilio.custodio@upc.edu

M. L. Calvache et al. (eds.), *Groundwater and Global Change in the Western Mediterranean Area*, Environmental Earth Sciences,
https://doi.org/10.1007/978-3-319-69356-9_11

Groundwater withdrawal from a coastal aquifer reduces the discharge to the sea. This is a delayed, slow process, in which the saline water wedge progresses landward, according to hydrological heterogeneities, and the mixing zone expands. When saline water, naturally or as the result of exploitation, exists below the wells and drains, saline water up-coning may cause serious salinity problems, as 2–3% of seawater made the abstracted water unfit for most uses.

All what has been said is well-known but not mastered by many non-specialized experts in hydrogeology and groundwater exploitation, who often do not realize the importance of variable density 3-D flow and the sometimes dominant role of local hydrogeological circumstances. Local circumstances are often poorly known as the needed intensity of study and survey is much greater than what is commonly done. For water authorities, coastal aquifer systems are often a small part or a much larger territory under their responsibility, so they pay special attention to them only when there is high population, human activity and intensively irrigated agriculture.

2 Spanish Coastal Mediterranean Aquifers. The SASMIE Project

Spain has a long coast with several kinds of aquifers and conditions. Coastal aquifer development started in the late 19th century in the neighborhood of Barcelona. Intensive groundwater abstraction was established in the 1940s in the Besós delta and in the 1960s in the Llobregat delta. The consequence was the salinization of part of the aquifers and the abandonment of important water supply wells. This was the starting point of detailed studies by the local water authority and the Public Works Geological Service of the Ministry of Public Works, with the collaboration of the Technical University of Catalonia. The studies were later extended to other parts of Catalonia, the Mediterranean area and the Balearic and Canary Islands. Shortly afterwards started the studies and surveys of the Geological Institute of Spain (IGME) in coastal areas. Much knowledge was gained at that time through dedicated studies and monitoring networks.

Serious seawater intrusion problems developed near Tarragona and in Mallorca and Eivissa (Ibiza), in the Balearic Islands, which soon were produced in many other places. This is mainly the result of uncontrolled private developments, for industrial and urban use in Catalonia, and mostly for irrigated agriculture and large tourist developments in other areas. In some moments, water for urban supply was so saline that it could be used only for sanitation or for industrial cooling.

Coastal problems were the subject of several specialized meetings, as the TIAC (Technology of Intrusion in Aquifers in the Coast), which included especially prepared summary reports, as that of Gómez Gómez et al. (2003). Spain hosted two of the international SWIM meetings (Salt Water Intrusion Meetings), in 1992 (Barcelona) and in 2002 (Cartagena). However, apart for a few detailed studies, most of the information about Spanish coastal aquifers was and is limited to descriptions and first surveys, as is also the case in most countries.

In order to put together and comment the existing experience in the Mediterranean area of peninsular Spain and the two archipelagos, the Balearic and the Canary islands, the SASMIE project was set. SASMIE is the acronym of *Salinización de las aguas subterráneas en los acuíferos costeros mediterráneos e insulares españoles* (*Groundwater salinization of Spanish Mediterranean and island coastal aquifers*). The goal is compare what is known, considering the environmental, legal, economic, managerial and social aspects and accomplishments. It is limited to what is available in written form or communicated through selected interviews and a questionnaire. The project has been carried out by the Technical University of Catalonia, with the support of Suez Solutions and overviewed by Cetaqua. The report was finished in June 2017, is being e-printed and is freely available (SASMIE 2017). In South-eastern Spain part of data come from MASE (2015).

Hereinafter, only the Mediterranean peninsular and Balearic coastal aquifers are considered. North of Barcelona, small recent alluvial and coastal plain aquifers dominate. Around Barcelona the main coastal formations are the Besós and especially the Llobregat deltas Plio-Quaternary alluvial and littoral formations. South of Barcelona, the Mesozoic carbonate and Miocene calcarenite formations dominate, down to Peníscola (Peñíscola), with the large Ebre (Ebro) delta in the middle. South of Peníscola, long, sometimes narrow Quaternary and Plio-Miocene plains dominate, down to Dénia. Southwards, in Alacant (Alicante) province, many coastal aquifers consist in highly yielding carbonate blocks embedded in a chaotic low-permeability mass associated to the fringe areas of the Alpine Betic-related orogeny, and the infilling sediments of some grabens. South of the important carbonate aquifers of the Campo de Dalías-Sierra de Gádor, coastal aquifers mostly reduce to small alluvial and coastal deposits, down to the the Campo de Gibraltar. In the Balearic Islands, Mesozoic carbonates and Plio-Miocene calcarenites dominate.

Deltas along the Spanish Mediterranean area have a similar sedimentary structure as in other Mediterranean and world deltas. They result from the fast sea level rise after the last glaciation (Custodio 2010). However, from the hydrogeological and seawater intrusion mechanisms point of view, there are important differences among them. The Ebre delta is a characteristic delta, with seawater in the confined deep aquifer due to the low elevation of the area. In the other Spanish Mediterranean deltas, seawater has been flushed out due to the high enough elevation of the surroundings and the deep aquifer outcrop or sub-outcrop of the deep late Pleistocene aquifer, which allows outflow. But again differences appear with respect to the Llobregat delta conditions, as in the La Tordera delta the intermediate aquitard has non-negligible permeability and in the Besós delta the smaller energy the old river was no able to fully excavate previous sediments. All this is the key to understand the natural circumstances and the rate of seawater intrusion under intensive exploitation.

Several large coastal karstic springs exist, which is common along the Mediterranean coast (Margat 2004), although not as large and more variable as the those existing in Southern France, Southern Italy, Croatia and Slovenia, some places in mainland and island Greece, Turkey and Syria-Lebanon, due to the smaller supply areas and the more arid climate. The most important ones in Spain are in the the Serra d'Irta, in Castelló (Peníscola, Badum and Alcossebre springs) and in the Benissa-Calp area, in Alacant (Moraig and Toix springs). Other coastal outflows are in the Garraf Massif, in Barcelona (La Falconera spring), in the Vandellós Massif, in Tarragona, and diverse points around Mallorca Island. In S'Almadrava spring, in Mallorca, brackish water outflows above sea level. No current use of these springs is foreseen.

The best developed studies correspond to the Llobregat delta, but also detailed studies on seawater intrusion conditions have been carried out in La Vall d'Uixó (Castelló) agricultural area and in the Andarax delta (Almería) for the supply of seawater though the aquifer to a large desalination plant. Especial mention deserve the studies of the economically important aquifers of the Campo de Dalías-Sierra de Gádor, one of the world largest concentrations of greenhouses. The most complete studies combine hydrodynamics with hydrogeochemistry and environmental isotopes, as complementary tools, and also combine recharge evaluation through soil water balance and the atmospheric chloride deposition balance (Alcalá and Custodio 2014). The contribution of University departments to these studies has been essential, mostly, but not exclusively, by the Technical University of Catalonia, the University of Castelló, the University of Alacant and the University of Almería, with valuable contributions from the University of Granada.

3 Environmental Considerations

Four main aspects should be considered: (1) Costal groundwater dependent wetlands in which salinity gradients and fluctuations play an important role, as well as the contribution of nitrate; this is the case of S'Albufera (Mallorca) and the Ebre delta lagoons and marshes; (2) The role of groundwater discharge in littoral marine water and the creation of especial salinity, chemical and nutrient conditions; (3) The effect of groundwater exploitation of coastal aquifers, which causes the reduction and desiccation of wetlands, changes of the salinity pattern, and decrease the continental water discharge to the sea; (4) The use of groundwater in desalination and debrackishing plants; this may alter continental and offshore littoral water conditions by groundwater level drawdown and adds the not well-solved problem of the disposal of return brines, which, in the case of brackish groundwater in several agricultural areas, may have high nitrate contents that affect biological conditions; this is currently the case of the Mar Menor lagoon (Murcia). Some attempts are being made for quantifying groundwater discharge in the coast. But results have to be used carefully, as continental outflow is less than total outflow, which incorporates recycled seawater.

4 Legal and Administrative Constraints

The water legislation in force in Spain does not directly consider coastal aquifers, as this is a too specific issue. Regulations are indirect and have to be deduced from rules referring to water quality and wetland preservation. The effect of continental groundwater disposal into the sea is not addressed at all. The water plan of the Júcar (Xùquer) Water District considers and estimate the order of magnitude of what is called "ecological" flow, which is the groundwater flow to be discharged in the coastal area in order to prevent, limit or restore salinity in the aquifer and in some coastal wetlands. Something similar is done in the water plan of Catalonia, but less detailed. These flows are subtracted from available groundwater flow before defining available groundwater resources and how they should be distributed to the different water demands.

According to the Water Act, since 1985 all waters are a public domain, but well owners holding private groundwater rights coming from earlier times are entitled to keep these rights, as did most of them. Before 1985, groundwater was a private affair and the public water administration was uninterested in them, except in special cases that led to special regulations in the Balearic Islands since the 1970s (in the Canary Islands since 1924). The incorporation into the public water authorities of responsibilities on groundwater management and affairs after 1985 was a hard task, as they were not prepared to, and especially due to the difficulties associated to coastal aquifers. So, studies and monitoring impaired. This situation has not recovered due to the economic crisis that started in 2009, although the technical capacity has improved.

The incorporation of the European Water Framework Directive of 2000 and the Groundwater Daughter Directive of 2006 moved the focus toward groundwater quality. This is indirectly related with coastal aquifer salinization. The administration has paid attention to characterizing aquifer good status but not to the causes and the solutions of salinization problems. Efforts have been mostly for making available good quality water from other sources, even if they are costly.

For groundwater administrative and management purposes, some simple indicators may help, as commented in Werner et al. (2013). This is really a difficult task still to be solved, if it is possible at all. Attempts in the Júcar Water District have not ended with a workable tool.

5 Economic Effects

Coastal aquifer salinization involves economic losses. They seem not quantified until present. These losses refer mostly to the cost of making available alternative freshwater sources, the increased cost of facilities, appliances and distribution networks for households, towns and factories due to enhanced corrosion and the associated water leakage, and the decreased productivity, crop shift, land abandonment and need of water blending in farming. This cost has been paid by water users, many of who do not receive the benefits of groundwater abstraction. The early abandonment of groundwater abstraction facilities is a further cost. Also there is the significant cost associated to impaired ecological services from coastal wetlands, although this is not quantified. Where groundwater is brackish, private small to medium-size reverse osmosis plants have proliferated, mostly for irrigation and hotels. This is an added cost to water. In the Campo de Cartagena, there are very numerous plants, perhaps up to 2000. The negative externalities associated to the saline return flows have to be added, although they are seldom considered such as the serious impairment of the Mar Menor coastal lagoon.

In some cases, the important water table drawdown in urban coastal areas allowed the use of the underground unsaturated space for facilities, storage and transport. The cease of groundwater abstraction due to salinization was accompanied by water table recovery and the subsequent inundation of these underground spaces. There is the cost of drainage by pumping (with possible increase of salinity problems to other users) or of the abandonment of non-amortized investments. Examples can be found in Barcelona area. No serious subsidence problems have been reported, but the sea shore retreat in the Llobregat delta is the compound result of subsidence due to deep groundwater abstraction with sea level rise and decreased sediment contribution by the river.

6 Management Action

Public management action to solve salinization problems in coastal areas depending on groundwater has been and is mostly structural. Non-structural action needs social complicity and cooperation, well designed plans, especially trained personnel, and a long time to mature them and being able to show results. Non-structural action through taxation has not been seriously designed and fiercely opposed. Whether positive economic results and incentives to correct problems are attained is unclear.

Most structural action has been directed to made available new sources of water. In the south-eastern area of Spain, this refers to the water transfer from inner areas and from Atlantic river basins through the Tajo-Segura and Negratín-Almanzora aqueducts. All along the coast, about 20 large seawater desalination plants have been constructed. Treated waste water reclamation for agriculture has also been made available. Reclamation has been a great success in the Segura water district, where river water quality recovery deserved recently an important international award. There is an important cost increase, especially for desalinated water. Some of the plants have a low use and this increases the cost of produced water. To foster the use of this water, the cost is subsidized in some areas, even non-covering the operation cost. In other cases, the cost is integrated into the operation of the water system, as in Barcelona Metropolitan area, where the desalination plan is mostly to increase water availability guarantee in droughts, when seawater intrusion has to be limited. There, to increase ground water availability in droughts and to reduce seawater intrusion, artificial recharge is practiced and a coastal water injection well barrier has been constructed. Injection water is intensively treated reclaimed waste water, with salinity reduction.

Coastal aquifers may be a source of brackish water for debrackshing and also a source of sediment and biologically free marine water to supply desalination plants. Good experience is available in Spain (Rodríguez-Estrella and Pulido-Bosch 2009).

7 Social Considerations

Sound coastal aquifer governance is still a poorly addressed issue, although noticeable steps have been done. Top-down action to form aquifer users associations to apply imposed management rules, as the “overexploitation” provisions in the Water Act have been mostly a failure, except when subsidies and large public investments are involved, paid by others. However, several bottom-up associations have been formed and are effective. They are worldwide pioneers. The first one started in the Lower Llobregat area, in 1975, even before the legal declaration of all water as a public domain. They are more easily formed in urban-industrial dominated areas than in rural areas, where people are more reluctant, even if farmers are currently well trained. One reason is the advantage that groundwater rights holders have relative to those that do not, in areas partly served with surface water and susceptible to scarcity during droughts. This is the case of the Campo de Cartagena. Until now, no groundwater users association have been formed in the Balearic Islands.

8 Future Role of Coastal Aquifers and Coming Changes

Past salinization problems have cast in many cases a poor appreciation of the important role of coastal aquifers. If carefully managed, they are not only a reliable permanent source of freshwater but also they contain a reserve for the dry season and those of larger size also to mitigate the serious Mediterranean area droughts. The preservation of what exists and the restoration of what has been degraded is a desirable goal, but if at a reasonable cost. Wells are restored in the Barcelona area, where old wells are again in operation, although with added chemical treatment.

The role of coastal aquifers is enhanced if available water resources will decrease due to global change and the common predictions related to climate change and their effects on groundwater (IAH 2012). In any case, integrated water resources management will be needed to confront water scarcity. Aquifers provide naturally the needed buffering storage. But this use needs adequate legal and administrative tools and the involvement of water users and the civil society. This is not duly considered in the European Water Framework Directive. Now, it is the right moment to incorporate improved consideration in the revision to be enacted by 2027 and to consider more adequately the specificities of the semiarid European

Mediterranean area especially in Spain. Good groundwater status could not be the best objective but the adequate status according to water resources optimization, compatible with the environment and the actualized economic values. So, aquifers should be classified according to uses and the expected social, global and climate changes.

Acknowledgements Most of the ideas and information come from the SASMIE project, carried out by the Technical University of Catalonia and supported by Suez Solutions under the overview of Cetaqua.

References

Alcalá FJ, Custodio E (2014) Spatial average aquifer recharge through atmospheric chloride mass balance and its uncertainty in continental Spain. Hydrol Proc 28:218–236

Bear J (2007) Seawater intrusion in coastal aquifers. III TIAC, Almería. SHAS, I, pp 3–22

Custodio E (1976) Relaciones agua dulce–agua salada en las regiones costeras. Hidrología Subterránea. In: E. Custodio, MR. Llamas (eds) Ediciones Omega. Barcelona. II–13, pp 1313–1389

Custodio E (2010) Coastal aquifers of Europe: an overview. Hydrogeol J 18:269–280

Custodio E, Bruggeman GA (1987) Groundwater problems in coastal áreas. Studies and reports in hydrology, vol 45, UNESCO, Paris, pp 1–576

Gómez Gómez JD, López Geta JA, Garrido Schneider E (2003) The state of seawater intrusion in Spain. II TIAC, Alicante, II, pp 169–186

IAH (2012) Climate changes effects on groundwater resources. A global synthesis of findings and recommendations. In: Treidel H, Martin-Bordes JL, Gurdak J (eds) Wallingford. IAH Book Series, UK, pp 1–398

Margat J (2004) Atlas de l'eau dans le Bassin Méditerrranéen. UNESCO/Plan Bleu, Paris, pp 1–46

MASE (2015) Aspectos hidrológicos, ambientales, económicos, sociales y éticos del consumo de reservas de agua subterránea en España: Minería del Agua Subterránea en España. Preparado por E. Custodio para UPC y AQUALOGY–Cetaqua, Barcelona, pp 1–730. http://h2ogeo.upc.edu/es/proyectos-empresa-ghs#d%C3%A9cada-2010

Post V, Abarca E (eds) (2010) Saltwater and freshwaters interactions in coastal aquifers. J Hydrol 19(1): 1–270

Rodríguez-Estrella T, Pulido-Bosch A (2009) Methodologies for abstraction from coastal aquifers for supplying desalination plants in the south–east of Spain. Desalination 249:1088–1098

SASMIE (2017) Salinización de las aguas subterráneas en los acuíferos costeros mediterráneos e insulares españoles. Preparado por E. Custodio para UPC y Suez Solutions–Cetaqua, Barcelona, pp 1–800 (in e–press)

Werner AD, Bakker M, Post VEA, Vandenbohede A, Lu C, Ataiee–ashtani B, Simmons, CT, Barry DA (2013) Seawater intrusion processes, investigation and management: recent advances and future challenges. Adv Water Resour 51: 3–26

Quantitative Impact of Climate Variations on Groundwater in Southern Italy

D. Ducci and M. Polemio

1 Introduction

In Italy, groundwater provides drinking water supply, supports agricultural and industrial activities, and contributes towards rivers and lakes. About 88% of the Italian's drinking water supply is from groundwater (Polemio et al. 2013). Groundwater resources are of different entity depending on the morphological and climate variedness. In southern Italy, some region, like the Campania region (Fig. 1), consists widely of carbonate mountains (limestone and dolomite), characterized by a well-developed karst, fed by very high recharge due to high rainfall and low temperature (Barberá and Andreo 2012). On the contrary, the flat Apulian region (Fig. 1), given the extreme scarceness of surface water and the lower recharge, valuable groundwater resources are exploited with increasing rates for domestic, irrigation, and industrial uses (Polemio 2016). The increase is particularly relevant during the numerous recent drought periods, on the basis a paradoxical management criterion.

Data from 1821 to 2003 of 126 rain gauges, 41 temperature gauges, 8 river discharge gauges and 239 wells, located in Southern Italy, have been analysed to characterize the effect of recent climate change on water resources availability, focusing on groundwater resources (Polemio and Casarano 2008).

D. Ducci (✉)
Department of Civil, Architectural and Environmental Engineering, University of Naples Federico II, Naples, Italy
e-mail: daniela@unina.it

M. Polemio
Research Institute for Geo-Hydrological Protection (CNR-IRPI), Via Amendola 122/I, Bari, Italy
e-mail: m.polemio@ba.irpi.cnr.it

M. L. Calvache et al. (eds.), *Groundwater and Global Change in the Western Mediterranean Area*, Environmental Earth Sciences,
https://doi.org/10.1007/978-3-319-69356-9_12

Fig. 1 Location of the study areas; *C* Campania; *P* Apulia

2 Data Analysis

A widespread decreasing trend of annual rainfall is observed over 97% of the southern Italy (Cotecchia et al. 2004; Polemio 2005). The spatial mean of trend ranges from −0.8 mm/year in Apulia (−10.1% of mean value in the study period), −2.44 mm/year (−17.5%) in Campania to −2.9 mm/year in Calabria (−22%). The decrease in rainfall is noticeable after 1980: the droughts of 1988–92 and 1999–2001 appear to be exceptional. On a seasonal basis, the decreasing trend is concentrated in winter; a slight positive trend is observed in summer, the arid season in which the increase is useless as it is transformed in actual evapotranspiration. Although the temperature trend is not everywhere significant and homogeneous, the temperature increase seems to prevail, especially from the eighties (Polemio and Lonigro 2015). Net rainfall, calculated as a function of monthly rainfall and temperature, shows a huge and generalized negative trend.

In Campania, the differences between the period 1951–1980 and the drier period 1981–1999 are evident (Ducci and Tranfaglia 2008). The mean annual rainfall data

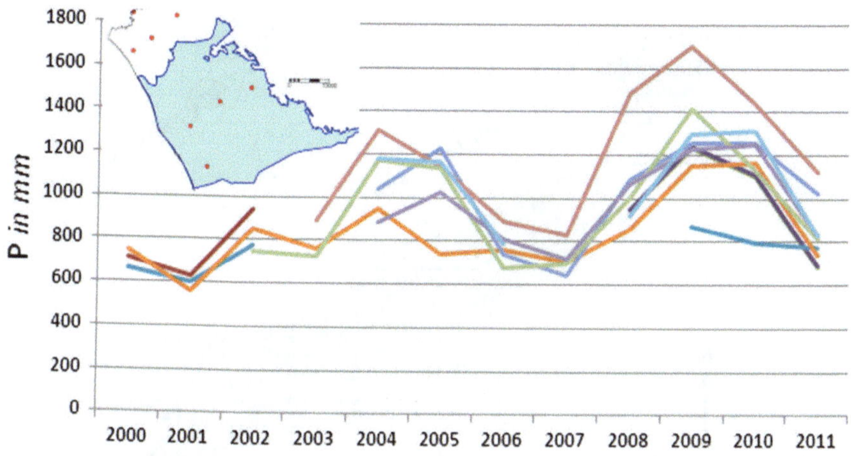

Fig. 2 Rainfall variation in rain gauge stations located in the Volturno-Regi Lagni plain groundwater body (P-VLTR GWB, Campania). On the upper left in light blue the P-VLTR GWB; in red the rain gauge stations and in black the coast of the Campania region (see Fig. 1)

show a decrease of about 15%. After 2001, the trend is increasing, but this trend is highly influenced by the period 2008–2011, very rainy, as shown in Fig. 2.

3 Piezometric Trend

The effects of recent climate variations on groundwater availability are evaluated considering 5 wide hydrogeological structures (HSs for sake of brevity), 4 in Apulia and 1, very large, in Campania. In each Apulian HS the shallow or outcropping aquifer is considered; three are constituted by carbonate rocks, one is porous, and all include coastal aquifers (Fig. 3).

The Apulian Tableland HS, hereinafter called Tavoliere HS, consists of a shallow and large porous aquifer within a conglomerate sandy-silty succession, less than sixty meters deep, with a clayey impermeable bottom (Polemio 2016). It is deep enough to allow seawater intrusion only in the vicinity of the coast. Groundwater is phreatic inland or far from the coast, in the recharge area, whereas it is confined in the remaining part of the aquifer; maximum piezometric levels reach 300 m asl.

Except for the Tavoliere, the Apulian region is characterized by the absence of rivers and the unavailability of surface water resources due to its karstic nature. Considerable groundwater resources are located in large and deep carbonate coastal aquifers as in the case of the Gargano (not considered in this study due to the low data availability), the Murgia and the Salentine Peninsula (Salento) HSs. The Murgia and Salento areas show some common features (Polemio 2016). They consist of large and deep carbonate aquifers, constituted mainly of limestone and

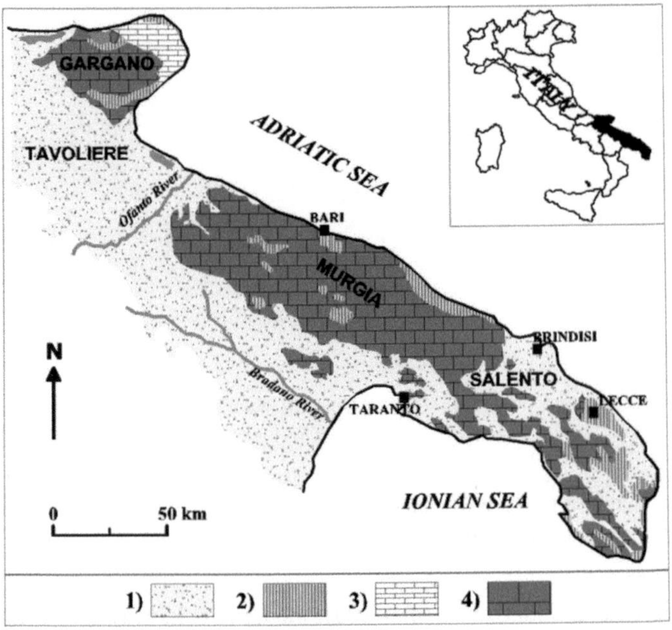

Fig. 3 Apulian schematic geological map and main hydrogeological structures (Polemio 2005, modified): (1) Recent clastic cover (Pliocene-Pleistocene); (2) Bioclastic carbonate rocks (Paleogene) and calcarenites (Miocene); (3) Scarp and basin chert-carbonate rocks (Upper Jurassic-Cretaceous); (4) Carbonate platform rocks (Upper Jurassic-Cretaceous)

dolomite rocks. Carbonate rocks are affected by karstic and fracturing phenomena, which occur also well below sea level, whereas intruded seawater underlies fresh groundwater owing to a difference in density. Confined groundwater is more widespread inland; groundwater is phreatic everywhere along a narrow coastline strip. The Maximum piezometric head is about 200 m asl in the Murgia area and 5 m asl in the Salento (Polemio 2005).

Data from fifty-eight wells or piezometric gauges are available for the three Apulian HSs, the Tavoliere, the Murgia and the Salento (Polemio and Dragone 2004) (Table 1) and for the P-VLTR GWB of Campania. The piezometric data sets regarding the Tavoliere are available for a minimum of 17 years and for a maximum of 55 years, covering a continuous period between 1929 and 1994 (Cotecchia et al. 2004). Continuous data are available from 1973 to 1978 for the Murgia and the Salento. Furthermore, sporadic recent data were collected in Apulia for the periods from 1995–1997 to from 2001–2003. The piezometric data set of Campania shows results coherent with the rest of the study area.

The spatial analysis is utilized to complete the trend analysis of piezometric data when sporadic but high density data are available.

The piezometric trend everywhere is decreasing; the continuous piezometric lowering has transformed many confined wells into phreatic wells; after that, the

Table 1 Piezometric data availability for each hydrogeological structure (HS) and straight line trend (AC, m/year). (1) The number of wells available for occasional years is higher and variable; (2) in the periods 1927–1940 and 1951–1984; (3) Determination not available due the characteristics of data set

HS/GWB	Well number	Data		Trend more probable at 2002 or 2003
		From	To	
Tavoliere	11	1929	2002	High decrease
Murgia	30	1965	2003	High decrease
Salento	17	1965	2003	Decrease
P-VLTR shallow	6	1926	1998	High decrease
P-VLTR main	100/200	1990	2003	High decrease

shallow groundwater of the Tavoliere is completely depleted in places. In terms of straight line trend, the trend everywhere is strongly negative, constituting a severe problem for groundwater discharge by wells (Table 1).

This trend was widespread confirmed by the spatial analysis of sporadic data. An almost affordable study case is due to Salento (Polemio and Casarano 2008), for which sporadic data from thirties up to 2010 are available (Fig. 4). Notwithstanding the spatial effect of the low density of measurement wells in some portion of the Salento, the worsening in the whole HS is self-evident.

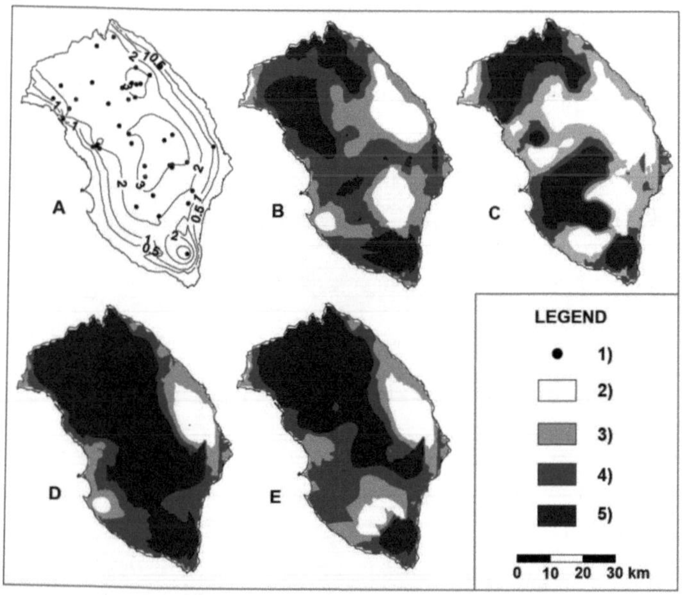

Fig. 4 Historical piezometric variations of Salento HS (**a**), Historic piezometric surface used as reference (1930, m asl); calculated Piezometric Variation (PV, m, positive values are increases) of 1976 (**b**), 1996 (**c**), 2003 (**d**), and 2010 (**e**). Legend: (1) wells; (2) PV > 0.5 m; (3) 0 m < VP < 0.5 m; (4) −0.5 m < VP < −0 m, 5) VP < −0.5 m

In Campania, the alluvial-pyroclastic groundwater body (GWB) of the "Volturno-Regi Lagni" plain (P-VLTR—1034 km^2) includes shallow aquifers, not continuous, constituted by alluvial and pyroclastic deposits, overlying the tuffs (Campanian Ignimbrite). This tuff confines or semi-cofines the main aquifer formed by alluvial, pyroclastic and marine porous sediments underlying the Campanian Ignimbrite (Corniello and Ducci 2014). The piezometric surface indicates a groundwater flow, departing from the neighboring limestone mountains, at east, and directed towards the Tyrrhenian sea, at west (Fig. 5).

At the foot of the carbonate mountains flowed out two high discharge springs (Qmed 1.3 m^3/s), located in the right lower corner in Fig. 5, almost exhausted over the last 20 years.

In P-VLTR groundwater body (Figs. 2 and 5) there are 6 wells of the shallow aquifer monitored since 1926 and until the end of nineties (Ducci and Onorati 1993). In these wells piezometric data make visible the drought period started in 1987 and finished in 1992: the mean lowering has been about 6 m and 2 wells were dried up. For the main aquifer two complete monitoring campaign were done in 1990 and 2003 (Fig. 5). The spatial comparison by GIS shows a mean lowering of 7.4 m, with peaks of 14 m. Afterwards, a campaign done in 2015 indicates almost the same levels of the year 2003 for the main aquifer, due to the constancy of the rainfall, as shown in Fig. 2.

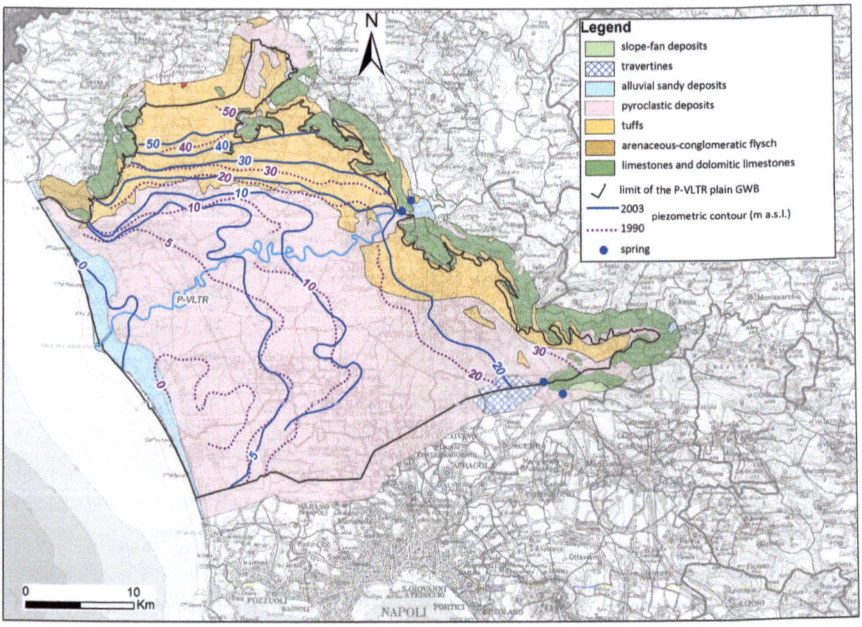

Fig. 5 Hydrogeological map of the P—VLTR GWB (Campania) highlightings the changes in groundwater levels of the main aquifer

4 Conclusion

The study deals with the climate variations recorded during many decades, observed in Southern Italy.

In Apulia the piezometric trend everywhere is decreasing; the continuous piezometric lowering has transformed in some parts the confined aquifer into phreatic and the shallow groundwater is depleted. The spatial mean of piezometric decreases almost 8 m over about 15 years.

In Campania, at the foot of the carbonate mountains some copious springs are almost exhausted over the last 20 years and the observation of piezometric data from 1990 and 2003 shows a mean lowering of more than 7 with peaks of 14 m.

The wide set of groundwater data, concerning piezometric and spring flow yield time series, shows the relevance of the overlapped effect of water demand and climate change, which is summarized by the widespread dramatic lowering of groundwater availability.

Acknowledgments We would like to show our gratitude to IAHS who provided permission to use the Figures/Table from Polemio et al. 2009.

References

Barberá JA, Andreo B (2012) Functioning of a karst aquifer from S Spain under highly variable climate conditions, deduced from hydrochemical records. Env Earth Sci 65(8):2337–2349

Corniello A, Ducci D (2014) Hydrogeochemical characterization of the main aquifer of the “Litorale Domizio-Agro Aversano NIPS” (Campania—southern Italy). J Geochem Explor 137:1–10

Cotecchia V, Casarano D, Polemio M (2004) Characterization of rainfall trend and drought periods in Southern Italy from 1821 to 2001. In Proceedings 1st italian-russian workshop “new trends in hydrology”, Rende (CS), 2004, Edibios, pp. 139–150

Ducci D, Onorati G (1993) Analisi di una lunga serie di dati piezometrici in Piana Campana. Atti 2 Convegno Nazionale di Geoidrologia—Quaderni di Tecniche di Protezione Ambientale—Protezione delle acque sotterranee, Pitagora, Bologna, vol 49, pp 339–357

Ducci D, Tranfaglia G (2008) The Effect of Climate Change on the Hydrogeological Resources in Campania Region (Italy). In Dragoni W (ed) Groundwater and climatic changes. Geological Society, London, Special Publications, 288, pp 25–38. doi:10.1144/SP288.3

Polemio M (2005) Seawater intrusion and groundwater quality in the Southern Italy region of Apulia: a multi-methodological approach to the protection. In: Maraga F, Arattano M (eds) Progress in surface and subsurface water studies at the plot and small basin scale, vol 77. Paris. UNESCO, IHP, pp 171–178

Polemio M (2016) Monitoring and management of Karstic coastal groundwater in a changing environment (Southern Italy): a review of a regional experience. Water 8(4):1–16

Polemio M, Casarano D (2008) Climate change, drought and groundwater availability in southern Italy. Geol Soc Spec Pub 288:39–51

Polemio M, Dragone V (2004) La siccità e la disponibilità di acque sotterranee in Puglia. Atti dei Convegni Lincei 204:187–193

Polemio M, Dragone V, Limoni PP (2009) The piezometric stress in the coastal aquifers of a karstic region. In: Taniguchi, M., Dausman, A., Howard, K., Polemio, M., and Lakshmanan, E., (eds) Sustainability of groundwater in highly stressed aquifers, Vol 329, IAHS, Apulia, Italy, p 138–144

Polemio M, Dragone V, Romanazzi A (2013) La risorsa idrica. Sfruttamento, depauperamento dei serbatoi sotterranei e utilizzo razionale nel caso della Calabria, In: Dramis, F., and Mottana, A., eds., L'acqua in Calabria: risorsa o problema? Roma, Aracne, p 2–29

Polemio M, Lonigro T (2015) Trends in climate, short-duration rainfall, and damaging hydrogeological events (Apulia, Southern Italy). Nat Hazards 75(1):515–540

Assessing the Uncertainties of the Water Budget in the Torrevieja Aquifer (Southeast Spain)

C. Duque, J. M. Gómez-Fontalva, J. M. Murillo and M. L. Calvache

1 Introduction

1.1 Water Problems in the Torrevieja Aquifer

The Torrevieja aquifer is located in the driest area of Europe where water resources are scarce and have a remarkable impact on society and economy and even political decisions are strongly linked to water management. As an example, the highest capacity desalination plant in Europe, with a yearly desalination capacity of 80 Mm^3, has been constructed in the proximity of this aquifer in response to the water deficit in the entire region. The growth of the economic activities of the area has produced an increasing requirement for water and studies for the quantification of groundwater resources (García-Arostegui et al. 2003), the geometry of the aquifer (Mediavilla et al. 2007; Tabares-Ródenas et al. 2009) and the hydrogeological properties of the aquifer (Rodes and Rodes 2003). In previous hydrogeological studies and reports (IGME 1990; Ramos-González et al. 2002; Rodríguez Estrella 2003; CHS 2004), the Torrevieja aquifer has been usually defined as a confined aquifer of 147 km^2 that outcrops in the northern border (16 km^2), whereas

C. Duque (✉)
Department of Geoscience, Aarhus University, Høegh-Guldbergs Gade 2, Aarhus 8000 Aarhus C, Denmark
e-mail: cduque@geo.au.dk

J. M. Gómez-Fontalva
Aljibe Consultores S.L., C/ José Luis Pérez Pujadas, 7, Bjo, Granada, Spain

J. M. Murillo
Instituto Geológico y Minero de España, C/ Rios Rosas 28, 28003 Madrid, Spain

M. L. Calvache
Department of Geodynamics, University of Granada, Granada, Spain

M. L. Calvache et al. (eds.), *Groundwater and Global Change in the Western Mediterranean Area*, Environmental Earth Sciences,
https://doi.org/10.1007/978-3-319-69356-9_13

the rest of the aquifer extension is covered with a confining unit composed of marls and silt (Mediavilla et al. 2007) (Fig. 1).

The annual pumping rates in the region are quantified as 5.50 Mm^3 (Ramos González et al. 2002), whereas the recharge induced by the rain and the irrigation return flow only adds 1.16 Mm^3 (Gómez-Fontalva and Calvache 2009) to the system. This indicates a clear deficit in the water budget that would not be sustainable for a long period given the size and the properties of the aquifer. The aquifer has been in this condition during the last 30 years, which shows that this imbalance should be reviewed. Based on this, two hypotheses are considered in this

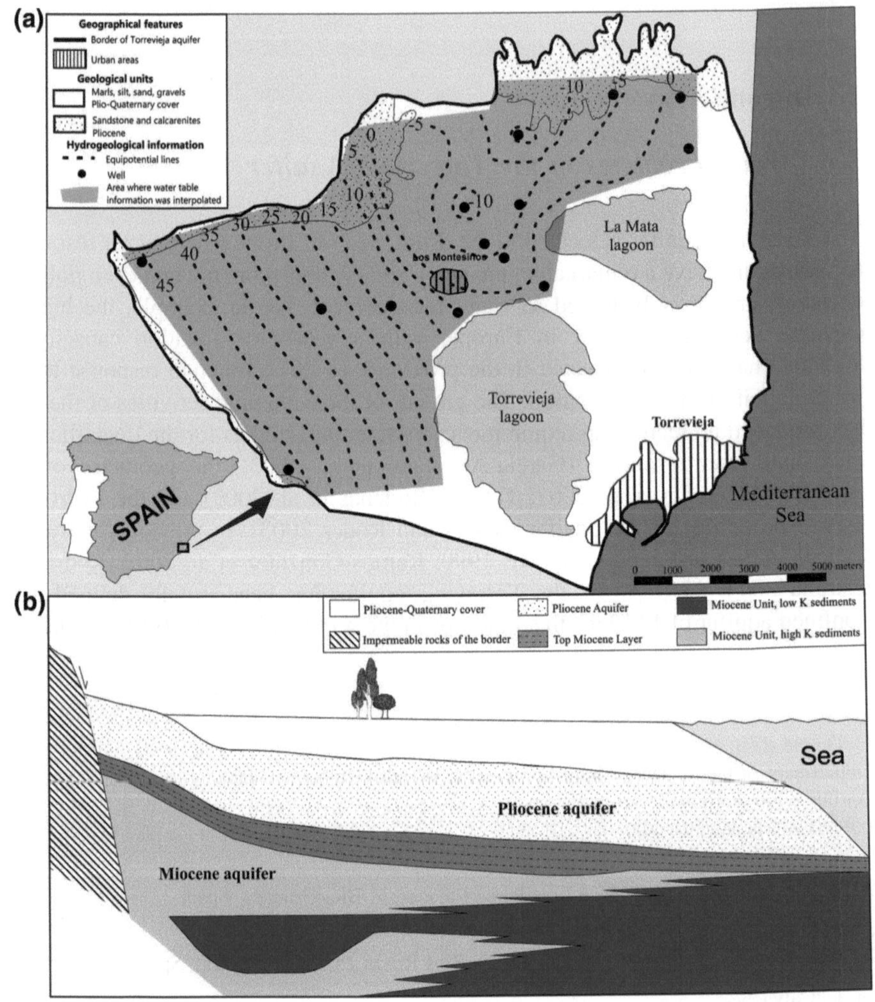

Fig. 1 **a** Torrevieja aquifer location, water table contour map and geological units in surface. **b** Geological cross section with idealized conceptual model from NW to SE

work: the impermeable properties of the top layer of the aquifer and the connection of the aquifer with a lower aquifer unit not very well-known in the study area. The hydrogeological information needed is scarce and the use of numerical models simulating groundwater flow is proposed as the tool for study the system complemented with hydrochemical data.

1.2 Methods

A set of 3D flow models was constructed using MODFLOW (McDonald and Harbaugh 1988), which solves the groundwater flow equation using modular finite differences. Only the Pliocene aquifer geometry and the upper unit on top were known and represented in the model. Initial steady state models were completed to decide which conceptual model would be more feasible. The calibration of the hydraulic conductivity and the transient simulations were accomplished after that. The model was designed on a regular grid of 304 rows by 370 columns at a spacing of 50 $\times$ 50 m, and two variable-thickness layers. In total, there were 22,4960 cells with their Y axis coinciding with the N-S direction. The top of layer 1 coincided with the topographic surface, and layer 2 represented the Pliocene aquifer. The morphology of both was irregular with changes in thickness and depth.

The hypothesis considered for the alternative conceptual models were:

Hypothesis 1: The upper Pliocene-Quaternary layer is not totally impermeable but permits certain inflows from irrigation return and precipitation. This option is based on the uncertainties about the hydraulic properties and the heterogeneity of these sediments.

Hypothesis 2: The extraction of water by the supply wells is partially assigned to a lower aquifer (Miocene) that decreases the outputs of the Pliocene aquifer. This possibility is proposed based on recent studies in stratigraphy (Soria et al. 2008; García-García et al. 2011) and the geometrical analysis of the lithological columns for assigning the layer from where the water is pumped to each well.

The hydrochemical dataset used was collected in several surveys from 2001 to 2003 (Rodes and Rodes 2003) by the Geological Survey of Spain (IGME). The sampling method consisted of the collection of 50 ml samples from purged wells or wells under exploitation. In many cases, these wells have long screens because their utility is to pump water for supply or irrigation, especially on the stretches with higher hydraulic conductivity, that result in samples being a mix of extensive sections of the aquifer. The information about the screen size and location was not always available. The water samples were analyzed for major ions (Ca^{2+}, Mg^{2+}, K^{+}, Cl^{-}, CO_3^{2-}, $SO4^{2-}$, NO_3^{-}) and for some minor ions (Br^{-}, Fe). Additionally, some parameters were measured in situ, such as pH, electrical conductivity and bicarbonates content by titration with HCl. The hydrochemical composition was compared with the assumed origins (based on stratigraphical criteria) to determine if the hydrochemical differences corresponded to different hydrogeological units. A hierarchical analysis based on the average cluster considering the Euclidean distance as a similarity measurement was accomplished using the values of Cl^-, SO_4^{2-}, HCO_3^-, NO_3^- and electrical conductivity. These are the parameters with higher variability and could thus better differentiate the samples. Each sample was assigned one of the three categories considered geometrically: Pliocene aquifer, Miocene aquifer or surficial aquifer for wells less than 10 m deep.

2 Objectives

Develop alternative conceptual models of aquifer functioning based on the study of the lithological columns of boreholes that can explain the imbalance of the aquifer.

Test the new hypotheses by applying simplified groundwater numerical models to decipher the feasibility of each case.

Verify the results with the assessment of hydrochemical characteristics of groundwater based on the data collected in previous surveys.

Analyze the hydrogeological status of the Torrevieja aquifer using the perspective of the new conceptual model.

3 Results

Based on the numerical model of Hypothesis 1, the recharge proceeding from the semipermeable top layer (Plio-quaternary) should be considered minor because of the large impact that would have the recharge over the water table in the areas where there was no pumping (the water table would exceed the topographic surface). Even if this layer could store a certain amount of water it would not have a relevant impact on the global water balance. In contrast, the possibility that a percentage of the water extraction came from a lower aquifer unit (Miocene aquifer) seems to be more plausible as a conceptual model. The implementation of the numerical model for Hypothesis 2 showed that the depth of the water abstraction was a very sensitive parameter and that the match between water table observations and model results present much lower errors compared with Hypothesis 1 (from 20–30 m to less than 5 m). This conceptual model greatly depended on the classification of the wells (from which layer is the water pumped), which was also connected with the stratigraphy model. There are uncertainties in the location of the well screens and in the contribution of each layer in cases when the wells had a long screen that penetrated both aquifer layers.

Hypothesis 2 was considered more adequate and a transient model of the year 2002 with monthly time steps was constructed using a percentage of the pumping abstracted from the upper Pliocene aquifer (25% of total) and without leakage from the top Quaternary layer. The pumping rates and recharge were assigned monthly based on the information collected in previous surveys. The model was manually calibrated by trial and error modifying the hydraulic conductivity and the specific storage in the range of the hydrogeological materials that were observed. The hydraulic conductivity of the Pliocene aquifer varied from 0.1 to 2.5 m d^{-1}. The homogeneity of the values was related to the spatial location of the observation wells, most of which were near the central area, which made the hydraulic conductivity closer to the borders more uncertain. The mean absolute error of the model for the 14 wells with monthly values was 1.56 m (Fig. 2).

The differentiation between the Pliocene aquifer and the Miocene aquifer established for the pumping could also be tested by analysis of their hydrochemical properties. Following the same geometrical criteria established for the flow model, the samples were plotted based on the SO_4^{2-} and Cl^- values that could distinguish between saltwater from the sea or from sediments because of the presence of gypsum or other sulfates. There was a chemical differentiation between the two groups designated geometrically in a high proportion, but there were still samples that were not possible to differentiate. In general, the Miocene aquifer groundwater had a higher concentration of Cl^- relative to SO_4^{2-}, possibly related to older saltwater intrusion events, the presence of connate water during the sedimentation process or ongoing processes that have not yet been deeply studied. The hierarchical analysis showed that 20 of 25 samples for the Miocene aquifer had similar hydrochemical properties, and 33 of 42 for the Pliocene aquifer (Fig. 3). A third group was detected where the properties of the groundwater were mixed and it was not possible to differentiate them based on the chemical properties of water. These were interpreted as wells with mixed water because of the location of the screen at

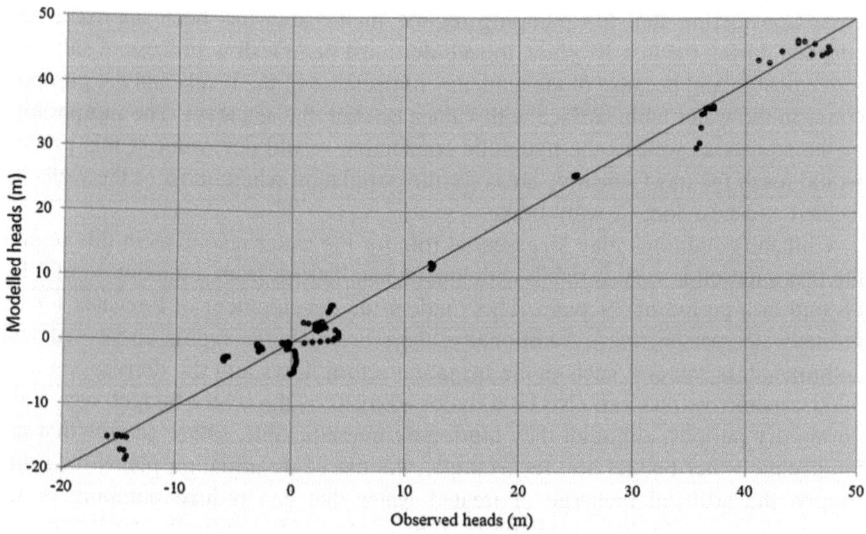

Fig. 2 Observations VS model results for the transient simulation of year 2002 including the modifications proposed based on the steady state models

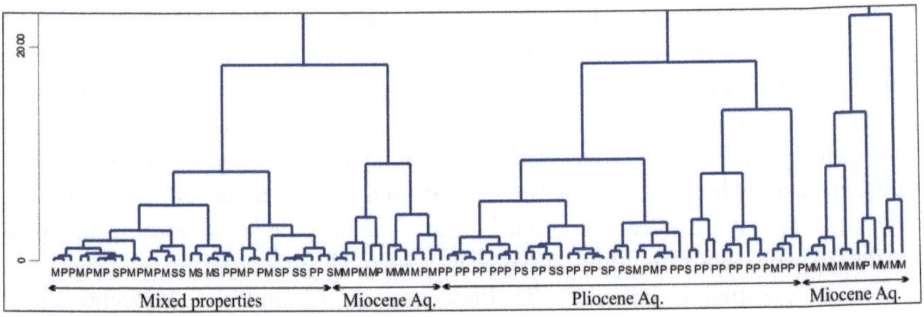

Fig. 3 Dendrogram for hierarchical cluster analysis for the 126 hydrochemical samples collected in the study area. M: Miocene aquifer, P: Pliocene aquifer, S: Surficial aquifer based on geometric criteria. The differentiation between the different aquifers was established based on geometrical data

an intermediate position or bad isolation during the drilling process. These results highlight the differences between the aquifers and support the geometrical classification initially considered.

4 Discussion

The Torrevieja aquifer presents an imbalance that can be explained based on numerical models and hydrochemistry analysis to the exploitation of different aquifers in the region. The increases of the outputs would lead to a compensation by seawater input assuming the connection of the Pliocene aquifer with the sea. The negative gradients farther inland indicate an ongoing saltwater encroachment process. Considering that the pumping regime in the area has been approximately constant during the last 30 years, the aquifer must be in a slow process of saltwater movement inland because of the tendency of decrease of the levels and the pumping cones in the water table surface with values beneath the sea level. The morphology of the aquifer as well as the hydraulic connection would determine if this process would reach the most sensitive areas for the population where most of the wells are located and how long it would take.

Climatic conditions play an essential role for the water resources in this region; the precipitation is very variable with annual oscillations from a maximum of 633–88 mm in a period of 39 years. This hinders the development of forecasts for the groundwater perspectives. Additionally, there are recharge inputs connected with anthropogenic sources, such as the irrigation return flows and the derivations from other catchments that can also be a major element of the water budget, especially during dry periods, although they are totally unpredictable. Other actions that can modify the water budget are, for example, the use of desalination plants for urban supply, the artificial recharge of treated water that can reduce pumping or the potential decrease in the pumping rate if the quality of the groundwater is reduced. This creates a very dynamic environment where it is challenging to establish strong guidelines for groundwater use and forecasting connected for example with climate change scenarios. Nonetheless, a trend of decreasing water resources can be detected in the current aquifer conditions both in the water balance of the model and in the water table elevation of the wells with longer series, even if the climatic conditions did not show a dryer tendency during the period 2002–2007.

The potential exploitation of the lower (Miocene) aquifer in the region can have a great impact for the aquifer management. In this study, it is assumed that most of the pumping wells were located in the lower aquifer, especially the ones with higher rates, whereas that ones with lower volumetric abstraction were pumping in the Pliocene aquifer. This highlights the relevance of this aquifer and opens new questions for further studies of recharge sources, reserves, the degree of exploitation and the connection with other nearby aquifers. Another possibility would be the merging between the Pliocene and Miocene aquifers into one bigger hydrogeological unit in the North and West sectors. In this case, the transfer between aquifers could act as a buffer when any of the aquifers has an excessive degree of exploitation or lack of recharge, although, at the same time, the contamination of one of them could directly affect both.

The differentiation between the Pliocene aquifer and the Miocene aquifer established for the pumping could also be tested by analysis of their hydrochemical properties. Following the same geometrical criteria established for the flow model, the samples were plotted based on the SO_4^{2-} and Cl^- values that could distinguish between saltwater from the sea or from sediments because of the presence of gypsum or other sulfates. There was a chemical differentiation between the two groups designated geometrically in a high proportion, but there were still samples that were not possible to differentiate. In general, the Miocene aquifer groundwater had a higher concentration of Cl^- relative to SO_4^{2-}, possibly related to older saltwater intrusion events, the presence of connate water during the sedimentation process or ongoing processes that have not yet been deeply studied. The hierarchical analysis showed that 20 of 25 samples for the Miocene aquifer had similar hydrochemical properties, and 33 of 42 for the Pliocene aquifer (Fig. 3). A third group was detected where the properties of the groundwater were mixed and it was not possible to differentiate them based on the chemical properties of water. These were interpreted as wells with mixed water because of the location of the screen at an intermediate position or bad isolation during the drilling process. These results highlight the differences between the aquifers and support the geometrical classification initially considered.

5 Conclusions

Two conceptual models were hypothesized to explain the imbalance in the water budget in the Torrevieja aquifer (South East Spain), a region with high demand for water and low resources. The assessment of the feasibility of each conceptual model was tested by the construction of groundwater numerical models and comparison with hydrochemical analyses of water samples. Based on this, it is proposed a conceptual model differentiating the pumping into two aquifer units: the one studied in this work and another underlying aquifer composed of Miocene sediments that would provide up to 75% of the water abstracted for human supply. The differentiation between the two aquifer units, Pliocene and Miocene, was supported also by the distinct hydrochemical signal of groundwater.

Acknowledgements This study was supported by project CGL2016-77503-R, the research group RNM-369 of Junta de Andalucía and the main author has received funding from the People Programme (Marie Curie Actions) of the European Union's Seventh Framework Programme FP7/2007–2013/ under REA grant agreement n° 624496.

References

CHS (2004) Asistencia técnica para el estudio de cuantificación del volumen anual de sobreexplotación de los acuíferos de la unidad hidrogeológica 07.48 terciario de Torrevieja y del acuífero 07.31.145 Cabo Roig. Technical report

García-Arostegui JL, Aragón-Rueda R, Hornero-Díaz J (2003) Situación de los acuíferos costeros de cabo Roig y Torrevieja (Alicante) en relación con la explotación intensiva de sus recursos hídricos. Technology of the seawater intrusion in coastal aquifers: mediterranean countries. IGME, Madrid, pp 445–452

García-García F, Corbí H, Soria JM, Viseras C (2011) Architecture analysis of a river flood-dominated delta during an overall sea-level rise (early Pliocene, SE Spain). Sed Geol 237:102–113

Gómez-Fontalva JM, Calvache ML (2009) Investigación y evaluación de procesos de intrusión marina mediante la aplicación de un modelo matemático de flujo: Acuífero costero de Torrevieja (Alicante). Technical report. University of Granada-Geological Survey of Spain—Diputacion Provincial de Alicante, 121p

IGME (1990) Estudio de la intrusión marina en acuíferos costeros de Murcia y Alicante. Geological Survey of Spain technical report

McDonald MG, Harbaugh AW (1988) A modular three-dimensional finite-difference ground-water flow model: techniques of water-resources investigations of the United States geological survey, Book 6, Chapter A1, 586 p

Mediavilla R, Murillo JM, Santisteban JI (2007) Stratigraphy in the mathematical modelization of aquifers. The case of Torrevieja underground reservoir (Alicante). Boletín Geológico y Minero 118:709–724

Ramos González G, Hornero Díaz J, Sánchez Guzmán J, Viñuales Guillén A (2002) Estudio de los recursos subterráneos de agua salobre en los acuíferos de Torrevieja y Cabo Roig. Technical report H7.02.02 Geological Survey of Spain, IGME

Rodes Amorós JJ, Rodes Martínez JJ (2003) Experiencia en la campaña piezométrica y de calidad en los acuíferos costeros de Torrevieja y cabo Roig enfocada a la ejecución de modelos tridimensionales de intrusión. Technology of the seawater intrusion in coastal aquifers: Mediterranean countries. IGME, Madrid, pp 667–671

Rodríguez Estrella T (2003) Situación de la intrusión marina en la cuenca del segura: evolución desde el TIAC'88. Technology of the seawater intrusion in coastal aquifers: Mediterranean countries. IGME, Madrid, pp 499–507

Soria JM, Caracuel JE, Corbí H, Dinarès-Turell J, Lancis C, Tent-Manclús JE, Viseras C, Yébenes A (2008) The Messinian-early Pliocene stratigraphic record in the southern Bajo Segura Basin (Betic Cordillera, Spain): implications for the Mediterranean salinity crisis. Sed Geol 203:267–288

Tabares-Ródenas P, Martínez Santos P, Martínez Alfaro PE (2009) 3D model of the Torrevieja Tertiary aquifer. Geometry of the aquifer. Boletín Geológico y Minero 120:53–60

Paleohydrogeological Model of the Groundwater Salinity in the Motril-Salobreña Aquifer

C. Duque, J. T. Olsen, J. P. Sánchez-Úbeda and M. L. Calvache

1 Introduction

The changes in groundwater systems can have both natural and anthropogenic origin such as sea-level rise, climate variations, excessive pumping or decrease in the recharge. These modifications present different time scales leading to hydrogeological variations that can impact in aquifers during hundreds or thousands of years (i.e. sea level rising or climate alterations) to fast modifications in a scale of years or decades (i.e. human intervention). If the changes are continuous, the groundwater might require a period to reach an equilibrium condition that, due to the slow flow of groundwater, can take a long period of time specially when considering the hydrochemical composition.

In coastal detrital aquifers, the equilibrium between saltwater and freshwater can be also affected by the variation in the position of the sea border. The aquifers composed by unconsolidated sediments can be exposed to the movement of the sediments and even, with a longer term perspective, can be connected to the genesis of the aquifer. In the Western Mediterranean areas, due to the combination of geological (high runoff associated with low infiltration of rain and slopes), biological (scarce vegetation) and climatic conditions (torrential rain), the transport of sediments

C. Duque (✉)
Department of Geoscience, Aarhus University, Høegh-Guldbergs Gade 2, Aarhus 8000 Aarhus C, Denmark
e-mail: cduque@geo.au.dk

J. T. Olsen
Norwegian Road Administration, Dreyfushammarn, 8012 Bodø, Norway

J. P. Sánchez-Úbeda
Departamento de Geodinámica, Facultad de Ciencias, Universidad de Granada, Av. Fuentenueva s/n, Granada, Spain

M. L. Calvache
Department of Geodynamics, University of Granada, Granada, Spain

M. L. Calvache et al. (eds.), *Groundwater and Global Change in the Western Mediterranean Area*, Environmental Earth Sciences,
https://doi.org/10.1007/978-3-319-69356-9_14

by rivers can produce the progradation of the shoreline distances of thousands of meters. This would have an impact in the saltwater wedge location but the differences between geological and hydrogeological scales of time can be very different.

The simulation of future scenarios due to climatic change includes the modification of the water inputs to aquifers. Additionally, in coastal areas, this would affect to the salinity distribution in the aquifer with a change in the position of the saline wedge. The use of numerical models, as for example, variable-density modelling requires an initial steady-state condition. However the movement of groundwater with different salinities can be a slow process and the assumption of steady state condition from the current groundwater status can be erroneous.

The historical modeling of landscape and the hydrogeology of an aquifer or paleo-hydrogeological modeling, could be used as a method to improve the definition of the initial conditions. This has been developed in previous studies as Lebbe et al. (2008) with the historical evolution of saltwater distribution in the mouth of an estuary for the last 500 years, or Delsman et al. (2014) showing the changes in saltwater intrusion in the Netherlands during the Holocene using variable-density simulations. The last one pointed out that within the 8500 years modeled period, the coastal groundwater distribution never reached an equilibrium with the contemporaneous boundary conditions.

The Motril-Salobreña aquifer is located at the South of Spain (Fig. 1) and changed its geometry in very recent history. The deforestation of mountain areas

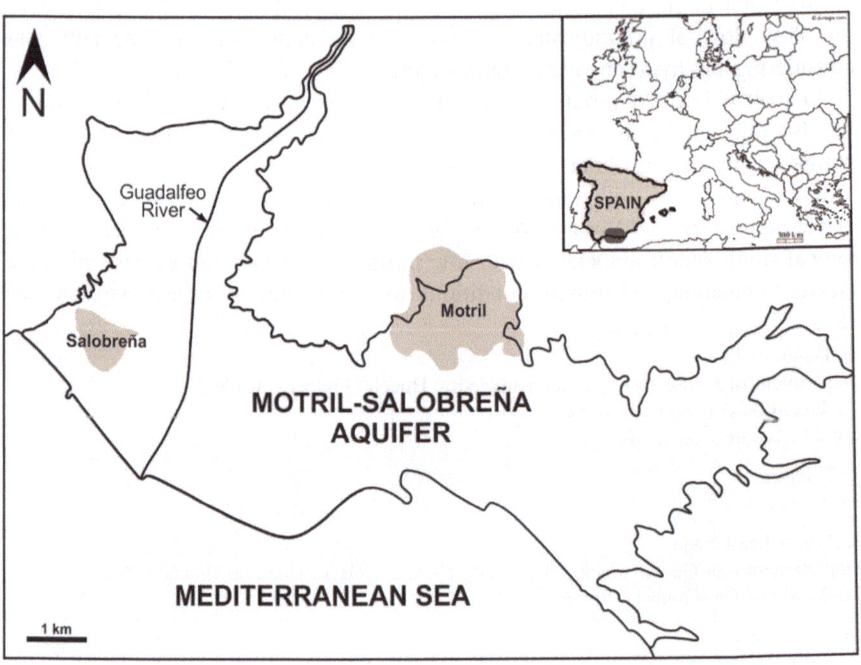

Fig. 1 Location of study area, boundaries of Motril-Salobreña aquifer and main urban areas

due to the colonization combined with the high slopes and the precipitation characteristics in Sierra Nevada, with annual strong storm events, increased the erosion rates.

It has been demonstrated that over the last 500 years the mean coastline advance of the aquifer is over 1200 m (Jabaloy-Sánchez et al. 2014); a progradation of over 2 m each year (Fig. 2). As a result, the location of the interaction sea-aquifer contact has been changing continuously and thereby modifying the natural equilibrium between saltwater and freshwater. Hence, the use of a paleo-hydrogeological model can be useful as a previous step to forecast future changes in the groundwater but also to explain the presence of seawater into the aquifer that could be related to old marine seawater trapped within sediments of low permeability (Duque et al. 2008) or the presence of preferential flow paths between sea and aquifer (Crespo 2012).

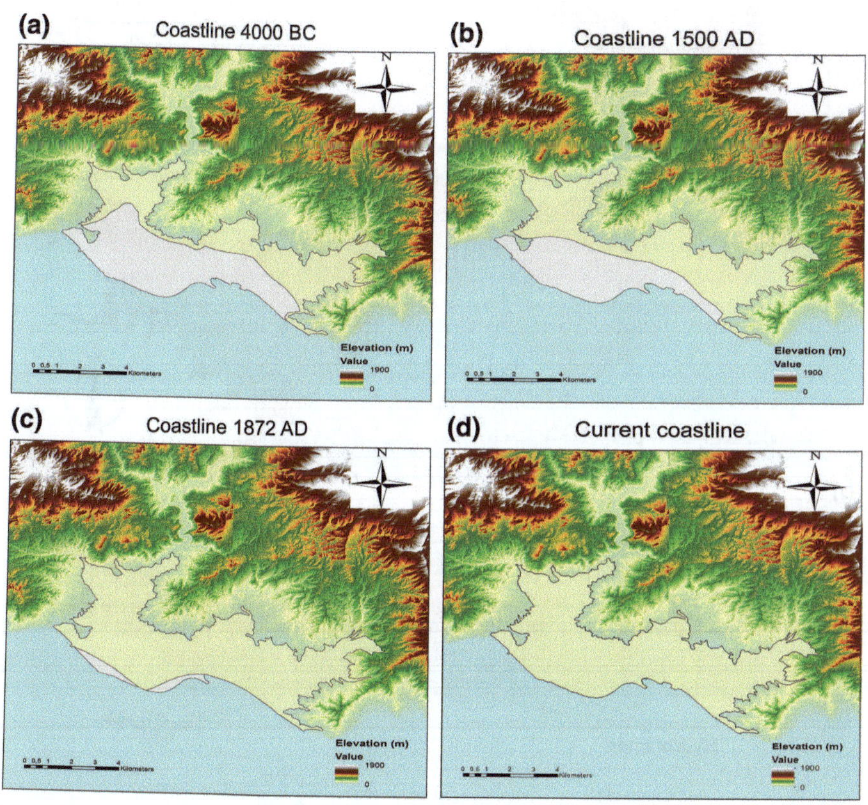

Fig. 2 Evolution of the coastline location in the Motril Salobreña aquifer for the last 6000 years. The *grey areas* indicate the areas covered by the sea

2 Methods

The study of the saltwater distribution in Motril-Salobreña aquifer is strongly linked to the aquifer properties. The aquifer presents differentiated sedimentological origins that would generate differences in hydraulic characteristics. The aquifer geometry was defined based on previous geophysical studies (Duque et al. 2008), the thickness of the aquifer increases from the borders of the aquifer towards the coastline where it reaches more than 250 m in depth (Fig. 3a). An analysis of 26 lithological columns together with a sedimentological interpretation of the different origin of the materials composing the aquifer was accomplished to establish different hydrofacies (Fig. 3b). For each of them it was assigned a range of hydraulic

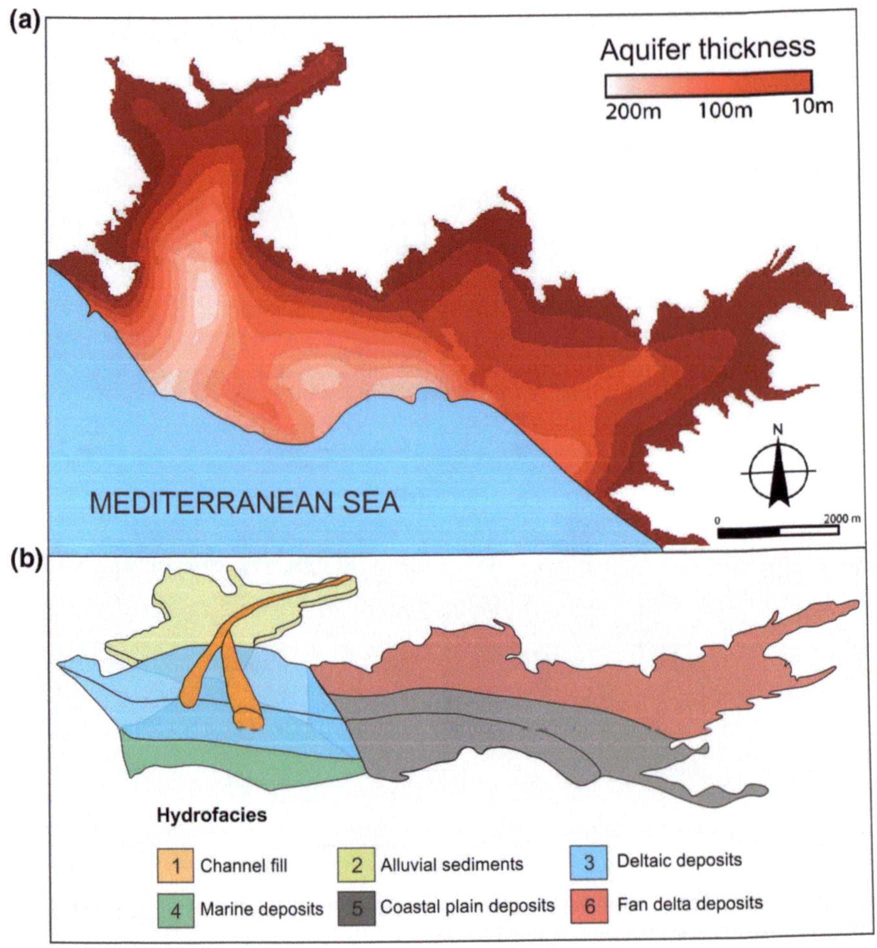

Fig. 3 **a** Motril-Salobreña aquifer thickness based on previous studies (Duque et al. 2008). **b** Hydrofacies within the Motril-Salobreña aquifer based on the sedimentological models

properties based on pumping test data, grain size analysis, bibliography and the review of previous studies using numerical models.

A numerical model with variable density (SEAWAT) (Guo and Langevin 2002) was constructed considering a database with hydrogeological measurements (groundwater heads and flow in the river and recharge from irrigation and precipitation) for the period 2001–2007. The grid used in the model has 16,921 cells with 100 m × 100 m × 20 m in the X, Y and Z dimensions respectively distributed over ten layers. A full description of the boundary conditions can be found in Olsen (2016). The model was calibrated using the pilot points technique with data from 2001 to 2004, and the values for each pilot point was constricted by the hydrofacies unit it was placed in. The model was validated using the data from 2004 to 2007. Once the model was considered valid, the boundary conditions were modified to reproduce the paleoscenarios inferred from the historical data and previous studies. The model was used to determine if the aquifer has reached an equilibrium condition and to study the flushing time for the different hydrogeological units. The model calibration was carried out matching measured groundwater head observations with the calculated heads from the model, and comparing the flow budget of the aquifer with obtained from previous studies (Calvache et al. 2009; Duque 2009). To quantify the agreement it was used the mean error (ME), mean absolute error (MAE) and root mean squared error (RMSE).

To implement the coastline changes in the Motril-Salobreña aquifer and study the salinity evolution of the aquifer the model simulation was divided in two time slices, each corresponding to distinct changes in the location of the coastline motivated by the historical reconstruction of the coastline (Jabaloy-Sánchez et al. 2014). The transitional movement of the coastline was simplified in 2 steps. The first time slice starts with the location of the sea border at 4000 BC as the first known reference point of the coastline (Mean coastline advance 0.09–0.15 m year^{-1}) until to 1500 AD. The second time slice corresponds to the change from 1500 AD to 2007 AD (Mean coastline advance 3.3 m year^{-1}). This division is made based on the acceleration of the changes happened in the last 500 years in comparison with the previous thousands of years that can also help to reach a stationary position of the saline wedge.

The boundary conditions of the aquifer for the past conditions (paleohydrogeological model) are unknown so several assumptions were needed: for the recharge it was considered the rain average and temperatures for the period 2001–2007 while the agriculture irrigation return distribution and quantity was assumed constant in the last 500 years with high fluxes entering the aquifer due to the dominant presence of sugar cane. This region has been under agricultural influence of the Arabs with flooding irrigation system to maintain the sugar cane with water derived from the Guadalfeo River. The pumping was eliminated excepting for the last part of the simulation when was technically possible and the current river characteristics for the period 2001–2007 were considered a good estimate of the river flow since it is included a dry period and a wet period and they were repeated in a loop. The hydrogeological properties were considered the same for all the time periods

Transport parameters used in the model were assigned based on the values estimated by Calvache et al. (2009) in a previous numerical model of the aquifer. They calibrated the parameters by matching the simulated saline wedge with geochemistry data from seven observation points in the aquifer. Depending on location, they estimated that longitudinal dispersivity ranges from 30 to 65 m, the ratio between longitudinal dispersivity and transverse dispersivity was 10 and the retardation coefficient was 1×10^{-7} m^3 kg^{-1}. For the diffusion coefficient it was used a value of 1×10^{-9} m^2 s^{-1} which corresponds to the diffusion coefficient of Cl^- and Na^+, the dominating ions in seawater. The only source of high salinity was the sea border, the rest of the inputs to the aquifer were considered freshwater.

3 Objectives

The objectives of this work are:

- Characterize the Motril-Salobreña aquifer sedimentary history to define both the sea border progradation and the hydraulic properties based on a sedimentary analysis.
- Develop a groundwater numerical model with variable density to define the impact of the changing position of the coastline.
- Analyze the time for the aquifer to reach steady state condition under big scale changes and define the flushing times for saltwater in the different hydrofacies.

4 Results

The sedimentological study of the aquifer showed spatial variations in lithology throughout the aquifer resulting from different depositional processes. It was distinguished four major depositional environments: (1) heterogeneous alluvial sediments in the northern parts of the aquifer with mostly gravel and sand, (2) deltaic deposit in the western part with mostly gravel and sand and some silt, (3) alluvial fans/coastal plain environment in the eastern parts with gravel, sand and silt and (4) marine deposition in the shallow parts, mostly clay and silt. From this the aquifer where delineated into 6 hydrofacies units (Fig. 3). Because of the heterogeneous grain composition observed within each depositional environment, each hydrofacies unit where assigned a possible range of hydraulic properties and the spatial distribution within each unit where estimated using pilot points (Fig. 4).

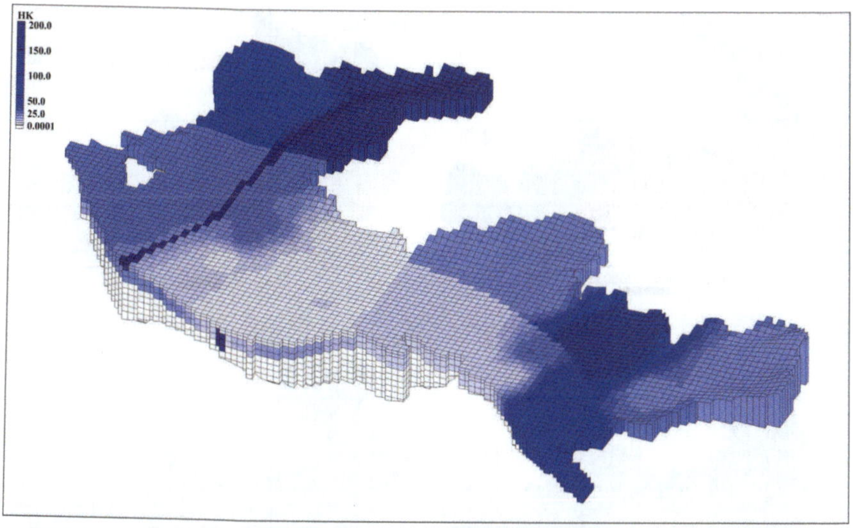

Fig. 4 Hydraulic conductivity distribution for the numerical model based on zones connected to the sedimentological study and the use of pilot points to include the natural variability in each of them

The calibration period of the model has a MAE of 0.54 m, ME of −0.03 m and RMSE of 0.77 and the validation period has MAE of 0.63, ME of 0.05 and RMSE of 0.83. The heads in the observation wells ranged from 35.71 to 0.49 m for the calibration period, and from 29.28 to 0.48 m for the validation period. Therefore, the RMSE relative to the observed head is 2.2% for the calibration period and 2.9% for the validation period.

The salinity distribution based on the paleohydrogeological model shows a fast movement of the saline wedge when the coast starts to prograde around year 1500 AD (Fig. 5). The change in the boundary conditions produce and immediate reaction with freshwater flushing the areas previously occupied with saltwater. This process is heterogeneous due to the changes in hydraulic properties and transport between the different hydrofacies but also due to the heterogeneity introduced in each unit. The area with higher hydraulic conductivity is along the channel of the Guadalfeo River and in 20 years the freshwater front moves several hundred meters towards the coastline while the central zone remains salty or slightly diluted probably due to the effect of freshwater recharge from the top of the aquifer. The top view of the model shows that in 50 years, all the saltwater in the areas with higher hydraulic conductivity can be washed and reached a stationary condition very similar to the current status of the aquifer. For the same period, in the central region the salinity concentration still can remain over 10–20 g/l at several hundred meters from the coastline.

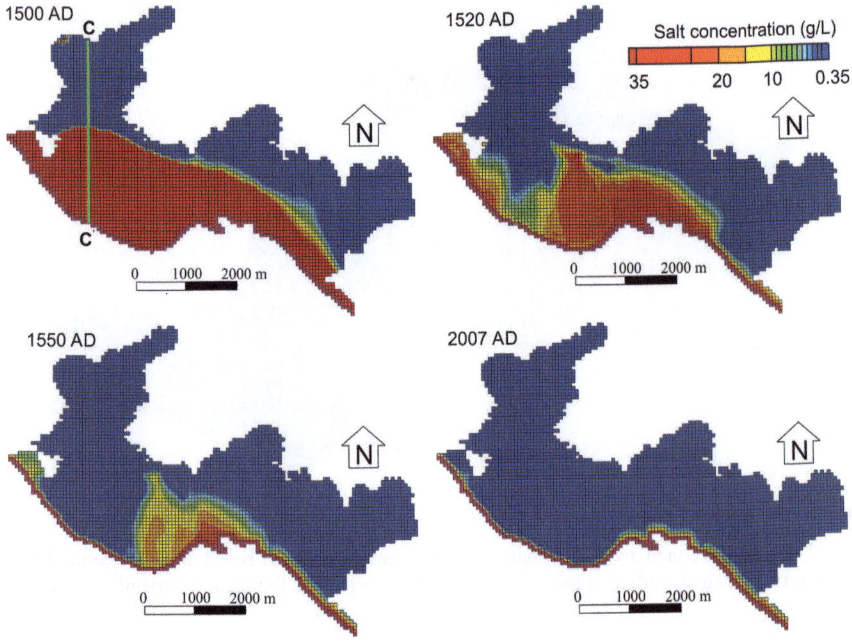

Fig. 5 Transient evolution of the salinity distribution in the Motril-Salobreña aquifer from 1500 AD to 2007

The paleohydrogeological model shows that there is a very distinct pattern in the reactions to the progradation of the sea border depending on the hydraulic properties of the aquifer. This is a good indicator that the aquifer can have a good recovery capability in case of potential saltwater intrusion processes but at the same time indicates that the aquifer can be very sensitive if the recharge is reduced. The area of the river can be flushed in a few decades while the central region can be more in the order of magnitude of 100 years.

The spatial distribution from the top view is not showing all the characteristics. The model results presented in cross section demonstrate also a differentiated pattern depending of the depth. The top part of the aquifer has a much faster reaction to the freshwater flushing that the deeper areas, and reaching a new equilibrium condition can take more than 300 years (Fig. 6). These results, considering the uncertainties in the boundary conditions indicate that some areas of the Motril-Salobreña aquifer could be still reaching an equilibrium derived from the changes in the coastline over the last 500 years. The distribution of sediments can be also critical and it can not be discarded that some geometrical configurations could hinder the washing of ancient saline water in the aquifer.

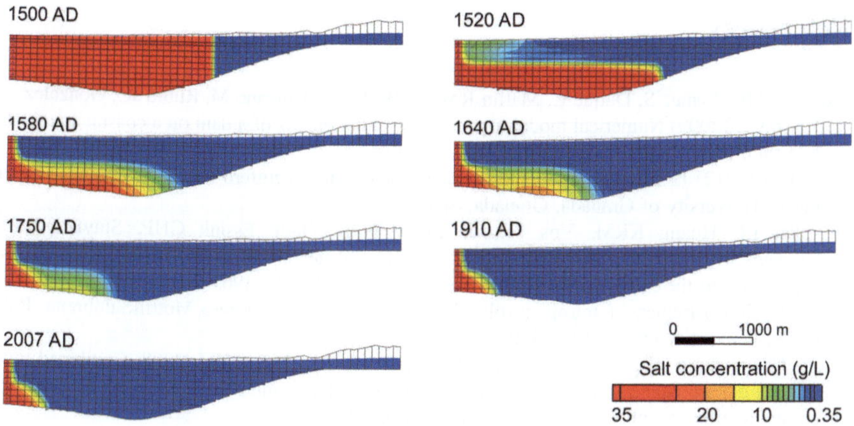

Fig. 6 Cross sectional view of the evolution of the salinity distribution along a S-N profile (in the *left* is located the sea border). Vertical exaggeration 5:1. The location of the cross section is presented in Fig. 5 C–C′

5 Conclusions

The paleohydrogeological model of the Motril-Salobreña aquifer indicates that most likely the aquifer is under steady state conditions in spite of the changes during the last centuries. However, it is possible the presence of saline water that has not been flushed depending on the geometric configuration of the different hydrogeological units or in case of presence of sediments with very low hydraulic conductivity. The flushing times for saltwater in the aquifer ranged from a few decades to a few hundred years depending on the zones that were considered and the hydraulic properties of them. There is a very distinct pattern of the aquifer with depth with very fast flushing in the top layers and much slower in the deeper zones. The areas where the flushing time is faster are also prone to have a quicker reaction if the recharge sources are reduced and generate saltwater intrusion problems due to future climatic changes or anthropogenic actions as intensive use of the water resources in the region.

Acknowledgements This study was supported by project CGL2016-77503-R, which was funded by the Ministerio de Economía y Competitividad (Government of Spain), by the research group RNM-369 of the Junta de Andalucía and the main author has received funding from the People Programme (Marie Curie Actions) of the European Union's Seventh Framework Programme FP7/2007-2013/ under REA grant agreement n° 624496.

References

Calvache ML, Ibanez S, Duque C, Martin Rosales W, Lopez Chicano M, Rubio JC, Gonzalez A, Viseras C (2009) Numerical modelling of the potential effects of a dam on a coastal aquifer in S. Spain. Hydrol Process 23:1268–1281

Crespo FJ (2012) Estudio de las fuentes de salinización en el acuífero Motril-Salobreña. Master thesis, University of Granada, Granada, Spain

Delsman JR, Huang KRM, Vos PC, de Louw PGB, Oude Essink GHP, Stuyfzand PJ, Bierkens MFP (2014) Paleo-modeling of coastal saltwater intrusion during the Holocene: an application to the Netherlands. Hydrol Earth Syst Sci 18:3891–3905

Duque C (2009) Influencia antrópica sobre la hidrogeología del acuífero Motril-Salobreña. PhD thesis, University of Granada, Granada, Spain

Duque C, Calvache ML, Pedrera A, Martin-Rosales W, Lopez-Chicano M (2008) Combined time domain electromagnetic soundings and gravimetry to determine marine intrusion in a detrital coastal aquifer (Southern Spain). J Hydrol 349:536–547

Guo W, Langevin CD (2002) User's guide to SEAWAT: a computer program for simulation of three-dimensional variable-density ground-water flow. Techniques of water-resources investigations of the U.S. Geological Survey, Book 6, Chapter A7, 77 p

Jabaloy-Sánchez A, Lobo FJ, Azor A, Martin-Rosales W, Perez Pena JV, Basrcenas P, Macias J, Fernandez Salas LM, Vazquez Vilchez M (2014) Six thousand years of coastline evolution in the Guadalfeo deltaic system (southern Iberian Peninsula). Geomorphology 206:374–391

Lebbe L, Jonckheere S, Vandenbohede A (2008) Modeling of historical evolution of salt water distribution in the phreatic aquifer in and around the silted up Zwin estuary mouth (Flanders, Belgium). In: Proceedings 20th salt water intrusion meeting, Naples, Italy, pp 141–144

Olsen JT (2016) Modeling the evolution of salinity in the Motril-Salobreña aquifer using a paleo-hydrogeological model. Master thesis, University of Oslo, Oslo, Norway

Numerical Groundwater Modelling as an Effective Tool for Management of the Deep Aquifer at the Dakhla Bay (South of Morocco)

S. M. El Kanti, B. El Mansouri, Y. Arjdal and A. Larabi

1 Introduction

Approximately one-third of the world's land area is arid or semi-arid. These regions face major challenges in the management of scarce freshwater resources under the pressures of population and the global change. Groundwater maybe the only source of water supply in such regions, and so needs water balance quantification to manage its limited resources. This study presents a groundwater flow model to assess optimal pumping rate under an expected increase in irrigation water requirements.

The Dakhla area, in the southern of the Morocco's Sahara, was selected in this study of groundwater resource quantification due to the fast growing need for utilization of non-renewable groundwater resources (Edoulati et al., 2013). Groundwater is mainly used for irrigation, domestic, and small industrial purposes. Under present climatic conditions, the rate of groundwater recharge is very low, so the groundwater resources will be exhausted in the near future if abstraction continues with present trends. Due to the increase in population and expansion of the irrigated area, groundwater management is under environmental constraints. Determining the availability of groundwater resources in the Dakhla area is done

S. M. El Kanti (✉) · B. El Mansouri
Geosciences Laboratory of Natural Resources,
Faculty of Sciences, University Ibn Tofail, Kenitra, Morocco
e-mail: elkanti01@gmail.com

Y. Arjdal
Applied Geology and Geo-Environment Laboratory,
University Ibn Zohr, Agadir, Morocco

A. Larabi
Laboratory LIMEN, Mohammadia School of Engineers,
University Mohammed V of Rabat, Rabat, Morocco

M. L. Calvache et al. (eds.), *Groundwater and Global Change in the Western Mediterranean Area*, Environmental Earth Sciences,
https://doi.org/10.1007/978-3-319-69356-9_15

using a simulation model in order to meet present and future water demands, and also to understand the physical behaviour of the Paleogene aquifer system in southern Morocco (Fig. 1). The Dakhla area lies within the Sahara in southern Morocco which is characterized by an arid climate zone; the aridity index is 0.9 (index De Martonne 1926).

This work describes the history of groundwater pumping from the Paleogene aquifer, the hydrogeological setting, the resulting negative impacts on groundwater levels and the various stages involved in the development of a three-dimensional (3D) simulation model of the interactive aquifer system followed by a prediction

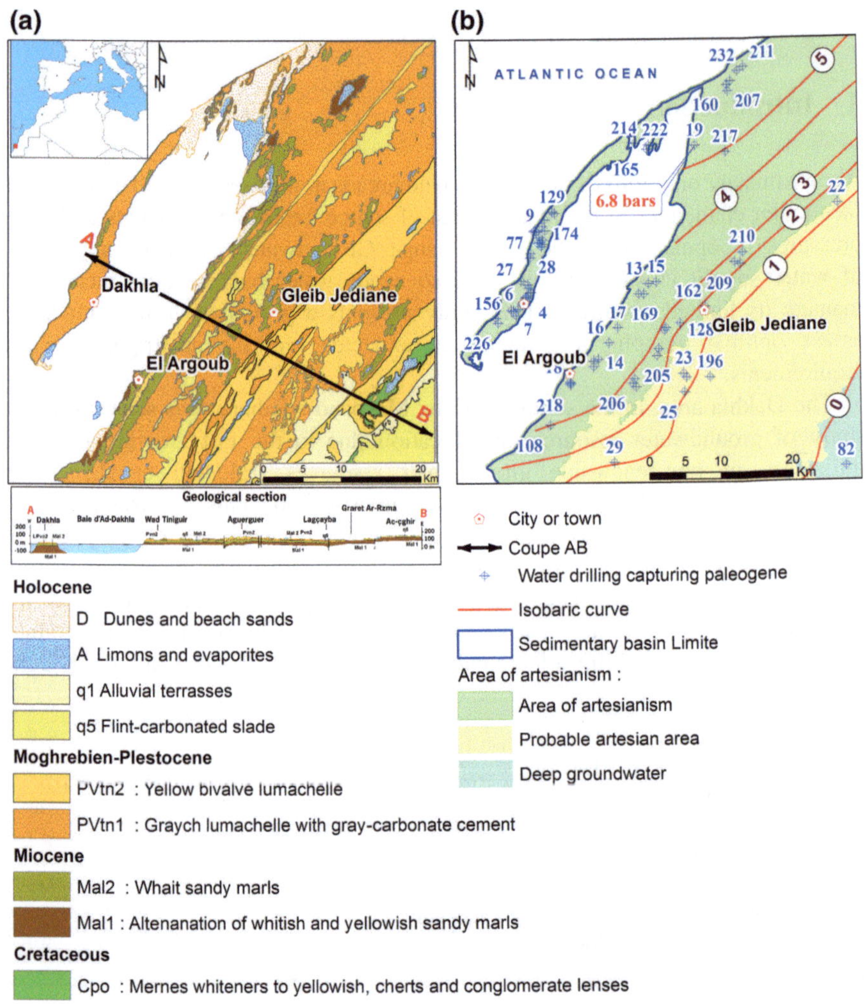

Fig. 1 **a** Location map of the Dakhla area, **b** distribution areas of artesian and isobaric curves map of Dakhla Bay (2011)

stage to evaluate the effects of long-term pumping on groundwater in the study area. Due to socio-economic growth over the last decade and due to consolidation of wells in Dakhla Bay, severe environmental problems have increased, with a steady decline in the piezometric surface area.

2 Objectives

The objective of this study is to find an optimal management solution to satisfy future water needs. This will be achieved through numerical groundwater flow modelling. The model results document a method to quantify the sustainable yield of a groundwater flow system. The transient calibration was used to simulate potential future water use scenarios, to guide the technical, regulatory and financial measures that constitute the contract of the Dakhla groundwater.

Note that there is no valuable predicted precipitation until 2025 that could be included in the scenarios.

3 Hydrogeological Background

The Paleogene aquifer is part of the Aaiun sedimentary basin (Labails and Olivet, 2009) in the Moroccan Sahara and attains a thickness of more than 250 m. It is represented by the formation of Puerto Rico (Rjimati and Zemmouri, 2002). The Dakhla area is located in the hyperarid region of Sahara. The groundwater resources of the study area flowing into a complex aquifer, which consists on (from bottom to top): a Lower Cretaceous formation contains a deep aquifer flowing into deposits of white sands and red sandy clays, a Paleogene formation containing a deep aquifer flowing in sandstones and sands, and a formation of Pliocene-Quaternary containing confined aquifer. The deep groundwater of the study area is fossil water and is non-renewable, confirmed by National Energy Center of Sciences and Nuclear Techniques (CNESTEN). More than 60 production wells were drilled within the area and the salinity varies between 1.8 and 2.8 g/l (Hydraulic Basin Agency). The area of Dakhla Bay exploited and explored by the electrical geophysical method (VES) were defined as a prerequisite to representing the geometric framework of the groundwater flow modelling.

The substratum characterized by marl is encountered at various depths ranging from +40 to −720 m relative to the mean sea level reference. The saturated thickness of the aquifer is highly variable ranging between 10 and 250 m. The lower boundary of the aquifer system is represented by the semipermeable sandy marl. The roof and the substratum of the aquifer of Paleogene shows a general decreases from East to West and North, and South to the North, except for a slightly lower area at Gleib Jediane in the center of the study area. The eastern boundary is represented by a discontinuity D1 representing the outcrop of the aquifer roof. The western boundary of the aquifer is unclear, because the formation continues under the sea (Fig. 2a).

Fig. 2 **a** Three-dimensional (3D) simulation model of aquifer of Paleogene, **b** head distribution of the pre-development calibrated steady simulation flow model 2005, **c** total head distribution of the calibrated transient flow model 2015, and **d**, **e** calculated versus observed head corresponds respectively to steady simulation flow model 2005 and transient flow model 2015

Consequently, the Dakhla Bay may have a considerable reserve of groundwater. All the wells in the study area are artesian (Fig. 1b); for instance the pressure at the well head no. 19 can reach 6.8 bars (Fig. 1b). Owing to low abstraction rates prior to 2005, the static water-level contour is considered to represent the steady-state conditions of the aquifer. The aquifer system is recharged from the east part of the study area, and naturally discharges to the west. Water-level monitoring from 1984 until 2015 was carried out in several observation boreholes.

Measurements in most of the boreholes record a decline of water level with time; evidently the heavy pumping of the aquifer has caused significant drawdown. The abstraction rate was relatively low until 2005; since then it has increased tremendously, which seemingly corresponds to the progressive population increase and the multiplication of agricultural projects.

The average increase of depth to water, measured from ground elevation, is about 15 m over the last 10 years, showing that the water level declined by 1.5 m $year^{-1}$. Evidently, the former remarkable decline of water level is due to the drilling of more than 60 production wells within the study area.

4 Results

4.1 Steady-State Conditions in 2005

The calculated boundary inflow and outflow in 2005 were estimated as 1.9×10^4 $m^3\ d^{-1}$. The water budget from the numerical model lies within the tolerable range of the previously estimated total budget. After calibration of the boundary conditions, the reducing head variance was adjusted by modification of the hydraulic conductivity distribution. The hydraulic conductivity has been divided into five zones and varied between 1.6×10^{-5} and 1×10^{-5} m s^{-1}, which are consistent with values calculated from pumping tests on wells in the study area, electric fields, and geological facies present.

For the inverse leakance field and after several simulations, the values are divided into five areas according to the geology, and the thickness of the layer aquitard, these values varied between 3.7×10^{-11} and 5.7×10^{-11} m s^{-1}. The total head varies between 30 and 90 m, while the RMS error is 5 m. It is evident that the total head of groundwater in the study area correlated well with the initial heads of 2005 (Fig. 2b).

4.2 Transient Conditions

Based on the pattern of the aquifer parameters obtained during steady-state calibration, the model was subjected to transient calibration for a period of 10 years between 2005 and 2015; the mean records hydraulic head to 2005, 2011 and 2015 are meanly valuable information used for model calibration, in addition to irregular information in time of some observation wells. During each stress period (year) all external stresses and boundary conditions were kept constant. The transient run was simulated from the calibrated steady flow model; the only modification is that of the quantity in flows through the eastern boundary, the flow imposed will therefore be 8500 $m^3\ d^{-1}$ as a calibrated value. The simulated piezometric levels at 2015 are shown in (Fig. 2c). On the calibrated transient model, the value of the homogeneous Specific Storage of $1 \times 10^{-5}\ m^{-1}$ was adjusted by the values $Ss1 = 5 \times 10^{-5}\ m^{-1}$; $Ss2 = 6 \times 10^{-5}\ m^{-1}$ and $Ss3 = 8 \times 10^{-6}\ m^{-1}$ the mean value are for different areas.

As for steady-state calibration, it is necessary to carry out transient-mode sensitivity tests. In this case, the analysis was carried out with respect to the storage coefficient and the recharge by inverse drainage. The results obtained by these sensitivity tests show that the model is always sensitive to recharge. The model verification was carried out with the final run of the transient calibration stage at the end of 2015; a scatterplot compares the simulated and observed.

5 Discussion

5.1 Model Predictions and Aquifer Potential

Unfortunately, human activities, such as groundwater extraction for irrigation, have resulted in excessive groundwater level declines and aquifer overdraft. A successfully calibrated and verified simulation model can be used for formulating various different water development schemes. The schemes can be compared in terms of their feasibility for efficient utilization of the available groundwater resources. The selected development schemes must, when implemented, meet the future water demand at a minimum cost commensurate with legal, organizational, political and environmental considerations. Three scenarios were considered to predict the consequences of future development in the model area. A planning period of 10 years (2015–2025) was used for the three scenarios, related to of a draft groundwater contract, the aim of which is to rationalize the exploitation of this water table, in a framework of consultation involving all the parties concerned. The starting conditions in each scenario were those obtained from the transient simulation at the end of 2015. These three scenarios are as follows:

Scenario I The first scenario assumes that the groundwater abstraction observed during 2015 continues without modification. The destocking of the Paleogene GW would be 13.8 Mm^3 in 2025 (Fig. 3a).
Scenario II Maintenance of current agricultural levies and simulation of future drinking water needs, with the commissioning of the purified wastewater treatment plant (WWTP). The destocking of the Paleogene GW would be 14.7 Mm^3 in 2025 (Fig. 3b).
Scenario III Increase in agricultural levies for irrigation of 150 ha per year up to 2022 and simulation of future drinking water requirements, The destocking of the Paleogene GW would be 20.7 Mm^3 in 2025 (Fig. 3c).

Scenarios I and II show a cone of depression with a maximum drawdown of about 10 m, the use of wastewater would reduce the impact of future drinking water requirements. The option of the Green Morocco Plan's strategy (www.agriculture.gov.ma) assumes an increase in cultivated areas of more than 200% (2000 acres), the model prediction shows the maximum drawdown in collecting drinking water area and the agricultural area, would be about 20 m for scenario III.

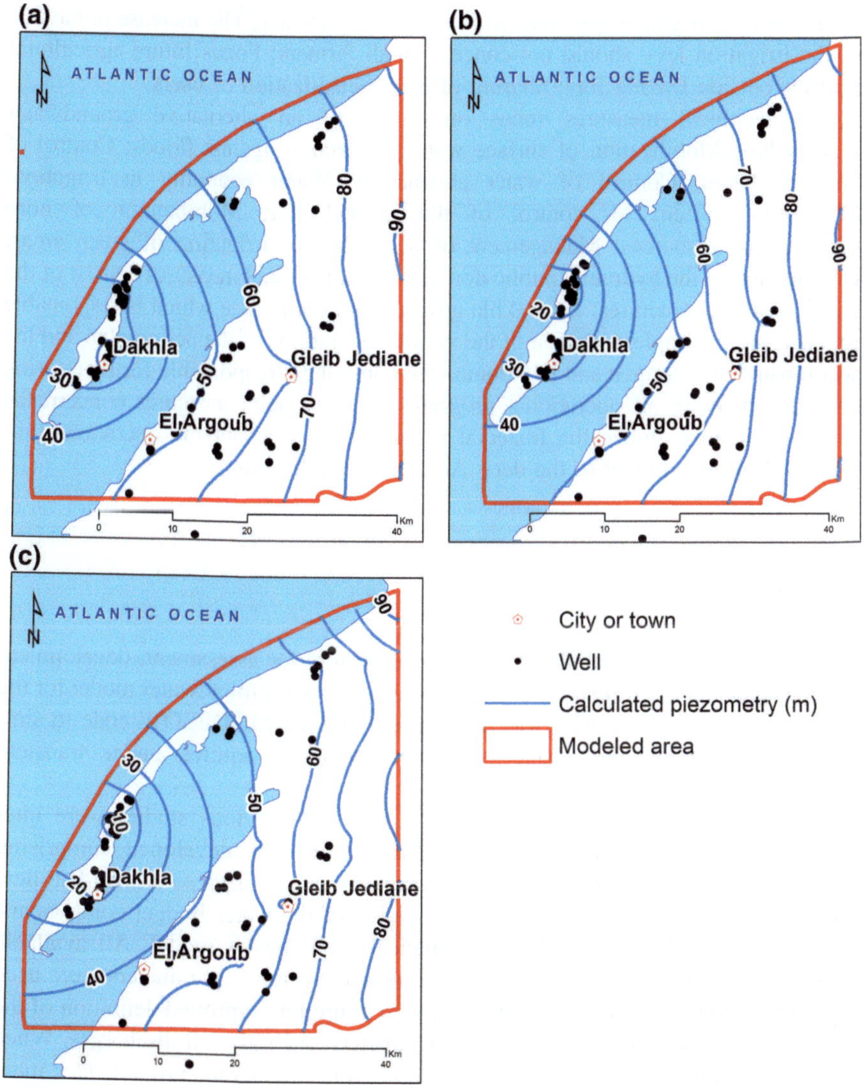

Fig. 3 Calibrated transient flow model in 2025 for: **a** scenario I, **b** scenario II, **c** scenario III

5.2 Groundwater Contract

Based on scenario III, a Dakhla groundwater contract is signed by the stakeholders to safeguard the groundwater and to develop the socio-economic activities of the region. The partners have agreed to respect the following principles which aim to safeguard the groundwater: extensions of agricultural areas will benefit primarily social and solidary agriculture; Use desalination of seawater to meet future water requirements; Irrigate green areas from treated wastewater; The increase in the rate of the irrigation levy should not concern small farmers; Focus future agricultural projects outside Dakhla Bay; Sharing of data; Sensitization of users.

The technical measures focus on: study of an alternative groundwater "Lagçayba"; Mobilization of surface water; Protection against floods; Control of volumes taken; Control of water abstraction; Water economy in irrigation; Monitoring system for control of the groundwater; Mobilization of non-conventional resources; Management of boreholes for irrigation of green areas; Delimitation of the hydraulic public domain. Organizational measures consist of the creation of two structures: the Dakhla groundwater committee which is responsible for ensuring the implementation of the various operations of the action plan and the accompanying measures and the monitoring committee Responsible for monitoring and implementing all operational programs. The financial measures concern the commitment to mobilize the financial resources to carry out the operations programmed in the contract of the deep Aquifer of Dakhla Bay.

6 Conclusion

Numerical groundwater modelling is a powerful tool for assessment, development and management of groundwater resources. A numerical groundwater model for the Dakhla arid area in Morocco was developed using the MODFLOW code to simulate local groundwater changes in the Paleogene aquifer under transient conditions.

Hydrogeological data from different sources and previous studies were integrated in the model. Initially a conceptual model was developed comprising identification of hydrogeological units, parameterization of these units, identification of the system boundaries and identification of the water budget components. These were transferred into the mathematical groundwater model. All available measurements of water level, well discharges and hydraulic parameters were used for model calibration. The model was optimized through improved definition of the boundary conditions and other relevant parameters, such as well discharges. When the model calibration and validation were complete, different groundwater management scenarios were evaluated to find an optimal management solution to satisfy future needs. Three different development schemes were assessed to test the Paleogene aquifer under long-term water pumping stresses. Scenario III, with a gradual increase in pumping for agricultural purposes over the next 10 years at 150 ha/year is selected after extensive consultation with all stakeholders in the water sector in the Dakhla region. A groundwater contract between all stakeholders was developed to save the underground resources and not compromise the socio-economic development of the region, particularly agricultural development. This contract, includes technical, organizational and financial measures, is concerted with all the stakeholders in order to govern the Paleogene water resources at the Dakhla Bay in southern of Morocco.

Acknowledgements The authors thank the members of the Hydraulic Basin Agency Sakia El Hamra and Oued Ed-Dahab, for their collaboration and for the documents that they make available to us.

References

Edoulati N, Boutaleb S, Bettar I, Ouchbani A (2013) Contributions of chemical and isotopic tools for understanding the groundwater modes recharge and flow in the lower cretaceous aquifer in the Moroccan Sahara. J Water Resour Prot 5:183–199

Labails C, Olivet J-L (2009) Crustal structure of the SW Moroccan margin from wide-angle and reflection seismic data (the Dakhla experiment). Part B—The tectonic heritage. Tectonophysics 468(1–4):83–97. doi:10.1016/j.tecto.2008.08.028

Rjimati E, Zemmouri A (2002) Geological map of Morocco at 1/100 000.Ad-Dakhla sheet, NF-28-XI-3. Notes Mem Geol Serv no 438bis. ISSN 0374-9789

Delineating the Aquifer Role in the Anthropogenic Fingerprint on the Groundwater-Dependent Ecosystem of the Biguglia Lagoon (Corsica, France)

M. Erostate, E. Garel, F. Huneau and V. Pasqualini

1 Introduction

The lagoons represent hydrosystems with major economic and ecologic interests particularly in the Mediterranean. The lagoon behaviour is intrinsically dependent on the freshwater quality and quantity brought to the lagoon in terms of control of salinity, nutrients fluxes and unfortunately pollution too. Groundwater is a freshwater component of first order to understand the global lagoon functioning. In this study we propose to clarify the aquifer behaviour in connection with the Biguglia lagoon and to estimate the role of the groundwater on the quality of the lagoon.

The Biguglia lagoon is the largest wetland of Corsica Island and is located south to the densely urbanized city of Bastia which is an important urban and commercial area of Corsica. To preserve the ecosystem, the Biguglia lagoon is recognised as a RAMSAR site (wetlands of international importance, especially as waterfowl habitat) since 1991.

2 Study Site

The Biguglia coastal watershed has a surface area of 182 km^2 with a maximum altitude of 1450 m a.s.l. The hills of the catchment area made up of lustrous schists (Forzoni et al. 2015) and the plain is composed of alluvial quaternary deposits (3–40 m of thickness), partly covered by the Biguglia lagoon (14.5 km^2), and corresponds to the main aquifer of the watershed. The lagoon is separated from the Tyrrhenian Sea by a sand bar and the seawater exchanges are controlled by a

M. Erostate · E. Garel · F. Huneau (✉) · V. Pasqualini
UMR CNRS SPE 6134, Hydrogeology Department, University of Corsica, Campus Grimaldi, BP 52, F-20250 Corte, France
e-mail: huneau@univ-corse.fr

M. L. Calvache et al. (eds.), *Groundwater and Global Change in the Western Mediterranean Area*, Environmental Earth Sciences,
https://doi.org/10.1007/978-3-319-69356-9_16

narrow channel located in the North part (Fig. 1). The catchment is drained by a perennial stream, the Bevincu River, and 3 temporary rivers (Pietre Turchine, Rasignani and Mormorana Rivers) which are only active during the rainfall periods (Huneau et al. 2016).

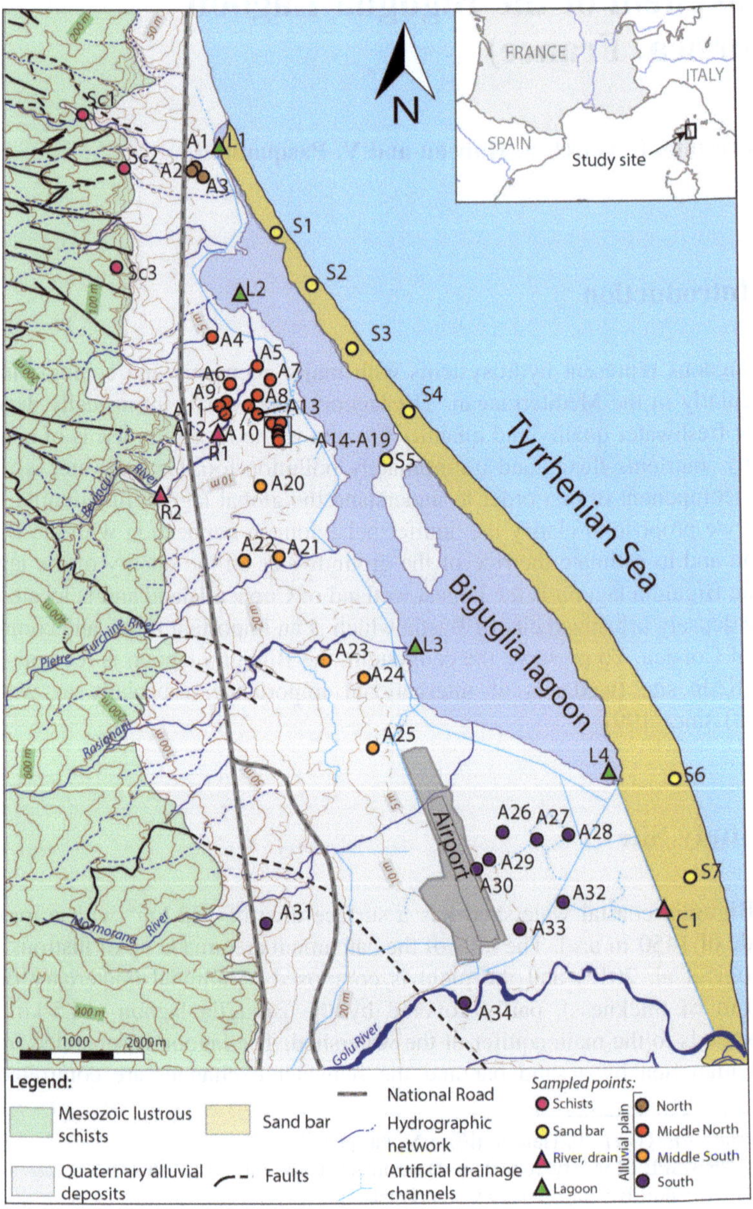

Fig. 1 Geological map of the study area and location of the sampling points (from Garel et al. 2016)

In the early XXth century, an artificial drainage network was realized to collect the subsurface groundwater on the western part of the plain and to use the land for the agriculture activities.

The Golu River borders the south of the alluvial plain and its channel is located 4 km south from the lagoon. Naturally, the Golu River is not integrated in the lagoon watershed, however, an active artificial canal links the river to the lagoon. Regarding the opening of the Golu narrow channel, the river water goes into the lagoon or inversely.

3 Methodology

The groundwater behaviour, at the catchment scale, was investigated by a multi-tracer approach (physico-chemical parameters, major ions, ^{18}O, ^{2}H and ^{3}H). In April and September 2015, during the high and low water period, 42 groundwater points, 6 river points, 1 drain point and 4 lagoon points were sampled (Fig. 1).

Most of the selected groundwater points correspond to private wells unused and often poorly maintained due to the derelict of the vineyards since 30–40 years. 10 wells (A12–A19 and A34) are pumped for the drinking water supply of Bastia city and its surroundings.

4 Results and Discussion

4.1 Spatio-Temporal Hydrodynamic

All the groundwater flow line directions converge to the lagoon and the hydraulic gradient calculated from the water table data ranges between 3 and 5‰ showing

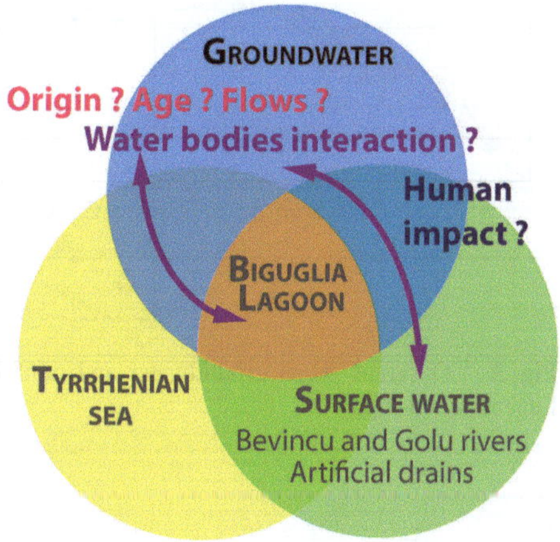

Fig. 2 Strategy of investigation for understanding the groundwater behaviour in connection with the Biguglia lagoon

favourable hydrodynamic behaviour. This result is in agreement with the conclusions of a previous hydrogeological study led in 2010 (BRGM 2010) on the same plain during also the high water period.

The seasonal variations of the groundwater level reach a maximum of 2 m, this variation is highly different from the Bevincu River flow variations. Indeed, the stream flows exist mainly during the autumn and winter period. And all along the rest of the year, very low flows go into the lagoon, while groundwater have, all over the year, a favourable flow towards the lagoon (Fig. 2).

4.2 Groundwater Origin

The isotopic signature (^{18}O, ^{2}H) shows 3 major origins of the groundwater recharge (Fig. 3):

- A recharge in altitude (headwater catchment of the Bevincu) for the groundwater located in the Northern part of the alluvial plain.
- A local recharge in the coastal plain for groundwater of the Southern part and of the sand bar. Their signatures are similar to the weighted mean isotopic rainfall signature of the Bastia rainfall station (monthly data since 2013).
- A seawater intrusion detected for the wells located in the southern plain near the lagoon.

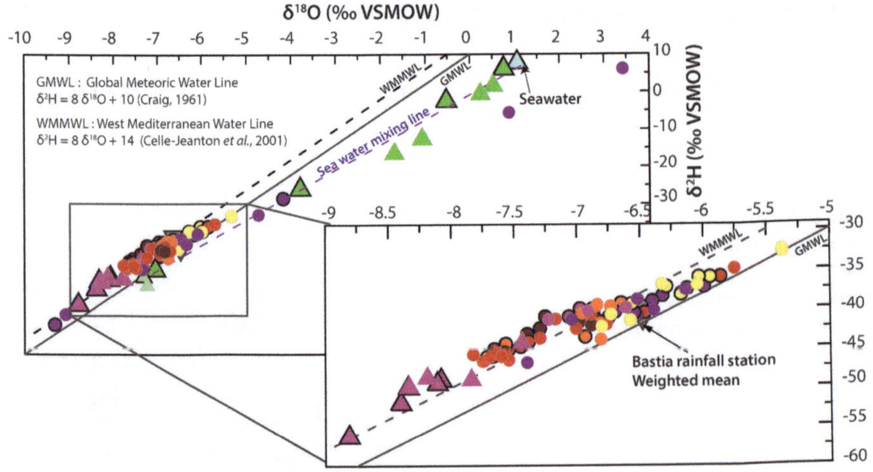

Fig. 3 ^{18}O versus ^{2}H in groundwater and rainfall at Bastia station (from Huneau et al. 2016). Legend as on Fig. 1

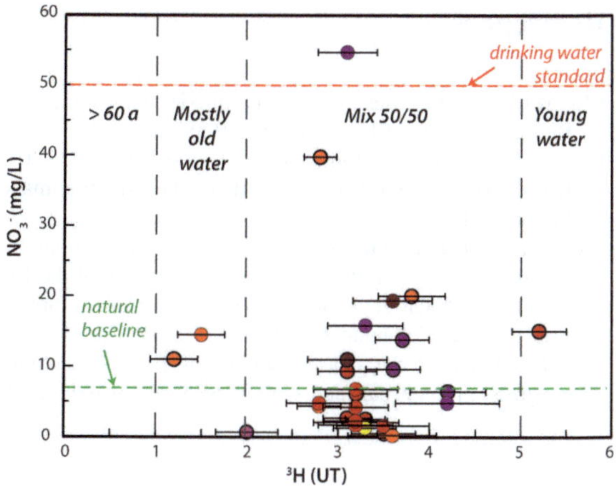

Fig. 4 Tritium versus nitrate on 15 groundwater samples from April 2015. Legend as on Fig. 1

4.3 *Long Term Anthropogenic Impact*

Most of groundwater samples show concentrations in nitrate above the natural groundwater baseline (5–7 mg/L) and in September A33 is above the drinking water limit of 50 mg/L (Fig. 4).

Tritium activities reveal that most of groundwater in the plain is a mixing in equal proportion between old and recent waters. One point, A24 close to the lagoon has very low activities (1.2–1.5 ± 0.3 UT) showing the presence of old water and the existence of inertial groundwater bodies.

5 Conclusion

The presence of old groundwater dated with ^{3}H indicates a current pollution of the groundwater due to former agricultural practice. The impact of the commercial gardening, which is a recent land activity, is not yet visible. However, decennial residence times of groundwater indicate in the near future, a potential threat to the drinking water supply quality as well as to the lagoon water quality by the current agricultural activities. The noticeable nitrate concentrations, above the drinking standards limits, and the trace elements content of the groundwater, compared to the low concentration in the surface water, reveals to be a major source of future potential disturbances for the lagoon and its dependent ecosystem.

Acknowledgements This study is supported by the CNRS Labex DRIIHM OHM "Littoral Mediterranéen" project.

References

BRGM (2010) Etude des interactions entre les eaux souterraines, les eaux de surface et l'étang de Biguglia – Rapport final. BRGM/RP-59068-FR

Forzoni A, Storms JEA, Reimann T, Moreau J, Jouet G (2015) Non-linear response of the Golo River system, Corsica, France, to Late Quaternary climatic and sea level variations. Quat Sci Rev 121:11–27

Garel E, Huneau F, Jaunat J, Celle-Jeanton H, Santoni S, Garrido M, Pasqualini V (2016) Groundwater role on the ecological preservation of the Biguglia lagoon (Corsica, France). In: 35th international geological congress, Cape Town, South Africa

Huneau F, Garel E, Jaunat J, Robert S, Fox D, Garrido M, Pasqualini V (2016) Land use and vulnerability of the coastal aquifer of Biguglia lagoon (Corsica, France). In: 43rd IAH congress, groundwater and society: 60 years of IAH, Le Corum, Montpellier, France

Change in the Hydrological Functioning of Sand Dune Ponds in Doñana National Park (Southern Andalusia, Spain)

A. Fernández-Ayuso, M. Rodríguez-Rodríguez and J. Benavente

1 Introduction

Doñana natural area is located in southwestern Spain, between the Guadalquivir River Estuary and the Atlantic Ocean. Its coastal area is constituted by aeolian sands, which form part of both actives and stabilized dunes (Fig. 1). The dunes play a crucial role in this study case, due to the fact that over 3,000 ponds are formed in rainy years (Green et al. 2016). These ponds are found close to the phreatic surface, which enables the existence of groundwater input to most of them.

The Doñana aquifer spreads over 2,600 km^2 in the provinces of Huelva and Seville (southwest Spain). This territory depends to a large extent on its groundwater for water supply to towns and crop irrigation. The natural area of Doñana is located over this aquifer, and it includes a protected area (National Park, Natural Park and Biological Reserve) of almost 1,080 km^2. Many of the ecosystems in Doñana depend on the groundwater, which are found at a very shallow depth over wide areas. The aquifers also feed numerous seasonal and permanent ponds and the main streams in the area (Olías and Rodriguez-Rodriguez 2013). The groundwater extraction from the Doñana area has caused a hydrological impact in some parts of the park, as different organizations have been warning for several years (Custodio et al. 2009). There is evidence that the water table fall, as a consequence of groundwater extraction, is causing changes in the vegetation, desiccation and a decrease in the pond water levels located near a coastal resort, as well as in the amount of water entering in streams (Serrano and Serrano 1996; Muñoz-Reinoso 2001; Serrano and Zunzunegui 2008; Custodio et al. 2009; Manzano et al. 2009).

A. Fernández-Ayuso (✉) · M. Rodríguez-Rodríguez
University Pablo de Olavide, Utrera Rd Km 1, 41013 Seville, Spain
e-mail: aferayu@upo.es

J. Benavente
University of Granada, Fuente Nueva Campus, 18071 Granada, Spain

M. L. Calvache et al. (eds.), *Groundwater and Global Change in the Western Mediterranean Area*, Environmental Earth Sciences,
https://doi.org/10.1007/978-3-319-69356-9_17

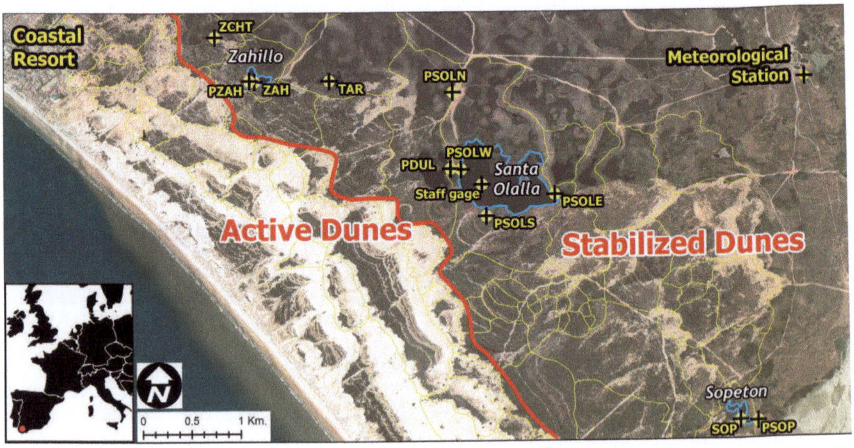

Fig. 1 Location of Zahíllo, Santa Olalla and Sopeton ponds and sensors on nearby piezometers. Palacio de Doñana meteorological station can also be seen. All of these ponds and sensors are located over the stabilized dunes. In the top left margin, a coastal resort (Matalascañas) is marked

Doñana aquifer is formed by two units: one confined and the other one unconfined. The studied ponds lie over the unconfined aquifer, classified by Vela (1984) as "Dune aquifer". It has a surface of 90 km^2, and extends from the Atlantic Ocean to the marshes. Its major components are permeable aeolian sands, with presence of estuarine clays and silts. An impervious bed of marine marls delineates the base of the aquifer. This heterogeneity in the aquifer composition makes difficult the establishment of the precise relationship between surface and groundwater.

The objective of this study is to reach a better characterization of the hydrological functioning of these sand dune ponds through the study of their daily water balances and hydrochemistry. The data period selected in this study was from January 2015 to December 2016 (two years).

2 Study Site

This research is focused on three ponds situated on Doñana Biological Reserve: Santa Olalla (SOL), Zahillo (ZAH) and Sopeton (SOP) whose morphometric characteristics are shown in Table 1. These ponds are located on the west side of the Park, on stabilized and vegetated dunes (Fig. 1). Large seasonal marshes are found on the east, which together with the mentioned ponds, configures unique wetlands endowed with several protecting status. The three ponds are located 2 km from the Atlantic Ocean.

In the study area the climate is Mediterranean, with high variability between the dry and the wet season. The annual rainfall, average of the years 2015 and 2016,

Table 1 Main morphometric indexes of the studied ponds

Name	AFS (ha)	WS (ha)	ER (mm/year)	Sal. (g/l)	Max. depth (m)	Min. Depth (m)	Height (m a. s.l.)
SOL	25	155	1684	3–16.6	2.19	1.21	2.45
ZAH	4.8	36	1538	0.3	0.38	0	9.08
SOP	2,3	38	1629	1.5	0.71	0	3.8

AFS Average flooded surface; *WS* Watershed; *ER* Evaporation rate; *Sal.* Salinity. Height refers to altitude of the bottom pond in meters above sea level

was 540 mm/year. The average daily temperature ranged from 5 °C in winter to 33 °C in summer.

3 Methodology

Submersible pressure transducers (Diver, Level-logger, and CTD-Decagon) were installed in each of the piezometers mentioned above and programmed to take three-hourly measurements of groundwater table. A Diver CTD-type sensor was also installed on the staff gage in the three ponds (ZAH, SOP and SOL). This sensor was programmed to measure the electrical conductivity, temperature and depth of water every three hours. Also, atmospheric pressure transducers were installed in close locations: a meteorological station and PSOLW (Fig. 1). Data were analyzed and instrumental errors, if found, were corrected.

In order to determine the variables from the water balance in Santa Olalla pond, a conceptual model developed previously by the authors Rodríguez-Rodríguez et al. (2016) was followed (Fig. 2). Input from rainfall in the pond (P) was calculated from the data of nearby meteorological stations (Almonte-Rocío and Palacio de Doñana). To calculate losses by direct evaporation from the water surface (E), the original Penman formula (1956) was used, which according to McMahon et al. (2013) is the most appropriate method for shallow ponds. Changes in the volume of water stored in the pond (ΔS) were determined from water level records. Surface run-off is negligible due to the high permeability of the aeolian sands and the absence of watercourses in the area. Consequently, Basin Discharge (BD) refers to the groundwater discharge from the hydrogeological basin. This component of the balance was estimated as the incognita of the balance (imbalance), using the water balance equation:

$$\boldsymbol{BD} = \boldsymbol{E} - \boldsymbol{P} \pm \Delta \boldsymbol{S}$$

The hypsometric curve of the lake basin, estimated by the equation developed by Hayashi and Van der Kamp (2000), was used in order to determine the pond surface at each time-step. The model in the three ponds was carried out in a period were no precipitation occurred (May-August 2016). Comparing the evolution of the

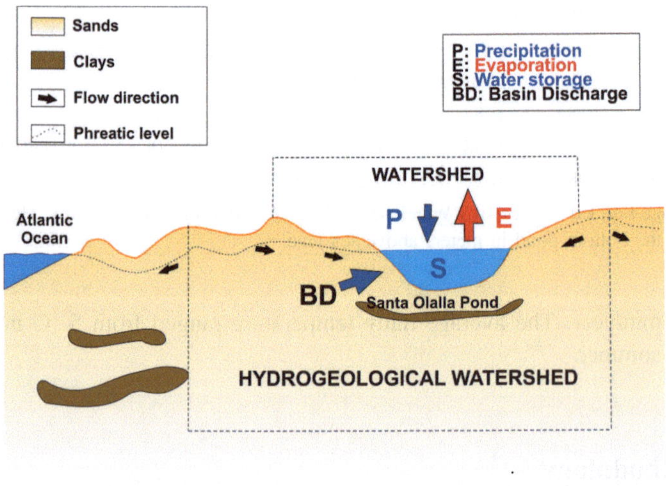

Fig. 2 Simplified hydrogeological conceptual model of Santa Olalla pond

evaporation from the pond surface and the actual measured level (*ΔS)* during the studied period—in mm/day—we estimated groundwater input or outputs to the aquifer.

The main components analysis of the water samples was performed with a DIONEX ICS-1000 ionic chromatograph. Bicarbonate analysis was made by colorimetry before a maximum period of five days since the sample was collected. Treatment of hydrochemical data has been made using AquaChem software (Waterloo Hydrogeologic) and Piper diagrams were plotted.

4 Results and Discussion

Results of the water balance in Santa Olalla pond, expressed as accumulated volume (m^3) are shown in Fig. 3, distinguishing between inputs (P and BD) and outputs (E) and the accumulated daily variation in storage. It has to be taken into account that the component BD has been has been calculated as the incognita of the equation. In this sense, this component of the equation accumulates all the residual errors of the rest of the estimated parameters (i.e., P, E and ΔS). A progressive increase was seen in this variable, thus, the water input into the pond by this means was constant. There were five moments in which higher values were observed. The output in this case was due to evaporation and, occasionally, to the discharge of water from the pond to the aquifer, as can be seen in the decreases in BD observed in late October, early November 2015, as well as in early May, November and December 2016. As it is shown in Table 2, the input due to direct rainfall during the

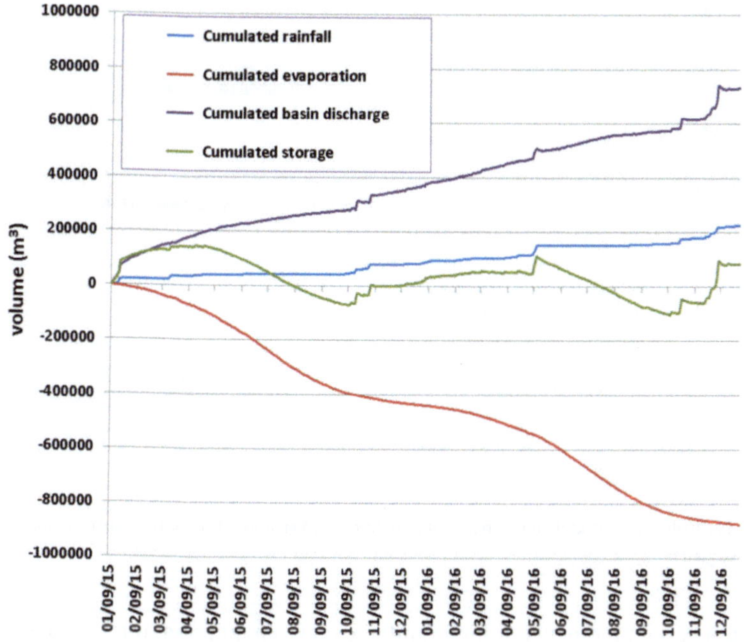

Fig. 3 Volumetric water balance at a daily scale in Santa Olalla pond during two years (January 2015–December 2016)

Table 2 Water balance in Santa Olalla pond (hm^3)

	Rainfall	Basin discharge	Evaporation	ΔS
Total (annual average)	0.11	0.37	0.44	0.04
Daily Max.	0.012	0.036	0.003	0.036
Daily Min.	0	−0.007	0.0002	−0.004
Daily average	0.0003	0.001	0.001	0.0001
Daily median	0	0.0006	0.001	−0.0007

study period was 866 mm (0.11 hm^3/year). The volume of BD, calculated from the balance was 0.37 hm^3/year, while the main loss of water was by evaporation, 3369 mm (equivalent to 0.44 hm^3/year). The increase in the storage (ΔS) was moderate (0.04 hm^3/year).

Even though the aquifer of the stabilized sand dunes is mainly formed by sands, there are also less permeable deposits of clays and organic silts at different depths, which contribute to the hydrological complexity of this system (Custodio et al. 2009). The formation of temporary ponds in certain areas is therefore conditioned, since they consist in local discharge sectors for the groundwater. It has been demonstrated in this study that the BD comes to 77% of the total inputs over the study period (groundwater discharge). The pond is only found to recharge the

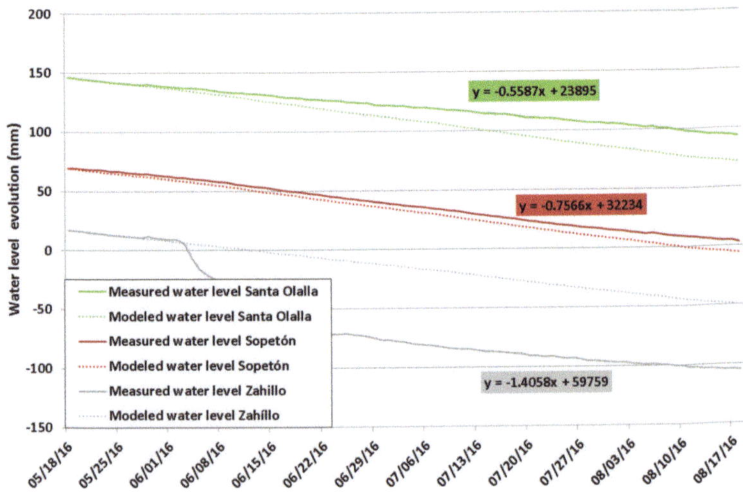

Fig. 4 Hydrological modelling carried out in SOL, ZAH and SOP ponds during summer 2016 (May–August)

aquifer at certain moments of high rainfall events. In addition, direct evaporation is responsible for the main loss of water from the system (>95%).

Another water balance, carried out in SOL, SOP and ZAH during the dry season of 2016, is shown in Fig. 4. The resulting Basin Discharge in SOL (2.4 mm/day) and SOP (1 mm/day), was positive, while in ZAH was negative (−0.9 mm/day). These values revealed the functioning of SOL and SOP as discharge ponds, where water input from the aquifer to the ponds exists. On the other hand, ZAH behaves as a recharge pond (i.e., the opposite process occurs). It is under study whether the currently hydrological functioning as recharge pond is an altered status caused by the pumping for water supply in a nearby coastal town, as it has been suggested previously (Serrano and Serrano 1996).

In Fig. 5, results of the hydrochemical analysis are represented on a Piper diagram. Firstly, Fig. 5a shows the surface water samples of ZAH, SOL and SOP. All the samples reveal that they are sodium-chloride type, especially in SOL, which registered a salinity between 3 g/l (December 2016) and 16.6 g/l (October 2016). This high salinity made SOL a brackish-water pond. This fact can be explained by the hydrological functioning of the system: inputs are from groundwater and rainwater of low salinity (<0.2 g/l) and outputs are by evaporation, so the salts remain in the system, gradually increasing the salinity of the pond (Lozano-Tomás 2007). SOP salinity ranged between 1.3 and 1.7 g/l and ZAH between 0.3 and 0.5 g/l. Secondly, Fig. 5b shows the hydrochemical facies of the groundwater samples. They are found to be mainly sodium chloride type, nevertheless with salinities much lower than the surface water. Those ranged between 0.15 g/l and 0.6 g/l, being in this case fresh water.

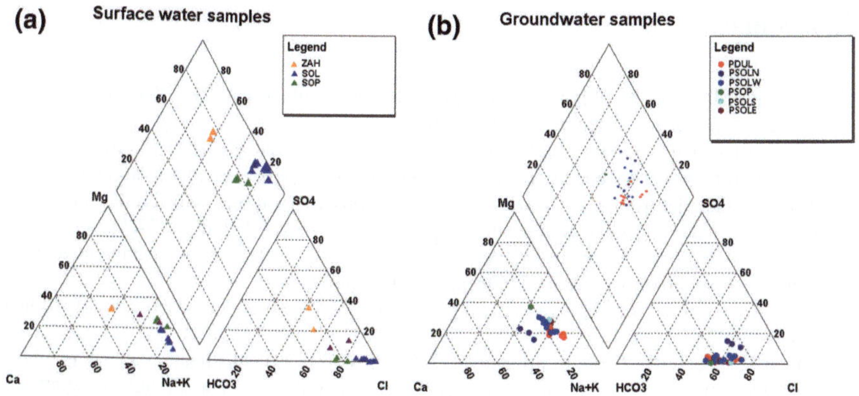

Fig. 5 Piper diagrams of **(a)** surface water samples and **(b)** groundwater samples. The size of the symbols in the diamond shaped diagram is related to its conductivity

5 Conclusions

The installation of sensors and the methodologies employed with the aim of determining the groundwater-surface water interaction, have allowed to improve the hydrological understanding of these aquatic systems over sand dunes, located on the southwest of Doñana National Park. Precipitations of the years 2015–2016 were slightly under the average. Thus, the water level of SOL registered low values of depth during summer 2015 and 2016 (min. depth 1.21 m). The results of the water balance in SOL revealed the strong dependence of its water inputs from the "Dune aquifer" (77%). The main water output from the system (95%) was caused by strong evaporation from the pond surface. Furthermore, the modelling carried out in SOL, SOP and ZAH ponds in the dry season of 2016, showed how the hydrological functioning of these three ponds seems to differ. While SOL registered water inputs of 2.4 mm/day and SOP of 1 mm/day, behaving as discharge ponds, ZAH functioning was the opposite: it recharged 0.9 mm/day of water to the aquifer during the study period. Finally, the hydrochemistry results highlight the crucial role of the evaporation on the high salinities in the surface water (max. value in SOL 16.6 g/l). The strong groundwater dependence of these ecosystems highlighted in this study, should be taken into account by the management authorities in order to avoid further alterations in the hydrological functioning of the ponds.

Acknowledgements The funding for this research comes from the Guadalquivir River Basin Authority (Project: Hydrologycal monitoring and modeling of the pond-aquifer interaction in the stabilized dunes in Doñana). Support provided by Carmen Díaz-Paniagua (EBD-CSIC) in field activities and data correction has been crucial. The technician Jose Manuel Bruque Carmona (Pablo de Olavide University) has helped in the elaboration of some of the figures.

References

Custodio E, Manzano M, Montes C (2009) Las aguas subterráneas en Doñana: Aspectos ecológicos y sociales. Agencia Andaluza del Agua

Green AJ, Bustamante J, Janss GFE, Fernández-Zamudio R, Díaz-Paniagua C (2016) Doñana wetlands (Spain). The wetland book: II: distribution, description and conservation, pp 1–14

Hayashi M, Van der Kamp G (2000) Simple equations to represent the volume–area–depth relations of shallow wetlands in small topographic depressions. J Hydrol 237(1):74–85

Lozano-Tomás E (2007) Las aguas subterráneas en los cotos de Doñana y su influencia en las lagunas. Dissertation, University Politécnica de Catalunya

Manzano M, Custodio E, Higueras H, Puig R, Soler A (2009) Boletín Geológico y Minero 120:377–392

McMahon TA, Peel MC, Lowe L, Srikanthan R, McVicar TR (2013) Estimating actual, potential, reference crop and pan evaporation using standard meteorological data: a pragmatic synthesis. Hydrol Earth Sys Sci 17(4):1331–1363

Munoz-Reinoso JC (2001) Vegetation changes and groundwater abstraction in SW Donana, Spain. J Hydrol 242(3):197–209

Olías-Alvarez M, Rodríguez-Rodríguez M (2013) Evolución de los niveles en la red de control piezométrica del acuífero Almonte-Marismas (Periodo 1994–2012). A. E. De Hidrogeólogos. X Simposio de Hidrogeología. Granada, 16–18 October 2013

Penman HL (1956) Estimating evaporation. Eos, Trans Am Geophys Union 37(1):43–50

Rodríguez-Rodríguez M, Fernández A, Moral F (2016) Hydrological regime and modeling of three ponds of the Mediterranean area (south of Córdoba, Spain). Hydrobiologia 782(1): 155–168

Serrano L, Serrano L (1996) Influence of groundwater exploitation for urban water supply on temporary ponds from the Doñana National Park (SW Spain). J Environ Manage 46(3): 229–238

Serrano L, Zunzunegui M (2008) The relevance of preserving temporary ponds during drought: hydrological and vegetation changes over a 16-year period in the Doñana National Park (southwest Spain). Aquat conserv mar freshw Ecosys 18(3):261–279

Vela A (1984) Estudio preliminar de la hidrogeología e hidrogeoquímica del sistema de dunas móviles y flecha litoral del Parque Nacional de Doñana. Dissertation, University Complutense de Madrid

Hydrogeochemical and Isotopic Investigations for Evaluation of the Impact of Climate Change on Groundwater Quality, a Case Study of the Plaine of Kasserine, Central Tunisia

I. Hassen, F. Hamzaoui-Azaza and R. Bouhlila

1 Introduction

In arid and semi-arid regions, water is fundamental and often a limiting factor for adequate livelihood. Located in an arid region in Central Tunisia, the Plaine of Kasserine aquifer is among the most important groundwater aquifer in the region of Kasserine.

Groundwater quality is a function of the chemical, physical, and biological characteristics of the resource. This latter is threatened by the consequences of climate change and human activities. Thus; changes in global climate are expected to affect the hydrological cycle, altering groundwater recharge, discharge and land use on groundwater systems (Dettinger and Earman 2007; Yagbasan 2016; Chenoweth et al. 2011).

In the last decades, many scientists focus their studies on understanding and investigating the climate change effect on components of groundwater system, including recharge, discharge, flow and storage, and its influence on groundwater quality (Alley et al. 2002; Kundzewicz et al. 2007; Dragon and Sukhija 2008; Ouysse et al. 2010; Green et al. 2011). Thse researches was focused on the effect of climate variability on interannual to multidecadal timescale in arid and semi arid regions, in coastal aquifers and even in rivers and estuaries (Burkett et al. 2002; Sukhija et al. 1998; Gurdak et al. 2007; Vandenbohede et al. 2008; Lambrakis and Kallergis 2001; Bates et al. 2008).

I. Hassen (✉) · R. Bouhlila
Laboratory of Modeling in Hydraulics and Environment (LMHE), National Engineering School of Tunis, University of Tunis El Manar, BP 37 Belvedere, 1002 Tunis, Tunisia
e-mail: imen.hassen@hotmail.fr

F. Hamzaoui-Azaza
Research Unit of Geochemistry and Environmental Geology, Department of Geology, Faculty of Mathematical, Physical and Natural Sciences, University of Tunis El Manar, Tunis, Tunisia

M. L. Calvache et al. (eds.), *Groundwater and Global Change in the Western Mediterranean Area*, Environmental Earth Sciences,
https://doi.org/10.1007/978-3-319-69356-9_18

As a direct consequence of warmer temperatures, the hydrologic cycle will undergo significant impact due to changes in the rates of precipitation and evaporation (Yagbasan 2016). In semi arid and arid regions, even small changes in precipitation may lead to large changes in recharge that may affect contaminant transport. Reduced groundwater recharge and increased pumping may disrupt the current balance of the freshwater/saline water boundary, resulting in saline water intrusion in coastal basins, and even inland aquifers (Chen et al. 2004; Grasby and Betcher 2002; Green et al. 2007; Sandstrom 1995; Alley 2001).

Consequently, it is of most importance to understand the hydrogeochemical mechanisms, processes and origin of the aquifer in order to assess possible threat of water quality due to excess of exploitation and climate change events.

To be a part of such studies, this paper deals with the study of the main geochemical mechanisms and origins controlling the plaine of Kasserine groundwater mineralization using Multivariate Statistical Analyses such as Principal Component Analyses (PCA) and Cluster Analyses (CA) and stable isotopes investigations. Review of these data may be used to better understand the system and to assess the success of current and/or need for further adaptation.

2 Study Area

The Plaine of Kasserine is located in arid region in Central Tunisia. It is surrounded by the Plateau of Kasserine in the south, Djebal Chambi in the west, Djebal Semama in the north and the region of Sbeitla in the east (Fig. 1). The Plaine of Kasserine is an unconfined aquifer filled up by thick Plio-Mio-Quaternary deposits. This aquifer is composed mainly by clayey sand inter-bedded with layers of clay and sand clay as well as conglomerates of gravels. The average thickness of this aquifer is about 150–200 m (Hassen et al. 2016b).

Groundwater recharge in the Plaine of Kasserine may occur from direct rainfall and from floods in the affluent of Oued Hatab and Derb, the main ouedi in this region. In the study area, there is a direct influence of climate change on variations in precipitation. The rainfall is characterized by irregularity in time and space. The hydrological year begins in September, when the rainy season starts. The temporal distribution of precipitation will determine the way and the time groundwater recharge and from rainfall and runoff in ouedi will occurs. Kasserine Station measured an annual average rainfall of 303 mm/year between 1932 and 2015, ranging from a minimum of 141.4 mm/year in 1996–1997 to a maximum of 756 mm/year in 1969–1970 (Fig. 1a). The average annual temperature is 17 °C with January the coldest month and July and August the warmest ones. The long-term trends for the Kasserine station shows significant decreasing trends in precipitation over a 83-year period (1932–2015). With high temperature and low rainfall, the potential evapotranspiration (ETP) reaches 2130 mm/year.

Understanding the profound consequences of climate change on precipitation trends in the long term is important for groundwater management studies.

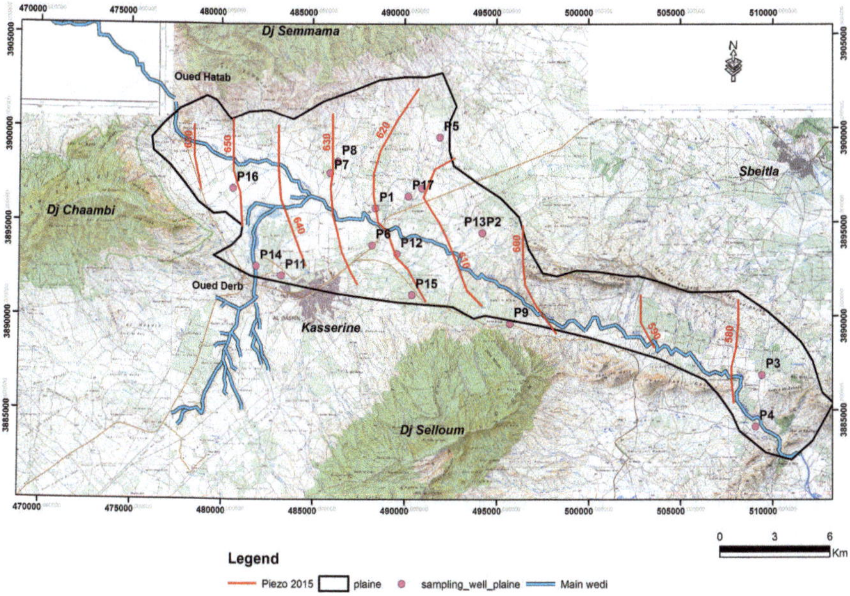

Fig. 1 Study area

In the Plaine of Kasserine, surface water resources are transported by Oued Derb and Oued Hatab. The dense hydrographic network of the region is characterized by perennial and non perennial water flows that occur only by flash floods in rainy seasons (Yangui et al. 2011; Hassen et al. 2016a). Along the Plaine of Kasserine, groundwater discharged to Oued Derb and Oued Hatab springs. Since major groundwater development increased, recharge rate decreased because of climatic change and discharge has been largely through pumping. Consequently, most of the springs are either no longer active or have a greatly reduced discharge such as Oued Derb and Oued Hatab springs (Fig. 2b).

2.1 Materials and Methods

Groundwater was sampled in 19 wells over the study area during Januray-February 2014 and 28 parameters including major, minor, and stable isotopes ($\partial\ ^2$H and ∂^{18}O) were analyzed. The water analyses were conducted in the laboratory of the Centre of Hydrogeology and Geothermic (CHYN) of the University of Neuchatel and in the laboratory of water–rock interaction in the Institute of Geological Sciences in Bern (Switzerland). According to Hounslow (1995), a charge balance error within ±5% is considered acceptable. Conventional graphical plots, Saturation Index (PHREEQC) and Multivariate Statistical Analyses including principal component analysis (PCA) and cluster analysis (CA) were used to

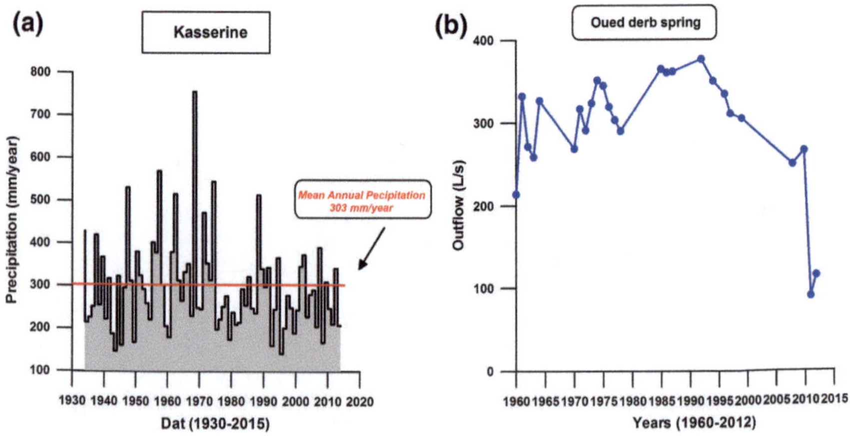

Fig. 2 Hydrological settings: **a** Rainfall in Kasserine meteorological stations (1930–2015). **b** Historical yield of Oued Derb spring (1930–2015)

determine the main factors controlling groundwater mineralization in the Plaine of Kasserine.

3 Results and Discussion

3.1 Hydrochemical Water Type

The Piper classification of Plaine of Kasserine indicates that the majority of samples have Sodium and Calcium dominant cations whereas the dominant anion is sulfate. As indicated in Piper diagram in Fig. 3a, two water facies may be distinguished: $Ca–HCO_3$ and $Na–SO_4$. The first group is mainly derived from recent recharge and calcite dissolution present in the limestone cretaceous of Dj Chaambi and Semama bordering the Plaine of Kasserine. Hence, recent recharge and weathering of carbonate minerals is the most important factors controlling groundwater chemistry (Hassen et al. 2016a). As for the second group, it may results from the replacement of calcium by sodium through cation exchange processes.

3.2 Saturation Index

The hydrochemical model PHREEQC (Parkhurst 1995) was used to determine the state of saturation of the groundwater of Plaine of Kasserine.

The result of the saturation Index in Fig. 3b shows that the groundwater samples of Plaine of Kasserine are undersaturation with respect to evaporates minerals such

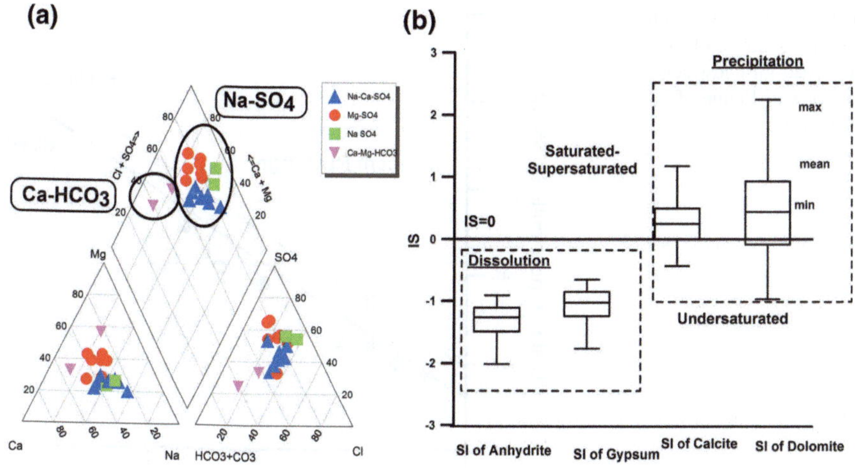

Fig. 3 **a** Piper diagram. **b** Saturation indices for several minerals

as anhydrite and gypsum while it is saturated to supersaturate with respect to carbonate minerals such as calcite and dolomite. Thus, precipitation of these minerals causes equilibrium in Ca^{2+} concentrations and leads to the dissolution of evaporite materials (gypsum and anhydrite) (Hamzaoui-Azaza 2011; Hassen et al. 2016a).

3.3 *Stable Isotopes*

The environment isotopes of oxygen-18 and hydrogen are excellent tracers for determining the origin of groundwater and are widely used for studying the groundwater recharge, migration pathways and mixing of waters from different sources (Faure 1986; Qian et al. 2013; Hamed and Dahri 2013). Only 9 water samples were analyzed for the stable isotopes. The groundwater samples of Plaine of Kasserine show that oxygen-18 varies from −5.33 to −7.35‰ with an average value of −6.69‰. As for the deuterium, it varies from −34.6 to −51.8‰ with an average value of −44.71‰. Stable isotope composition of the analyzed groundwater samples is plotted in Fig. 4 in the ∂^2H–$\partial^{18}O$ space with respect to global and regional meteoric water lines. These groundwater samples are located between the global meteoric water line (GMWL) ($\partial^2H = 8\times\partial^{18}O + 10$) (Craig 1961) and regional meteoric water line (RMWL) ($\partial^2H = 8\times\partial^{18}O + 13.5$) (Zouari et al. 2005) established for Sfax City. The Fig. 4 shows two distinct groups: the first group containing the majority of the water samples is characterized by lower $\partial^{18}O$ and ∂^2H values is interpreted as recharged ensured by rainwater infiltration the water sample P5 seems to have more positive values that may be due to local low-altitude infiltration. Otherwise, only one sample (P9) is plotted on the RMWL. The P9 well

Fig. 4 ∂^2H–$\partial^{18}O$ stable isotope diagram of the analyzed groundwater samples of Plaine of Kasserine

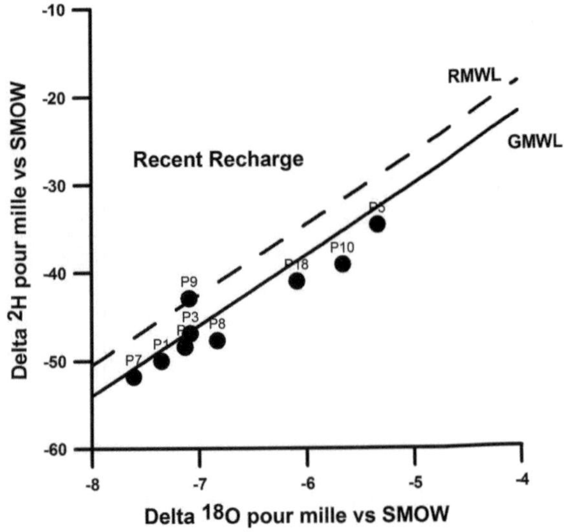

is located approximately near Dejabl Selloum. Such localization may indicate shallow water that infiltrated exclusively in the apparently higher altitude lime-stone's of Dj Selloum. The apparently low salinities of the samples seem to support this interpretation. Furthermore this isotopes classification enhanced the water type classification presented in Piper diagram.

Hence, the stable isotopes investigation highlighted the main origin of the Plaine of Kasserine groundwater, which is from diffuse recharge.

3.4 *Multivariate Statistical Analysis*

3.4.1 ACP

In this study, the PCA is performed on 10 variables (TDS, Na^+, Mg^{2+}, K^+, Ca^{2+}, Cl, HCO_3^-, SO_4^{2-}, pH, and T°). The first two components account 71%. The first component is characterized by a high positive correlation between TDS and Na^+, Mg^{2+}, Ca^{2+}, Cl^-, SO_4^{2-} and K^+, it may be defined as the "Salinity" component". This component is related to the dissolution of evaporates deposit such as gypsum and water-rock interaction. The source of sodium is the process of cationic exchange in clay formation in the Plio-Mio-Quaternary deposits. As for the high positive loading in Ca, it suggests the importance of dissolution of carbonate rocks in the study area. While the second component presents 21% of the total variance and characterized by high correlation between, temperature, ph and HCO_3^- Therefore component 2 is called "alkaline" component.

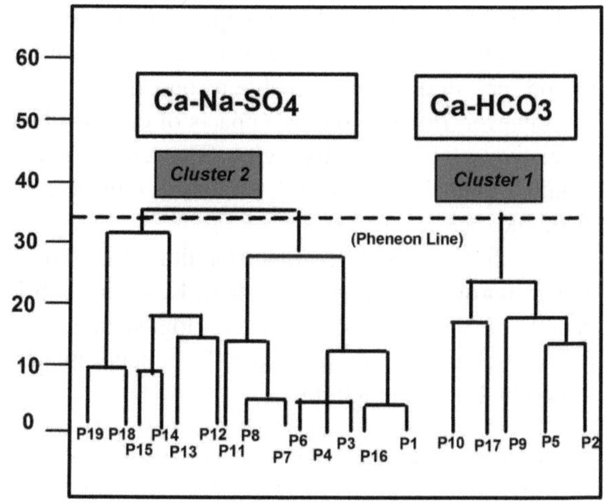

Fig. 5 Dendrogram of Q-mode HCA including all water samples in the Plaine of Kasserine aquifer

3.4.2 Cluster Analysis

Cluster Analysis (CA) was performed in Ward's mode with the Euclidean distance as a measure of similarity of samples and using the physico-chemical parameters of the groundwater samples of Plaine of Kasserine. The result of CA was indicated in the dendogram in Fig. 5. This dendogram exhibits two main clusters identifying the hydrochemical facies of the Plaine of Kasserine; the Ca–HCO_3 and Na–SO_4 as mentioned in Piper diagram in Fig. 3a.

4 Conclusion

In many areas, aquifers provide an important source of freshwater supply. Maintaining water quality in these aquifers is essential for the communities and farming activities.

In arid and semi arid regions, increased evapotranspiration may lead to groundwater salinization and deterioration of groundwater quality. Hence, climate change and hydrological variability may affect the quality of groundwater available for use in a groundwater dependent system.

In the arid region of the Plaine of Kasserine, nineteen groundwater samples were studied based on multivariate statistical analyses including PCA and CA investigations to highlight the main processes controlling the groundwater mineralization of this aquifer.

These geochemical investigation show that dedolomization process generated by dissolution of evaporates minerals and calcite dissolution and cationic exchange reaction, may be among the main mechanisms controlling groundwater mineralization of Plaine of Kasserine aquifer.

In the other hand, stable isotopes analyses indicated that direct recharge represent the main origin of the Plaine of Kasserine aquifer. Hence, precipitations ensure the preservation of this valuable resource.

However, research on the impacts of climate change on the groundwater system is relatively limited in this region. The reasons may be that long historical data are required to analyze the characteristics of climate change. These data are not always available.

Therefore, for a better understanding of the correlation between climate change and groundwater quality of Plaine of Kasserine aquifer, future investigations will be conducted using historical meteorological data and numerical modelling to test different climatic change scenario for a better management of water demand in the Plaine of Kasserine aquifer.

Acknowledgements This study was supported by the LMHE-ENIT and CILIUM Project partially funded by the Swiss government. The authors warmly thank Dr. Ellen Milnes and Dr. Pierre Perrochet from the laboratory of the Centre of Hydrogeology and Geothermic (CHYN) in Neuchatel in Switzerland for their contribution to the analyses of the major and trace elements. We thank also Dr. Waber Niklaus from the laboratory of rock–water interaction in the Institute of Geological Sciences in Bern in Switzerland for his contribution to the isotopic analyses. The authors gratefully thank the National Society of Drinking Water in Tunisia (SONEDE), the Resources Water Direction of Tunis (DGRE) and the Regional Direction of Agriculture and Water Resources of Kasserine (CRDA Kasserine).

References

Alley WM (2001) Groundwater and climate. Ground Water 39:161

Alley WM, Healy RW, LaBaugh JW, Reilly TE (2002) Flow and storage in groundwater systems. Science 296:1985–1990

Bates B, Kundzewicz ZW, Wu S, Palutikof JP (2008) Climate change and water, Technical paper VI of the Intergovernmental Panel on Climate Change. Intergovernmental Panel on Climate Change Secretariat, Geneva, 210 pp

Burkett VR, Zilkoski DB, Hart DA (2002) Sea-level rise and subsidence: implications for flooding in New Orleans, Louisiana, U.S. geological survey subsidence interest group conference. U.S. Geological Survey, Galveston, pp 63–71

Chen Z, Grasby SE, Osadetz KG (2004) Relation between climate variability and groundwater levels in the upper carbonate aquifer, southern Manitoba. Can J Hydrol 290(1–2):43–62

Chenoweth C, Hadjinicolaou P, Bruggeman A, Lelieveld J, Levin Z, Lange MA, Xoplaki E, Hadjikakou M (2011) Impact of climate change on the water resources of the eastern Mediterranean and Middle East region: modeled 21st century changes and implications. Water Resour Res 47:W06506. doi:10.1029/2010WR010269,2011

Craig H (1961) Isotopic variation in meteoric waters. Science 133:1702–1703

Dettinger MD, Earman S (2007) Western ground water and climate change—pivotal to supply sustainability or vulnerable in its own right? Ground Water 4(1):4–5

Dragoni W, Sukhija BS (2008) Climate change and groundwater: a short review, Geological Society special publication. Geological Society, London, pp 1–12

Faure G (1986) Principles of isotope geology. Wiley, New York

Grasby SE, Betcher RN (2002) Regional hydrogeochemistry of the carbonate rock aquifer, southern Manitoba. Can J Earth Sci 39:1053–1063

Green TR, Bates BC, Charles SP, Fleming PM (2007) Physically based simulation of potential effects of carbon dioxide-altered climates on groundwater recharge. Vadose Zone J 6(3):597–609
Green TR, Taniguchi M, Kooi H, Gurdak JJ, Allen DM, Hiscock KM, Treidel H, Aureli A (2011) Beneath the surface of global change: impacts of climate change on groundwater. J Hydrol 405 (3–4):532–560
Gurdak JJ, Hanson RT, McMahon PB, Bruce BW, McCray JE, Thyne GD, Reedy RC (2007) Climate variability controls on unsaturated water and chemical movement, High Plains aquifer, USA. Vadose Zone J 6(3):533–547
Hamed Y, Dahri F (2013) Hydro-geochemical and isotopic composition of groundwater, with emphasis on sources of salinity, in the aquifer system in northwestern Tunisia. J Afr Earth Sci 83:10–24
Hamzaoui-Azaza F (2011) Géochimie et modélisation des nappes de Zeuss-Koutine, des grés du Trias et du Miocène de Jorf-Jerba-Zarzis. PhD Thesis, Faculty of Science of Tunis, 260p
Hamzaoui-Azaza F, Ameur M, Bouhlila R, Gueddari M (2012) Geochemical Characterization of groundwater in a Miocene aquifer, southeastern Tunisia. Environ Eng Geosci 18:159–174
Hassen I, Hamzaoui-Azaza F, Bouhlila R (2016a) Application of multivariate statistical analysis and hydrochemical and isotopic investigations for evaluation of groundwater quality and its suitability for drinking and agriculture purposes: case of Oum Ali-Thelepte aquifer, central Tunisia. Environmental Monitoring and Assessment, pp 188–135. doi:10.1007/s10661-016-5124-7
Hassen I, Gibson H, Hamzaoui-Azaza F, Negro F, Rachid K, Bouhlila R (2016b) 3D geological modeling of the Kasserine Aquifer System, Central Tunisia: new insights into aquifer geometry and interconnections for a better assessment of groundwater resources. J Hydrol (2016) doi:10.1016/j.jhydrol.2016.05.034
Hounslow AW (1995) Water quality data. Analysis and interpretation. Lewis, New York
Kundzewicz ZW, Mata LJ, Arnell NW et al (2007) Freshwater resources and their management. In: Parry ML, Canziani OF, Palutikof JP, van der Linden PJ, Hanson CE (eds) Climate change 2007: impacts adaptation and vulnerability. Cambridge University Press, Cambridge, pp 173–210
Lambrakis N, Kallergis G (2001) Reaction of subsurface coastal aquifers to climate and land us changes in Greece: modelling of groundwater refreshening patterns under natural recharge conditions. J Hydrol 245(1–4):19–31
Ouysse S, Laftouhi NE, Tajeddine K (2010) Impacts of climate variability on the water resources in the Draa basin (Morocco): analysis of the rainfall regime and groundwater recharge. In: Taniguchi M, Holman IP (eds) Groundwater response to changing climate, International Association of Hydrogeologists selected paper. CRC Press/Taylor & Francis Group, London, pp 27–48
Parkhurst DL (1995) User's guide to PHREEQC – A computer program for speciation, reaction-path, advectivetransport, and inverse geochemical calculations. U.S. Geological Survey Water Resources Investigations Report. 95–4227
Qian H, Li P, Wu J, Zhou Y (2013) Isotopic characteristics of precipitation, surface and ground waters in the Yinchuan Plain, Northwest China. Environ Earth Sci 70(1):57–70. doi:10.1007/s12665-012-2103-3
Sandstrom K (1995) Modeling the effects of rainfall variability on groundwater recharge in semiarid Tanzania. Nord Hydrol 26:313–330
Sukhija BS, Reddy DV, Nagabhushanam P (1998) Isotopic fingerprints of paleoclimates during the last 30,000 years in deep confined groundwaters of southern India. Quatern Res 50(3):252–260
Vandenbohede A, Luyten K, Lebbe L (2008) Effects of global change on heterogeneous coastal aquifers: a case study in Belgium. J Coast Res 24(2 Suppl B):160–170

Yagbasan O (2016) Impacts of climate change on groundwater recharge in Küçük Menderes River Basin in Western Turkey. Geodin Acta 28(3):209–222. doi:10.1080/09853111.2015.1121802

Yangui H, Zouari K, Rozanski K (2011) Hydrochemical and isotopic study of groundwater inWadi El Hechim-Garaa Hamra basin, central Tunisia. Environ Earth Sci 66:1359–1370

Zouari K, Trabelsi R, Abid K (2005) Chemical and isotopic composition of rain water of station of Sfax. Tunisia. In: Workshop on chemical and isotopic composition of Rain Water of some Arab countries Beyrouth, Liban 25–26 December 2005

Groundwater Recharge Assessment Using WetSpass: Case Study of the Sidi Marzoug-Sbiba Aquifer (Tunisia)

F. Jarraya Horriche, N. Mgaidi, N. Ghouili and M. Zammouri

1 Introduction

The groundwater resources assessment is often based on the overall estimation of the natural average infiltration of rainfall per year. However, water infiltration process varies in space and time according to several factors such as climatic and hydrological conditions, hydrogeologic characteristics, land use, etc. Ignoring the variation of these factors can lead to uncertainty in the calculation of the groundwater recharge.

It is obvious that in situ measurement of water infiltration is the best way to assess the whole infiltration. However, this can't be managed for a large area. Thus, indirect tools are usually used to assess groundwater recharge at spatial and temporal scales. WetSpass, a physical-based model integrated with the ArcView GIS, has been largely used to assess the groundwater recharge (Tilahun and Merkel 2009; Park et al. 2014; Ghouili et al. 2017). It allows also the assessment of evapotranspiration and surface runoff.

2 Materials and Methodology

The Sidi Marzoug-Sbiba basin is located in the center of Tunisia, in the north-east of the governorate of Kasserine (Fig. 1). The aquifer system is formed by three juxtaposed reservoirs contained respectively in limestone and sandstone

F. Jarraya Horriche (✉) · N. Ghouili
Georesources Laboratory, Centre of Research and Water Technologies, Technopole Borj Cedria, BP273, 8020 Soliman, Tunisia
e-mail: faten.horriche@topnet.tn

N. Mgaidi · N. Ghouili · M. Zammouri
Faculty of Sciences of Tunis, University of Tunis El Manar, 2092 Tunis, Tunisia

M. L. Calvache et al. (eds.), *Groundwater and Global Change in the Western Mediterranean Area*, Environmental Earth Sciences,
https://doi.org/10.1007/978-3-319-69356-9_19

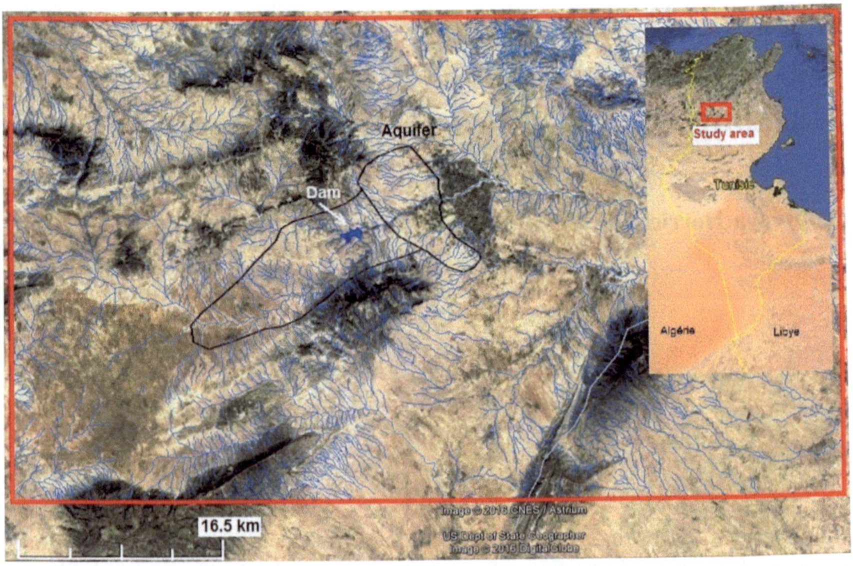

Fig. 1 Location of the study area (2016 Google Earth background)

(Amouri 2005). The basin is characterized by series of faults which are at the origin of fracturing and the emergence of several natural springs. The groundwater exploitation has increased during the last decades following the economic development in the region, mainly for irrigation needs. This increase caused piezometric drawdown especially in the sandstone aquifer.

The study is carried out for average data considering the period 1980–1985 and including wet and dry seasons. The rainfall average is 183 mm for the dry season and 223 mm for the wet season. The study area is characterized by slightly dense forests, agriculture crops and several types of natural vegetations. The agricultural activities are mainly vegetable crops, cereals and arboriculture orchards. The general texture of the soil is essentially clayey and sandy. The highest elevation of the soil surface (1000–1300 m) is located in the bordering areas of the aquifer (NW and SE).

The WetSpass (Water and Energy Transfer between Soil, Plants and Atmosphere under quasi-Steady State) model is integrated with the ArcView GIS (Batelaan and De Smedt 2007). It calculates evapotranspiration, runoff and recharge for a vegetated fraction of a raster cell, per season or per year and accordingly takes into account the spatial variability of the water balance equation components (Eq. 1):

$$P = I + S_v + T_v + R_v[L] \tag{1}$$

Where P, the precipitation; I, the interception; S_v, the surface runoff; T_v, the actual transpiration; and R_v, the groundwater recharge. The total evapotranspiration ET is the sum of I, T_v and the evaporation from the bare soil.

Table 1 Average climatic data input

	Wet season	Dry season	Year
Precipitation (mm)	223.2	183.1	406.3
Temperature (°C)	10.5	17.5	14.0
Wind velocity (m/s)	37.0	29.5	33.3
PET (mm)	194.7	419.9	681.1

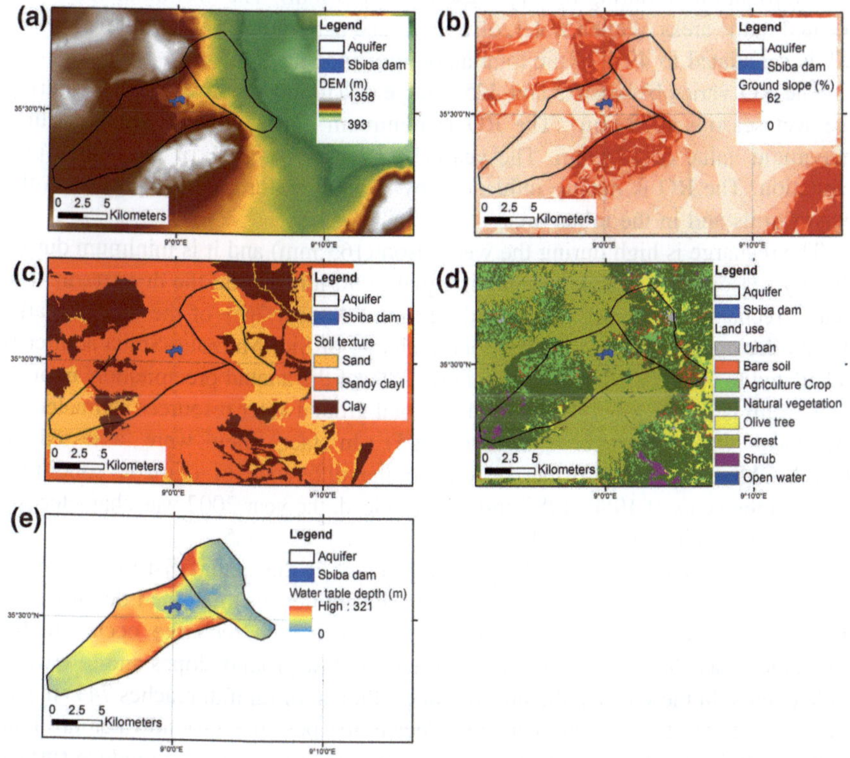

Fig. 2 Grid data input: **a** Digital Elevation Model. **b** Ground slope. **c** Soil texture. **d** Land use. **e** Water table depth

The input data for WetSpass include the ground level and slopes deduced from digital elevation model, the soil texture, the land use, the precipitation, the temperature, the potential evapotranspiration (PET), the wind velocity and the depth of water table. All the data is introduced as a regular grid with cell size equal to 100 × 100 m. The water table depth is deduced from the groundwater level measured in the monitoring network of the Sidi Marzoug-Sbiba aquifer during the years 1980–1985 and gathered from the national water resource agency (DGRE). Monthly measurements of climatic data collected from the national meteorological institute of Tunisia (INM) are used to calculate the average values for the wet and dry seasons, during the same period (Table 1). The input grid maps are presented in the Fig. 2.

3 Results and Discussion

The results of WetSpass model are presented as several grid maps for seasonal and yearly scales. The main calculated maps concern the actual evapotranspiration (ET), the surface water runoff (RO) and the groundwater recharge (RE) (Fig. 3). The yearly ET varies between 134 and 615 mm; it is low during the wet season (58–195 mm) and high during the dry season (53–420 mm). The maximum values are located in the areas of open water, olive trees and agriculture crops. The total yearly ET is calculated to 49% of the total amount of precipitation.

The RO varies between 2 and 325 mm/year. The maximum RO occurs during the wet seasons (178 mm). The RO is minimum during the dry season with a maximum value of 147 mm. The yearly RO represents 23% of the annual precipitation. The RO is mainly influenced by the soil texture; it is important in the clayey soils and in the urban areas.

The recharge is high during the wet season (165 mm) and it is minimum during the dry season; the maximum value is equal to 90 mm during the dry season. The yearly recharge varies between 0 and 222 mm. Almost 76% of the total recharge occurs during the wet season. The total recharge of Sidi Marzoug-Sbiba aquifer is calculated to 14.8 MCM/year representing 28% of the annual precipitation amount. This result is a little bit higher than of the total groundwater resources estimated by the local water manager (Amouri 2005). In fact, they were assessed to 11.9 MCM/year. This gap can be due to the difference of climatic conditions between the years of 1980–1985 and 2002. Indeed, the year 2002 was characterized by a lower precipitation than those of the period 1980–1985.

The second major result of the present research is the spatial distribution of the recharge. The maximum recharge occurs in the aquifer upstream and downstream, in the limestone and the sandstone outcrops. The high recharge occurs in the favorable areas characterized by permeable soils, low ground slopes and low water table depths. In these areas, the infiltration coefficient of rainfall reaches 74% in wet season (Fig. 4). These areas can be selected for possible groundwater artificial recharge. Indeed, artificial barriers can be managed in these areas to reduce surface water runoff and accordingly increase the groundwater recharge process during the wet seasons. The dam located in the center of the studied area (Fig. 1) can be used for planned operations of artificial recharge.

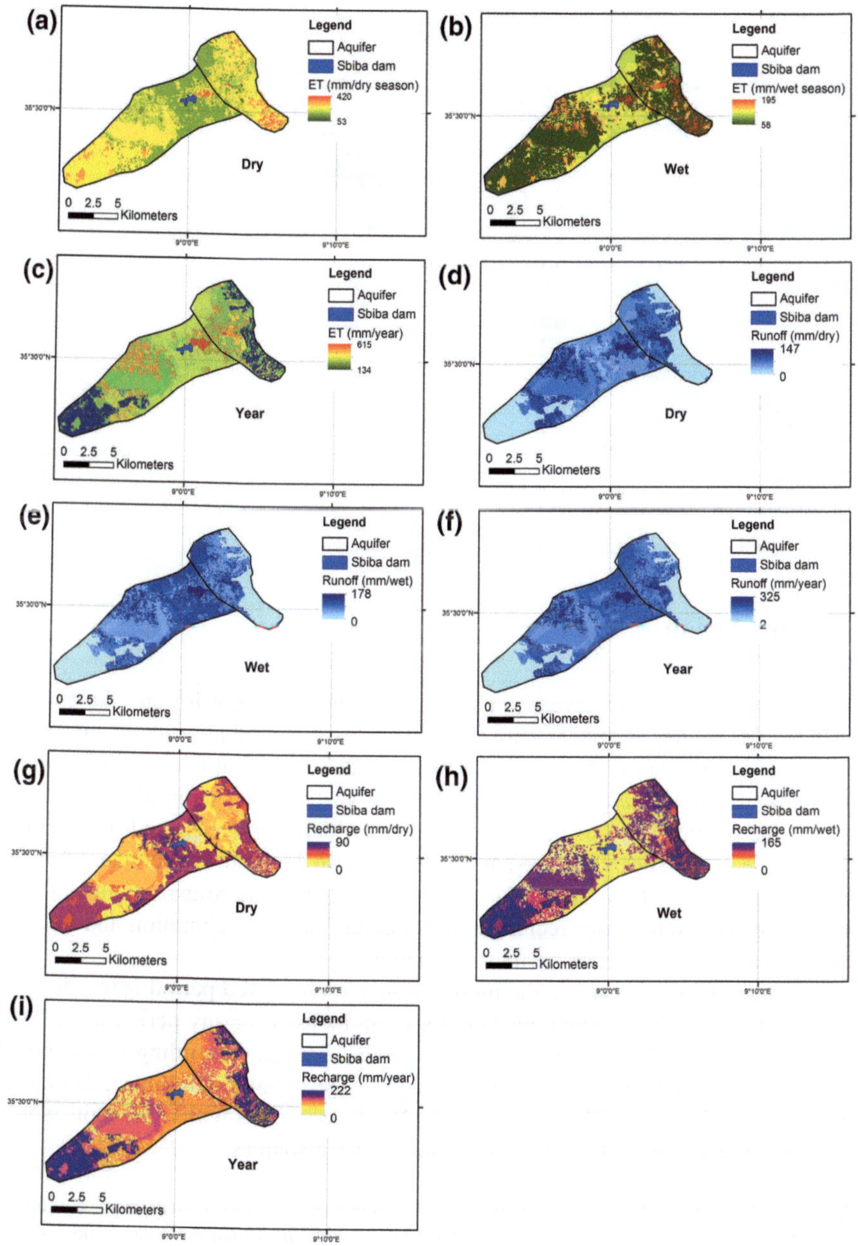

Fig. 3 WetSpass simulated results in dry and wet seasons and yearly: **a–c** Evapotranspiration. **d–f** Runoff. **g–i** Aquifer recharge

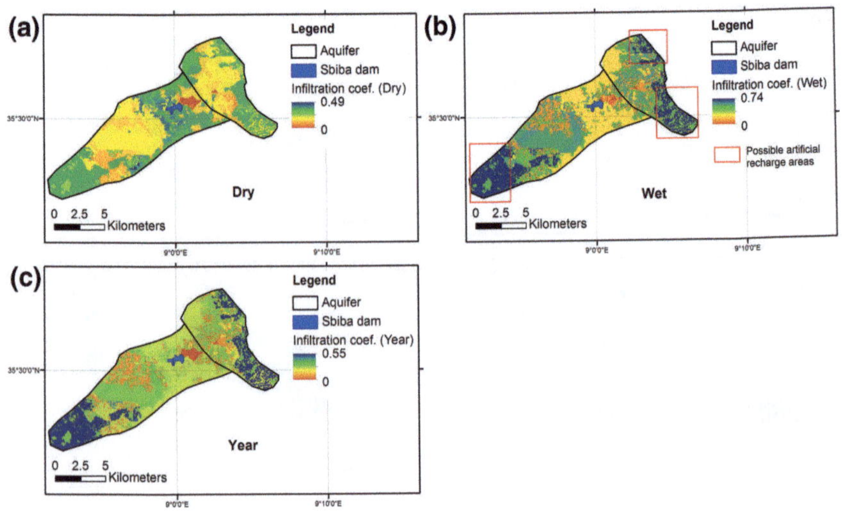

Fig. 4 Infiltration coefficient and possible artificial recharge areas

4 Conclusion

The assessment of groundwater recharge is an important tool for an efficient and sustainable aquifer management since it helps to evaluate the groundwater resources and to plan further wells for groundwater extraction without inducing an overexploitation of the aquifer. In addition, the spatial distribution of the recharge allows the identification of the favorable areas which can be selected for possible artificial recharge sites. The present results can also be used as accurate input for next groundwater flow modeling of the Sidi Marzoug-Sbiba aquifer. Good knowledge of groundwater recharge allows better model's calibration and accordingly reliable results of perspective simulations.

WetSpass model is used in the present research for a fixed period between 1980 and 1985. However, the same methodology can be used for any period if we have the requested data. We can also use it for perspective state according to the global changes that can undergo the region, such as the change of climate, land use, groundwater depth, etc. Thus, coupling the WetSpass results with other tools would be a useful decision making tool for groundwater resources.

Acknowledgements The authors would like to acknowledge water management authorities "Direction Générale des Ressources en Eau DGRE", mainly Mr. Mohamed Ayadi, and "Institut National de Météorologie INM" for their supports and data access.

References

Amouri M (2005) Etude hydrogéologique et possibilité de recharge artificielle du système aquifère de Sbiba. Direction Générale des Ressources en Eau, Tunis, Tunisia

Batelaan O, De Smedt F (2007) GIS-base recharge estimation by coupling surface-subsurface water balances. J Hydrol 337:337–355

Ghouili N, Jarraya Horriche F, Zammouri M, Benabdallah S, Farhat B (2017) Coupling WetSpass and MODFLOW for groundwater recharge assessment: case study of the Takelsa multilayer aquifer (Northeastern Tunisia). Geosci J 21(5):791–805

Park C, Seo J, Lee J, Ha K, Koo MH (2014) A distributed water balance approach to groundwater recharge estimation for Jeju Volcanic Island, Korea. Geosci J 18:193–207

Tilahun K, Merkel JB (2009) Estimation of groundwater recharge using a GIS-based distributed water balance model in Dire Dawa, Ethiopia. Hydrogeol J 17:1443–1457

Application of GALDIT Index to Assess the Intrinsic Vulnerability of Coastal Aquifer to Seawater Intrusion Case of the Ghiss-Nekor Aquifer (North East of Morocco)

T. Kouz, H. Cherkaoui Dekkaki, S. Mansour, M. Hassani Zerrouk and T. Mourabit

1 Introduction

The vulnerability of groundwater to seawater intrusion (SI) can be defined as a sensitivity of groundwater quality to imposed groundwater pumping, or sea-level rise, or both, in the coastal zone. This sensitivity is determined by the intrinsic characteristics of the aquifer. Indeed, the SI is one of the most widespread salinization mechanisms affecting water quality of coastal aquifers (Custodio 2010). It is an active process leading to the breakdown of the hydrodynamic equilibrium between fresh water and seawater. One of the most common reasons for this discrepancy is the overexploitation of aquifers, exacerbated by the ever-growing population of coastal areas, but also due to a natural deterioration in water recharge or a rise in the level of marine waters.

In the present report, the vulnerability of the Ghiss-Nekor aquifer to the SI is assessed using the GALDIT approach developed by (Chachadi and Lobo Ferreira 2001) within the framework of COASTIN Euro—Indian project and then changed in 2005 by (Chachadi et al. 2002). This approach consists of the integration of six parameters considered most relevant, coupled with the Geographical Information System (GIS).

The Ghiss-Nekor aquifer is one of the most main groundwater aquifers in the Moroccan Mediterranean. It covers about 100 km^2 and has a significant role in satisfying water needs for agriculture and potable water supply in the city of Al Hoceima (Fig. 1). It is an unconfined aquifer, circulating in plio-quaternary alluviums composed of detrital sediments, sand, gravel and silt, crossed by several

T. Kouz (✉) · H. Cherkaoui Dekkaki · S. Mansour · M. Hassani Zerrouk · T. Mourabit
Laboratory of Research and Development in Engineering Sciences (LRDSI), Department of Geology, Faculty of Sciences and Techniques of al Hoceima, University Mohammed First, BP 34 Ajdir, 32003 Al Hoceima, Morocco
e-mail: T.kouzmaster@gmail.com

M. L. Calvache et al. (eds.), *Groundwater and Global Change in the Western Mediterranean Area*, Environmental Earth Sciences,
https://doi.org/10.1007/978-3-319-69356-9_20

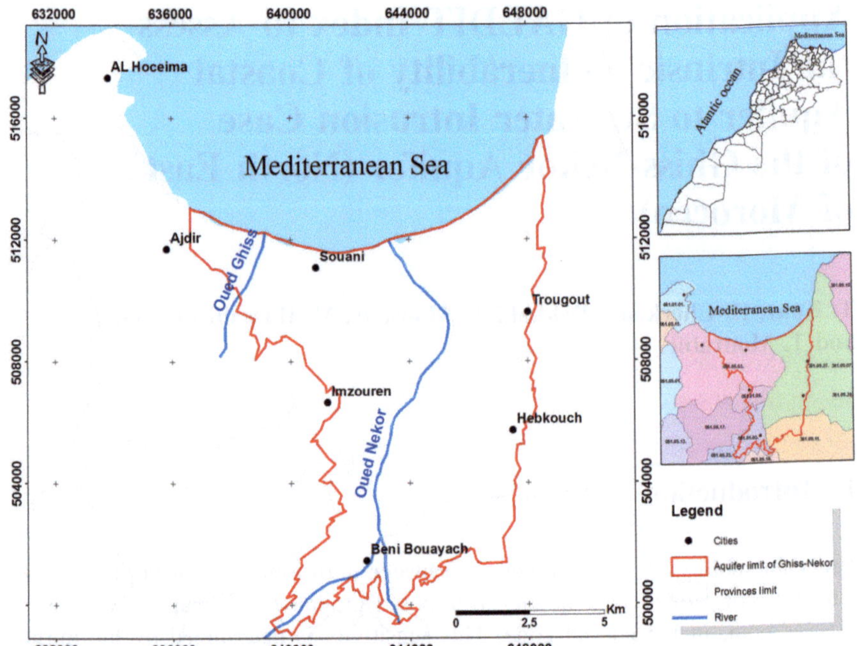

Fig. 1 Location of the study area

discontinuous clay layers. The thickness of the aquifer varies between 5 and 430 m (Fig. 1). The thinnest areas are located in the south, northeast and northwest of the alluvial plain (Salhi 2008).

2 Materials and Methods

2.1 *Principle of Application of the GALDIT Approach*

The GALDIT approach is based on the system of weighted classes. It is a qualitative method based on six parameters:

(G) Groundwater Occurrence. Thus, vulnerability to SI depends on the type of aquifer that can be confined, semi-confined or unconfined. As expected the unconfined aquifer is the most exposed to seawater intrusion due to the impermeable roof compared to the confined or semi-confined aquifer. However, in the case of overexploitation, excessive pumping can cause a cone of depression and exploits seawater and therefore generates a salt water wedge.

(A) Aquifer Hydraulic Conductivity. For the same operating rate, the forward speed of the fresh salt water interface is closely related to the hydraulic conductivity.

(L) Height of Groundwater Level below Sea Level. This parameter determines the hydraulic pressure capable of pushing the fresh salt water interface. The level of the groundwater in relationship to the mean sea level is a very important factor in controlling the advancement of the salt water wedge and in assessing the SI in the coastal aquifer; it expresses the capacity of the water table to reduce the penetration of saltwater into the sea (Ghyben-Herzberg model). As a result, the lowest values increase the vulnerability to possible seawater intrusion. The sea level rise also contributes to the decrease of groundwater height level above sea level, which leads to the SI.

(D) Distance from the Shore. This distance from the coast is measured perpendicularly to the shore. The impact of saline intrusion decreases as it moves away from the coast.

(I) Impact of existing status of seawater intrusion in the area. This parameter is represented by the spatial variation of the ratio ($Cl^-/HCO_3^- + CO_3^-$). Chlorine is the dominant ion in marine waters while it is present in poor quantities in fresh waters. The bicarbonate ion, on the other hand, is present in large quantities in fresh waters and less in marine waters. This is why (Revelle 1941) proposes this report as a criterion for identifying seawater intrusion areas, and was taken up by (Chachadi et al. 2002). We also consider that this parameter of the present state of saline intrusion is an important factor of vulnerability in our study.

(T) Thickness of the aquifer. The more this thickness is important, the more the extension of the SI is important

For the development of the index maps relating to the 6 parameters: GALDIT as well as the final map of the groundwater vulnerability to the seawater intrusion, we have followed the protocol below:

- Delimitation of the study area, on which the operations of combination will be carried out;
- Calculation of the partial indices relative to the different parameters retained, at the level of each pixel of the studied zone. In this stage, an independent mapping must be drawn up for each selected parameter before combining them.

Table 1 below shows the different partial indices, in a more flexible way according to (Chachadi et al. 2002).

- Automatic calculation of the vulnerability global index (Iv) by additive combination of the index maps relating to the different parameters selected, at the level of each pixel of the study area according to the Eq. 1.

Table 1 Matrix of correspondence and digitization of partial indices (I_i)

Parameters	Weight	Classes and ranges			
		Very low	Low	Medium	High
		2, 5	5	7.5	10
G	1	Bounded aquifer	Leaky confined	Unconfined	Confined
A	3	<5	5–10	10–40	>40
L	4	>2	1.5–2	1–1, 5	<1
D	4	>1000	1000–750	750–500	<500
I	1	<1	1–1.5	1.5–2	>2
T	2	<5	5–7.5	7.5–10	>10

$$\mathrm{Iv} = \sum_{i=1}^{6} \{\mathrm{Wi} * \mathrm{Ni}\} / \sum_{i=1}^{6} \mathrm{Wi} \quad (1)$$

Where: Wi is the weight assigned (1–4) to each parameter "i" and N_i is the rating given to each parameter (2.5–10).

The result is an interval between a lower limit < 5 and a top limit, greater than 7.5.

The subdivisions selected are:

Class > 7.5 ⇔ Area of high vulnerability;

Class [5–7.5] ⇔ Area moderately vulnerable;

Class < 5 ⇔ Area weakly vulnerable.

The maximum value of the index I_V indicates a high vulnerability to seawater intrusion.

- Cartographic evaluation of the degree of groundwater vulnerability by zoning of areas with the same classes of variation.

2.2 Data Used

In order to develop the final map of the intrinsic vulnerability of the Ghiss-Nekor ground water, the following data and documents are used:

Map of the spatial distribution of parameter "G", developed from the outline of the Ghiss-Nekor aquifer delimited and supplied by the Loukkos Hydraulic Basin Agency (ABHL);

Map of the spatial distribution of parameter "A" based on the interpolation of the hydraulic conductivity values compiled from the drill records implanted in the study area and archived by the ABHL;

Map of the spatial distribution of the parameter "L", developed according to the interpolation of the data of the piezometric program carried out in april 2017;

Map of the spatial distribution of parameter "D", developed by direct application on GIS;

Map of the spatial distribution of the parameter "I", developed by interpolation of the water quality data (2017) of the aquifer archived at the level of the laboratory of the ABHL;

Map of the spatial distribution of the parameter "T", established from the interpolation of the values of the asynchronous data of the thickness of the aquifer, compiled from the records of the boreholes implanted in the study area and archived at the level of the ABHL.

3 Results

3.1 Spatial Distribution of Individual Vulnerabilities Regarding GALDIT Parameters

3.1.1 Parameter G, Type of Aquifer

The studies carried out and the geological framework of the aquifer of Ghiss-Nekor, consider the water table of Ghiss-Nekor as non-confined. The classification assigned to the aquifer of Ghiss-Nekor is therefore 7.5.

3.1.2 Parameter A, Hydraulic Conductivity of the Aquifer

The calculation of this parameter shows values fluctuating from 0.58 to 102 m/d and the mapping of this indicator illustrates that the values above 40 m/d characterize the upstream part of the aquifer, the adjacent zones to the banks of the Ghiss wadi and the central part of the aquifer.

3.1.3 Parameter L, the Aquifer Indicator Below Sea Level

In order to assess the influence of sea-level rise on the vulnerability of the Ghiss-Nekor aquifer to saline intrusion, two scenarios are chosen

- The scenario 1 is the reference scenario, representing the current state of mean sea level;
- The scenario 2 corresponds to an assumption of sea-level rise of 1 m. This increase is likely to be based on the Intergovernmental Panel on Climate Change (IPCC) prediction on a worldwide scale (Church et al. 2011; Jevrejeva et al. 2016).

3.1.4 Parameter D, the Distance to the Shoreline

Distances in relation to the coastline and watercourses have been calculated. Accordingly, the distances from the coastal shoreline have been classified into four classes: distance less than 500, 500–750, 750–1000 m and distance greater than 1000 m.

The mapping of this parameter for the two scenarios demonstrates clearly the role that the presence of wadis can assume in the diffusion of marine saline waters inside the reservoir of the coastal aquifer and thus accentuate the salinization of the waters of this aquifer reservoir.

3.1.5 Parameter I, the Impact of the Current State of Seawater Intrusion

Taking into account the value attributed to this parameter (weight I = 1) for the Ghiss-Nekor aquifer, the ratio $Cl^-/HCO_3^- + CO_3^-$ is calculated for all wells by keeping the same adopted rows by Chachadi et al. (2002).

3.1.6 Parameter T, Thickness of the Ghiss-Nekor Aquifer

The average thickness of the alluvial layers which form the Ghiss-Nekor aquifer is about 240 m and can reach a maximum of 430 m particularly towards the centers of Souani and Imzouren neighboring. The low thicknesses are noted upstream of the Ghiss-Nekor plain and at the level of the two north-eastern and northwestern points (Salhi et al. 2008). Almost to all of the Ghiss-Nekor aquifer was assigned a value of 10.

3.1.7 The GALDIT Index Under Current and Future Sea-Level Conditions

The calculation of the notes for each criterion of the acronym GALDIT was used to assess the spatial distribution of the degree of groundwater vulnerability to pollution. The final calculation of the GALDIT index was carried out automatically by the GIS used. The final result is a vulnerability map, which supports the six criteria. It reflects the vulnerability of the hydrogeological system to seawater intrusion by gathering the global indices by classes.

The overall indices obtained are distributed in three classes corresponding to the degrees of vulnerability fluctuating from "Low", "Moderate" and "High" Differently distributed for both scenarii 2017 and 2100 (Figs. 2 and 3).

- For a current state of the sea level.
- For a hypothesis of the rise of the sea level of 1 m by 2100 (SLR-1 m).

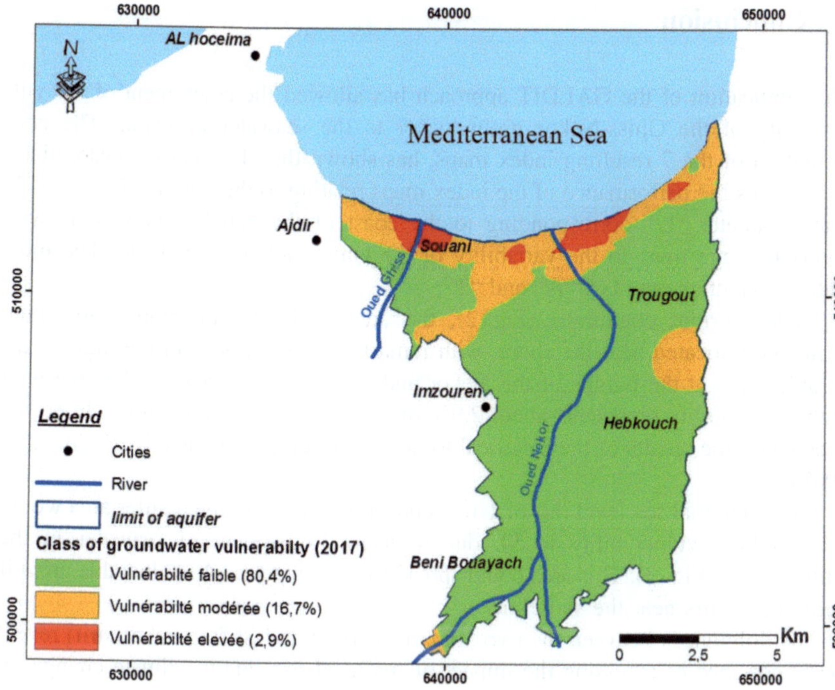

Fig. 2 Vulnerability of the Ghiss-Nekor aquifer to seawater intrusion—Current Conditions

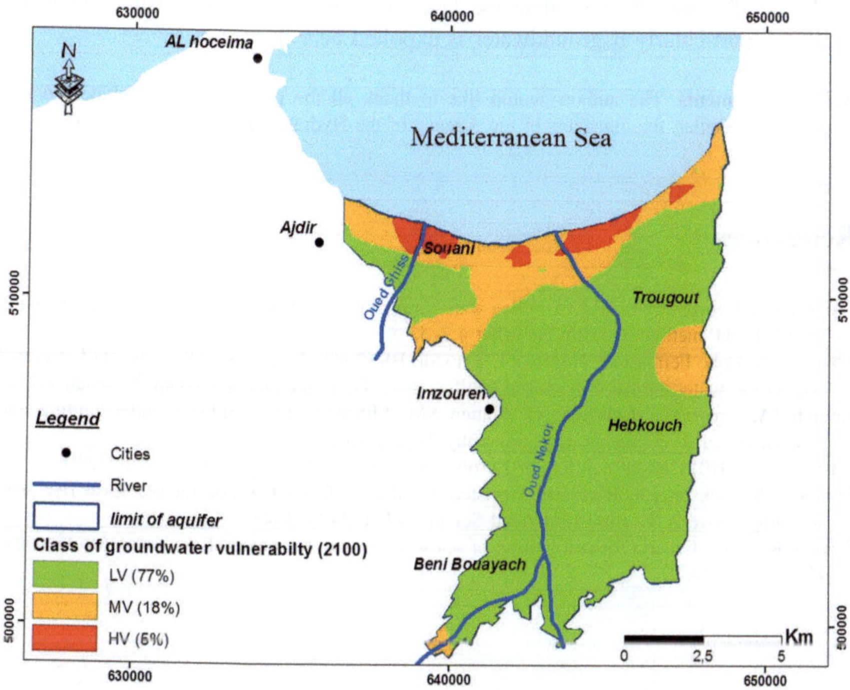

Fig. 3 Vulnerability of the Ghiss-Nekor aquifer to seawater intrusion—With a sea level rise of 1 m

4 Conclusion

The application of the GALDIT approach has allowed the assessment of the vulnerability of the Ghiss-Nekor groundwater to the seawater intrusion. The confrontation of the 7 resulting index maps, has shown that the final map essentially recurrences the performance of the index maps relating to the criteria "L" and "D". The parameter "L", corresponding to the coast of the aquifer below sea level, contributes the most to the variability of the Ghiss-Nekor water table. The most static parameters are "G", "A" and "T".

Under current conditions (Fig. 2), the area of high vulnerability to saline intrusion is located near the shore, with remarkable progress from the highly vulnerable areas at the bumps of the Nekor and Ghiss wadi or rivers. The area with high vulnerability represents about 2.9% of the total area of the aquifer. In accordance with the results of the electrical tomography survey obtained by (Salhi et al. 2008).

For a potential sea-level rise of 1 m, approximately 5% of the aquifer area would show a high vulnerability to SI, due to lateral movement of seawater in the Ghiss-Nekor (Fig. 3). This area could spread to other moderately vulnerable areas if pumping occurs near the shoreline.

The differences between the two scenarii (current and SLR-1 m scénarii) reveal the importance of assessing the impact of sea-level rise due to climate change on coastal aquifers.

It is also essential to consider these results in the conditions of overexploitation of the aquifer, which may affect the level of groundwater by causing salt water intrusion, particularly if groundwater is expoited near the coast.

Acknowledgements The authors would like to thank all the parties who contributed to this project. In particular, the managers of the Agency of the Hydraulic Basin of the Loukkos.

References

Chachadi G, Lobo Ferreira JP (2001) Sea water intrusion vulnerability mapping of aquifers using the GALDIT method. Coastin Newslett 4:7–5

Chachadi G, Lobo Ferreira JP, Noronha L, Choudri BS (2002) Assessing of the impact of sea-level rise on salt water intrusion in coastal aquifers using GALDIT model. Coastin Newslett 7:7–32

Church JA, Gregory JM, White NJ, Platten SM, Mitrovica JX (2011) A understanding and projecting sea level change. Oceanography 24(2):130–143

Custodio E (2010) Coastal aquifers of Europe: an overview. Hydrogeol J 18(1):269–280

Jevrejeva S, Jackson LP, Riva RE, Grinsted A, Moore JC (2016) Coastal sea level rise with warming above 2 °C. Proc Natl Acad Sci 113(47):13342–13347

Revelle R (1941) Criteria for recognition of sea water in ground waters. Eos, Trans Am Geophys Union 22:593–597

Salhi A (2008) Géophysique, hydrogéologie et cartographie de la vulnérabilité et du risque de pollution de l'aquifère de Ghis-Nekor. PhD thesis, Université Abdelmalek Essaadi, Tétouan, Maroc

Salhi A, Stitou El Messari J, Benabdelouahab S, El Gettafi M, Diaz Y, Himi M, Casas A (2008) Cartografía comparativa de la vulnerabilidad del acuífero costero de Ghis-Nekor (Marruecos), revue Geogactea 44:167–170

Impacts of Global Changes on Groundwater Resources in North-East Tunisia: The Case of the Grombalia Phreatic Aquifer

F. Lachaal, S. Chargui, R. B. Messaoud, A. Chekirbane, M. Tsujimura, A. Mlayah, S. Massuel and C. Leduc

1 Introduction

A sustainable groundwater use requires an integrated assessment of the different water resources (Palazzo and Brozovic 2014; Wu et al. 2016). But this task is often hard to achieve because of the complexity of groundwater systems and their connections with surface waters. The latter ones are affected by the highly variable distribution of rainfall in space and over time, exchanges with the river network, return of excess irrigation, infiltration of waste waters, and artificial recharge. Natural and anthropogenic processes combine and lead to a large range of modifications in quantity and quality of groundwater (Lachaal et al. 2014). Some cases are paradoxical like situations of rising water table in semi-arid areas where a groundwater depletion is generally expected as a consequence of the more intense water exploitation. Large areas may then become unproductive because of the water-logging and increased salinization (e.g. Lachaal et al. 2016).

An example is the Grombalia aquifer, in NE Tunisia. Its overexploitation during the last century led to notable modifications in the piezometric levels as well as in the water quality (Gaaloul et al. 2014; Chenini et al. 2015). As a response, the local authorities have raised the need since 1984 for providing supplementary water through water transfer network from the Medjerda and Ichkeul basins located in the North-West of Tunisia. These resources have been mainly used for irrigation and

F. Lachaal (✉) · S. Chargui · R. B. Messaoud · A. Chekirbane · A. Mlayah
Georesources Laboratory, Water Research and Technology Centre, Borj Cedria Ecopark, PO Box 273, Soliman 8020, Tunisia
e-mail: lachaalfethi@yahoo.fr; fethi.lachaal@certe.rnrt.tn

M. Tsujimura
Faculty of Life and Environmental Sciences, University of Tsukuba, 1-1-1 Tennodai, Tsukuba, Ibaraki 305-8577, Japan

R. B. Messaoud · S. Massuel · C. Leduc
IRD, UMR G-EAU, BP 5095, 34196 Montpellier Cedex 5 France

M. L. Calvache et al. (eds.), *Groundwater and Global Change in the Western Mediterranean Area*, Environmental Earth Sciences,
https://doi.org/10.1007/978-3-319-69356-9_21

drinking water. Previous chemical studies showed the degradation of Grombalia groundwater quality (Ben Moussa et al. 2009). A high salinization was observed in the aquifer due to both natural processes, including dissolution of geological rocks such as halite and gypsum (Charfi et al. 2013; Tlili-Zrelli et al. 2013), and to human activities mainly related to industrial and irrigation development (Ben Moussa et al. 2009, 2010; Re et al. 2017). Socio-economic problems related to the poor water management were observed in the basin (Tringali et al. 2017).

Therefore, our objectives were to update the quantitative and qualitative assessment of groundwater in the Grombalia region, and to identify the main drivers of changes and their impacts.

2 Study Site Description

The Grombalia basin is 40 km away from Tunis and 20 km from Hammamet (Fig. 1). It extends over an area of about 100 km^2. It is bounded by Jbel Korbous on the Northeast, by Jbel Abdelrrahman and Takelsa syncline on the East, and by the Mediterranean Sea on the Northwest (Figs. 1a, b).The geological features, mainly the normal faults of Borj cedria and Hammamet, have shaped the Grombalia basin into a NW-SE oriented graben (Fig. 1b). The Beglia (Miocene) and Segui Formations (Mio-Plio-Quaternary) are characterized by strong subsidence with variable depths (Chihi 1995; Hadj Sassi et al. 2006) (Fig. 1c).

The groundwater flow system understudy is composed of shallow and deep aquifer systems. The aquifers are separated by a 15 m-thick clay unit (Ben Moussa et al. 2010) while the shallow aquifer is formed by the Quaternary sediments of

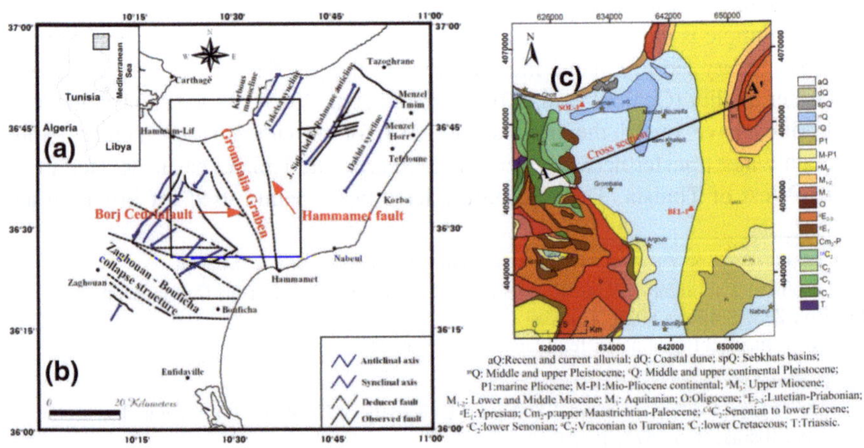

Fig. 1 Geological map of the Cap-Bon region: **a** Geographical location. **b** Structural map of the Cap-Bon region (Mzali et al. 2016), and **c** Geological map of the study area (Ben Haj Ali et al. 1985)

alluvium, sands, and sandy clays reaching 50 m depth. The deep aquifer is composed by the Mio-Plio-Quaternary (Segui Formtion), Miocene (Beglia Formation), and Oligocene series. The Mio-Plio-Quaternary is characterized by lithological and geometric complexities and it is formed by intercalation of sand, sandy clay and clay deposits.

3 Materials and Methods

This work exploits 147 water samples collected during four campaigns on March 2013, November 2013, April 2014, and April 2015, simultaneously with piezometric observations. In situ measurements concern temperature (T), electrical conductivity (EC), and pH. A particular attention was paid to preserve the collected water samples for further chemical analyses. Sulfate (SO_4^{2-}) concentration was measured using the gravimetric method. Chloride (Cl^-) was analyzed using titration technique according to the Mohr method. Bicarbonate (HCO_3^-) and Carbonate (CO_3^{2-}) were determined by titration with sulphuric acid. Major cations Calcium (Ca^{2+}), Sodium (Na^+), Magnesium (Mg^{2+}), and Potassium (K^+) were analyzed by atomic absorption spectrometer. Stable isotopes ($\delta\ ^{18}O$ and δ^2H) of 27 groundwater samples taken in November 2013 were analyzed using an isotopic liquid water and vapour water analyzer (Picarro L1102-i) in the Faculty of Life and Environmental Sciences, Tsukuba University, Japan. In addition to the piezometric data collected in 2013–2015 period, we used similar historic piezometric data covering the 1973–2013 period. These data were provided by the local representatives of the Ministry of Agriculture in Nabeul (CRDA, Nabeul).

4 Results and Discussion

4.1 *Grombalia Water Supply Strategy*

For a longtime the Grombalia aquifer system has been considered as an important water resource with a good water quality. It has been used for several purposes including drinking water for the local population, agricultural and industrial development in the region. In order to increase the local available water resources, the government constructed some dams. The principals reservoirs are Bezigh dam built in 1954, with 6.5 Mm^3 $year^{-1}$ capacity located upstream of Menzel Bou Zelfa city, the Tahouna dam built in 1967 with 0.96 Mm^3 $year^{-1}$ capacity, and El Masri dam built in 1968, with 6.9 Mm^3 $year^{-1}$ capacity and located upstream of Bou Argoub city.

Since 1973, the artificial recharge techniques were applied in groundwater piezometric drawdown area located in Bou Argoub and East aquifer side area. Later on, since 1984 the Ministry of Agriculture has transferred water from the

North-West Tunisia (Medjerda and Ichkeul basins) to fill the gap between the demand and the availability of water needed for citrus crops irrigation in the Menzel Bou Zelfa, Beni Khalled, and Soliman regions. The current management of the water resources in the Grombalia region relies on water supply from various sources including both shallow and deep aquifers, local dams and water from the North-West.

4.2 *Impacts of Global Changes on Groundwater*

4.2.1 Piezometric Perturbation of Grombalia Shallow Aquifer

We have drawn the piezometric map of April 2015 using interpolation of 73 observation wells (Fig. 2). The map reveals that the measured water level varies between 1.9 m near the El Maleh Sebkhat to 60 m in upstream in Bou Argoub region with a general water flow from the South-East to North-West. Figure 3 shows the difference in the observed groundwater level between 1968 and 2015 as calculated by simply subtracting the water level in the later year from the former one. The piezometric difference map shows the presence of two piezometric behaviours.

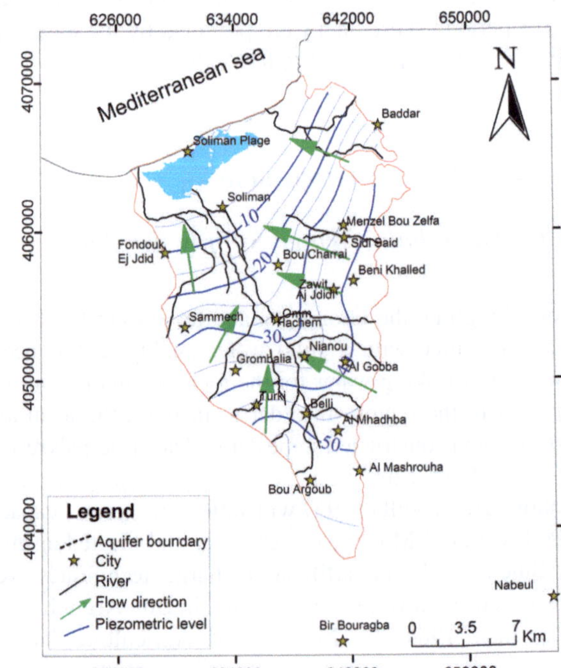

Fig. 2 Piezometric head of shallow Grombalia aquifer in April 2015

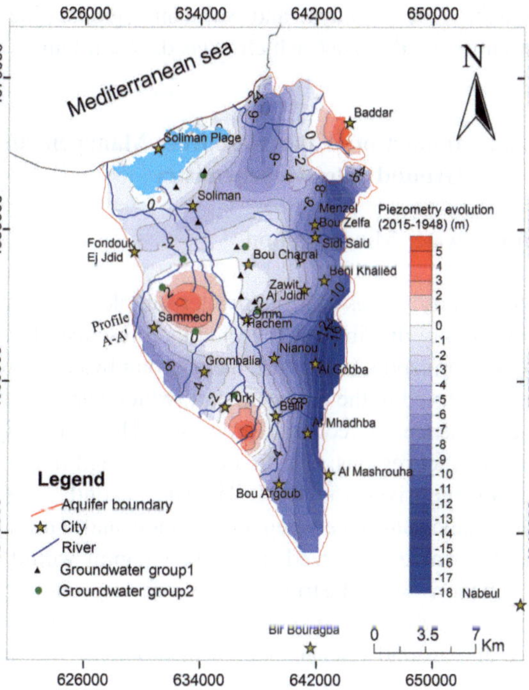

Fig. 3 Piezometric difference between 2015 and 1948 of Grombalia shallow aquifer

Groundwater Piezometric Drawdown

The piezometric drawdown is located in the aquifer upstream (Bou Argoub region) and in its boundaries. The piezometric deacrease is coused by the intensively use of shallow aquifer. This area is not equipped by the public irrigation system and the North-West water are not used for irrigation. However, the farmers use shallow and deep aquifers for irrigation. The expensive cost of drilling deep wells promote the use of shallow aquifer.

Groundwater Piezometric Rise

The shallow groundwater in the central part is caracterized by an increase in water level (Fig. 3). Since 1992, the groundwater has risen-up to within two meters of the ground surface, causing agricultural, environmental, and economic problems linked to suffocation of plants, trees and crops. The field observations and farmers interview during 2013–2015 period show that in recent years, the groundwater rise exceeded the root zone of trees. Therefore whole fields of citrus fruits are asphyxiated. The water level depth map shows that the groundwater rise is observed in the aquifer centre, between Beni Khalled, Menzel Bou zelfa, Grombalia, and Soliman cities. It is concentrated in Zaouiet Jedidi, Omm Hashem, Beni Khalled,

Henchir Bou Charaya, and Soliman regions. The piezometric profile shows the extention of this zone which exceeds 7 km lengh.

4.2.2 Impact of Water Resources Management in Shallow Groundwater Geochemistry

Groundwater Mineralization

The salinity values of groundwater samples range from 1.54 to 5.81 g L^{-1} (Fig. 4). The salinity distribution revealed the presence of salinization gradient from East to West, corresponding to the main groundwater flow direction. The lowest salinity was observed in the upstream part, which varied between 1.54 and 2.5 g L^{-1} reveal the influence of recharge processes. However, high salinities were observed in two distinct regions: The first one, is located in the northwestern of the basin close to Soltane river (5.81 g L^{-1}). The second, is situated in the central part of Grombalia Basin between the Nianou and Turki towns (4.6 g L^{-1}). The higher salinity values observed in the first region would are attributed to the high contamination by industrial wastewater, especially, in Soliman region. The high salinity is observed also in the central aquifer which coincides with the piezometric rise and groundwater management problem. The high salinity can be explained by

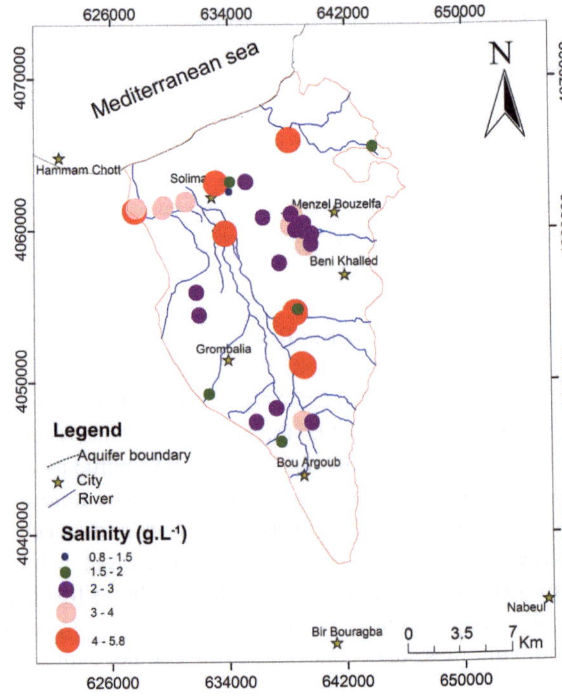

Fig. 4 Spatial distribution of the salinity in the Grombalia aquifer in April 2015

the influence of water irrigation return and high nitrate concentration in this area (Tringali et al. 2017; Re et al. 2017).

Hydrogeochemical Water Types

Major ions of Grombalia groundwater are plotted on the Piper diagram, in order to distinguish the different water facies (Fig. 5). Nitrate concentration was taken into account because of its abundance in the groundwater (Lachaal et al. 2016).

According to the Piepr diagramme (Fig. 5), groundwater presents tree water types: Na–Cl facies, Mixed Ca–Mg–Cl facies, and Mg–Cl facies.

- Na–Cl water type characterizes Grombalia SGW and it was found in wells located mainly along the coastal region and near Sebkhet El Malah, which corresponds to discharge zone.
- A mixed Ca–Mg–Cl water type characterizes shallow and deep aquifers and it was identified in the artificial recharge zone, as well as, in the south part of the

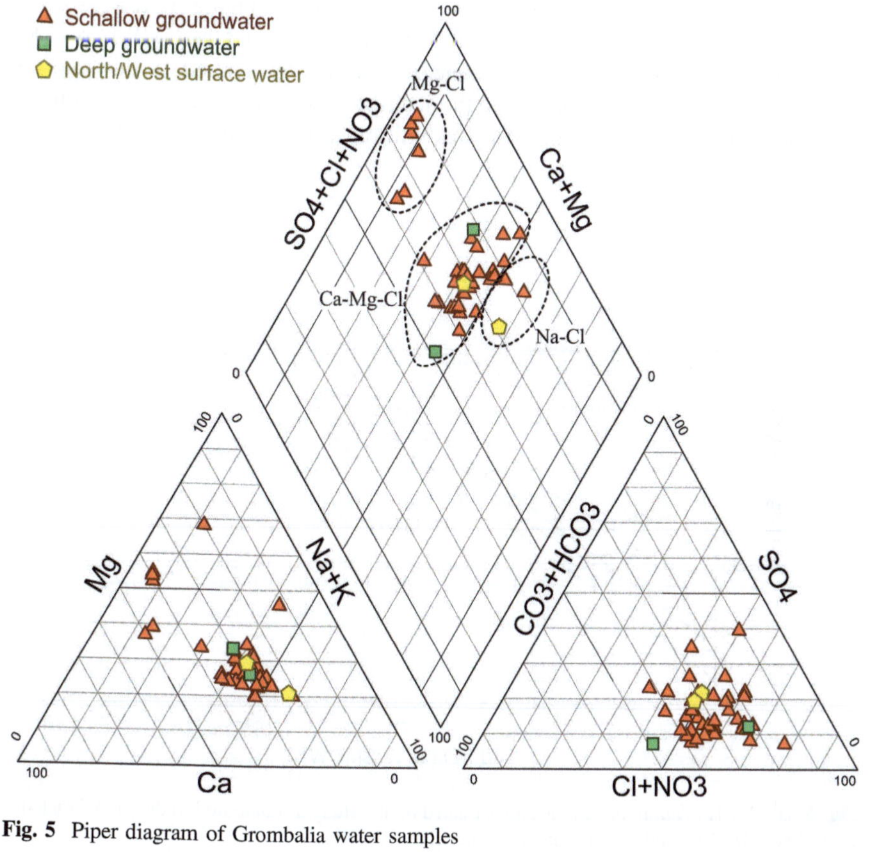

Fig. 5 Piper diagram of Grombalia water samples

aquifer. This facies shows no dominant cation, which signifies the combination of the two natural processes: dissolution of evaporatic minerals and the cation exchange or mixing process.

- Mg–Cl water type, which characterizes the wells situated in the central part of the aquifer. Cl and Mg were the dominant anion and cation of this facies, respectively. This water type may indicate the influence of the both, natural process probably the dissolution of evaporatic minerals (Halite) and the ion exchange.

4.2.3 Isotopic Data and Shallow Groundwater Rising Origin

The oxygen and hydrogen isotopes ($\delta^{18}O$ and $\delta^{2}H$) of water samples, the Global Meteoric Water Line (GMWL), and the Local Meteoric Water Line of Tunis-Carthage (RMWL) are plotted in Fig. 6. The $\delta^{18}O$ and $\delta^{2}H$ contents of the sampled groundwater vary from −5.28 to −3.65 ‰ and from −29.29 to −21.27 ‰, respectively. The plot shows two water groups which are described as follow:

- Group 1: This water group contains 7 water samples. Figure 6 presents a significant correlation between first studies group samples and the GMWL, proving that the origin of these waters is meteoric and non-evaporated. In this case the rainwater was rapidly infiltrated to the saturated zone. The groundwater recharge is principally from rainwater that was infiltrated directly to the aquifer.

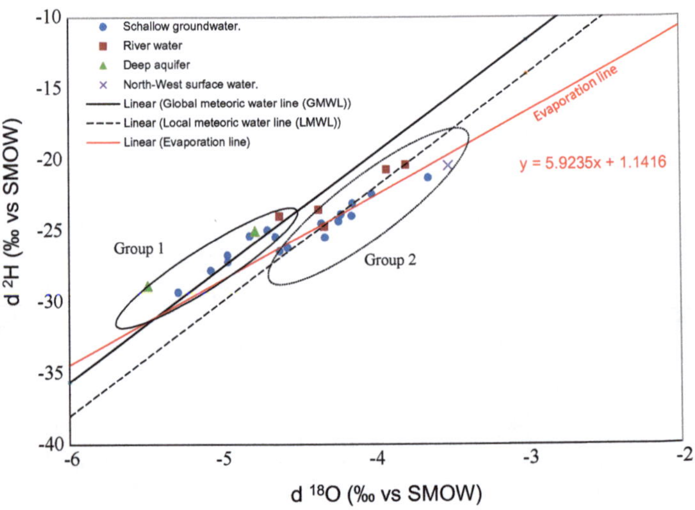

Fig. 6 $\delta^{18}O/\delta^{2}H$ diagram of samples investigated in this study as compared to the global meteoric water line (GMWL) and Local meteoric water line (LMWL)

– Group 2: The second group contains 10 water samples. Which are placed below the GMWL and correlate with the RMWL. This water is resulting from the infiltration of evaporated water. This group is falling a line reflecting the evaporation effect evaporation (Fig. 6). The slope of the evaporation line is 5.9235 that are coinciding with that expected for evaporation in a semi-arid climate (Clark and Fritz 1997). This water group can be related to the return-flow of irrigation that represents in this case an important origin to the groundwater recharge. This group is located in the central zone of the aquifer (Om Hachem, Henchir Bou Charraya, and Menzel Bou Zelfa) where the irrigation activities are very developed. These areas coincide with the SIP and PuIP that are irrigated from North-West water resources and deep aquifer.

5 Conclusion

As consequence of poor water resources management, the groundwater flow perturbation was recorded in the central part of the Grombalia aquifer where a piezometric rise was registered. The increase of water level in the central and downstream parts beyond 2 m depth from the surface led to many agricultural, environmental, and economic issues in the region. Additional threats related to the increase in water salinity were found in this area and highlighted in this study. The piezometric rise is correlating with the increase of deep aquifer abstraction and the use of North-West water resources.

The isotopic analyses show two main groundwater origins which are natural infiltration of rainfall and irrigation return-flows which resulted in the piezometric level rise in the central and downstream regions.

The water resources management in the Grombalia region in the recent years led to the increase in water levels and salinity of groundwater in the central part of the aquifer. Therefore, the need for a new integrated water management plan in the region is apparent while taking into account all the water resources origins and users.

References

Ben Haj Ali M, Jedoui Y, Dali T, Ben Salem H, Memmi L (1985) Carte géologique de la Tunisie au 1/500.000. Office National des Mines, Service Géologique National, Tunis

Ben Moussa A, Zouari K, Oueslati N (2009) Geochemical study of groundwater mineralization in the Grombalia shallow aquifer, north-eastern Tunisia: implication of irrigation and industrial waste water accounting. Environ Geol 58:555–566

Ben Moussa, A, Bel Haj Salem, S, Zouari, K, Jlassi, F (2010) Hydrochemical and isotopic investigation of the groundwater composition of an alluvial aquifer, Cap Bon Peninsula, Tunisia. Carbonates Evaporites 25:161–176

Charfi S, Zouari K, Feki S, Mamic E (2013) Study of variation in groundwater quality in a coastal aquifer in north-eastern Tunisia using multivariate factor analysis. Quatern Int 302:199–209
Chenini I, Zghibi A, Kouzana L (2015) Hydrogeological investigations and groundwater vulnerability assessment and mapping for groundwater resource protection and management: state of the art and a case study. J Afr Earth Sci 109:11–26
Chihi L (1995) Les fossés néogènes à quaternaires de la Tunisie et de la Mer pélagienne : leur étude structurale et leur signification dans le cadre géodynamique de la méditerranée centrale. Thesis of "Doctorat d'Etat", University of Tunis II, Tunisia
Clark ID, Fritz P (1997) Environmental isotopes in hydrogeology. Lewis, New York
Gaaloul N, Candela L, Chebil A, Soussi A, Tamoh K (2014) Groundwater flow simulation at the Grombalia phreatic aquifer (Cap Bon, Northeastern Tunisia). Desalination Water Treat 52:10–12
Hadj Sassi M, Zouari H, Jallouli C (2006) Contribution de la gravimétrie et de la sismique réfléxion pour une nouvelle interprétation géodynamique des fossés d'effondrement en Tunisie: exemple du fossé de Grombalia. Comptes Rendus Géoscience 338:751–756
Lachaal F, Chekirbane A, Mlayah A, Hjiri B, Tarhouni J (2014) A multi-tracer approach to understand the hydrogeochemical functioning of a coastal aquifer located in NE Tunisia. IAHS Publ 363:191–196
Lachaal F, Chekirbane A, Chargui S, Sellami H, Tsujimura M, Hezzi H, Faycel J, Mlayah A (2016) Water resources management strategies and its implications on hydrodynamic and hydrochemical changes of costal groundwater: case of Grombalia shallow aquifer, NE Tunisia. J Afr Earth Sci 124:171–188
Mzali H, Harbi A, Alyahyaoui S (2016) Geometry characterization and depth determination of deep water reservoirs: the case of the Jebel Sidi Abd Er Rahmane region (Cap Bon, northeastern Tunisia). Arab J Geosci 9:295
Palazzo A, Brozovic N (2014) The role of groundwater trading in spatial water management. Agric Water Manag 145:50–60
Re V, Sacchi E, Kammoun S, Tringali C, Trabelsi R, Zouari K, Daniele S (2017) Integrated socio-hydrogeological approach to tackle nitrate contamination in groundwater resources. The case of Grombalia Basin (Tunisia). Sci Total Environ 593:664–676
Tlili-Zrelli B, hamzaoui-Azaza F, Gueddari M, Bouhlila R (2013) Geochemistry and quality assessment of groundwater using graphical and multivariate statistical methods. A case study: Grombalia phreatic aquifer (Northeastern Tunisia). Arab J Geosci 6:3545–3561
Tringali C, Re V, Siciliano G, Chkir N, Tuci C. Zouari K (2017) Insights and participatory actions driven by a socio-hydrogeological approach for groundwater management: the Grombalia Basin case study (Tunisia). Hydrogeol J 1–15
Wu X, Zheng Y, Wu B, Tian Y, Han F, Zheng C (2016) Optimizing conjunctive use of surface water and groundwater for irrigation to address human-nature water conflicts: A surrogate modeling approach. Agric Water Manag 163:380–392

Groundwater Resources Scarcity in Souss-Massa Region and Alternative Solutions for Sustainable Agricultural Development

I. Mansir, L. Bouchaou, R. Choukr-allah, B. Chebli and M. El Otmani

1 Introduction

The world will have to face a global water deficit of 40% by 2030, write experts from the United Nations World Water Assessment Program. Three-quarters of the Arab countries inhabitants live below the water scarcity threshold of 1000 m^3/year, and almost half are in an extreme situation with less than 500 m^3/year in Egypt and Libya in particular. Not only the undeveloped countries are affected. The West American, some provinces of China, Mexico, Southeast Australia, India and the Southern Mediterranean also suffer from water stress (WWDR 2016). As for Morocco, it faces a water shortage linked to an increased drought in the last two decades. If Moroccan population growth continues, consumption will have to be reduces from 830 m^3/capita/year in 1990 to 411 m^3/capita/year in 2020 (ABHSM 2006).

The hydrological basin of Souss-Massa suffers from depletion of its water resources. Pressures on the available water resources due to the irrigated agricultural land expansion, urban and industrial growth have led to the overexploitation of groundwater. The decline rhythm of the water table is increasing at a worrying rate. Moreover, this situation is exacerbated by the long and severe droughts and the certain risks of the water quality deterioration linked to the anthropogenic activities

I. Mansir (✉) · B. Chebli
Laboratory of Mechanics, Processes Energy and Environment, National School of Applied Sciences, BP 1136, Agadir, Morocco
e-mail: iman.mansir@gmail.com

L. Bouchaou
Laboratory of Geology and Geo-Environment, Faculty of Sciences, Ibn Zohr University, BP 8106, Agadir, Morocco

R. Choukr-allah · M. El Otmani
Laboratory of Salinity and Plant Nutrition, Hassan II Institute of Agronomy and Veterinary Medicine, BP 773, Agadir, Morocco

M. L. Calvache et al. (eds.), *Groundwater and Global Change in the Western Mediterranean Area*, Environmental Earth Sciences,
https://doi.org/10.1007/978-3-319-69356-9_22

(Malki et al. 2016) and to the marine intrusion observed in several places of the coastal areas (Tagma et al. 2009). Reducing quantities and degrading water quality in the Souss-Massa region is expected to become more acute in the near future. Therefore, the use of non-conventional waters has become a necessity. The aim of this study is to summarize the causes and effects of groundwater vulnerability, their impact on the agricultural and socio-economic sectors and some solutions to reduce water crisis in the area.

2 Study Area

The Souss-Massa region is located in southern-west of Morocco (Fig. 1). This zone is characterized by a semi-arid to sub-desert climate. Rainfall in the region shows great spatial and temporal variability, with an average of 250–300 mm/year in the plain and about 500–600 mm/year in the mountains (Bouchaou et al. 2012). Temperatures are warm; the annual average can reach up to 20 °C. Agriculture is a strategic sector for the economy in the Souss-Massa region. It employs more than 51% of the working population and contributes to 13% of the regional Gross Domestic Product (GDP). Although, this region is intensively cultivated and ensures a leading position at national level, especially in the production of citrus

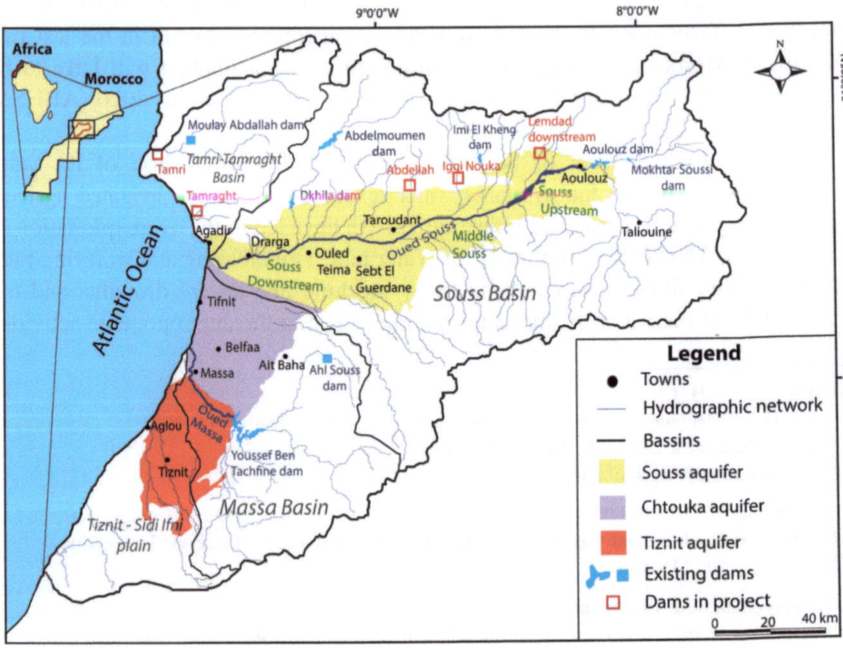

Fig. 1 Souss-Massa region

and vegetables (more than 50% of the national export volume) (Haddouch et al. 2016). In regards to water resources, Souss-Massa region has made enormous efforts to save water. Groundwater is obtained from Souss, Chtouka and Tiznit aquifers (Fig. 1). The hydrogeological basin of the Souss is the most important in southern Morocco. In addition, the large-scale hydraulic works carried out allow the exploitation of surface water resources for irrigation, the supply of drinking water to agglomerations in Agadir and the artificial recharge of the Souss aquifer.

3 Methodology

In order to assess the vulnerability of groundwater to over-exploitation and pollution in the Souss-Massa region, several studies have been carried out. For that, we used the data collected from the Souss-Massa hydraulic basin agency and from Laboratory of Geology and Geo-environment (LAGAGE) of Ibn Zohr University (Seif-Ennasr et al. 2016; Malki et al. 2016). Annual sampling campaigns, piezometric level calculations, physicochemical and bacteriological analyses were carried out at the level of the Souss-Massa hydraulic basins in order to monitor the quality and quantity of water. The groundwater are subjected to several analyzes (electrical conductivity, Chloride, Nitrate, Ammonia nitrogen, fecal coliforms and oxidizable matters). Using these data, we were able to draw groundwater quality assessment maps as well as curves to assess the piezometric level of water tables.

4 Results

4.1 Groundwater Variation (Quantitative Aspect)

The piezometers monitoring in the Souss-Massa plain allow us to study the evolution of the water level in different parts of the aquifers (Fig. 2a). Souss upstream shows a gradual and accentuated depletion since 2000s. The return of normal precipitation at the end of 2011 resulted in a relative increase in groundwater level (Fig. 2b). From 2013, a sharp drop was observed. The influence of intensive pumping and the lack of precipitation would explain this decline. In the Middle Souss, the decrease was moderate in the early 2000s which can be explained by the supply of irrigation water by Abdelmoumen dam and also it is due to the change in the irrigation system from gravity irrigation to drip irrigation, which means an optimization of the groundwater consumption. In the Souss downstream, the groundwater level declined in a rapid way during 2005. In 2011, the groundwater level increased due to the significant precipitation in the region. Within irrigated areas in the southern sector of Chtouka, piezometers generally show small

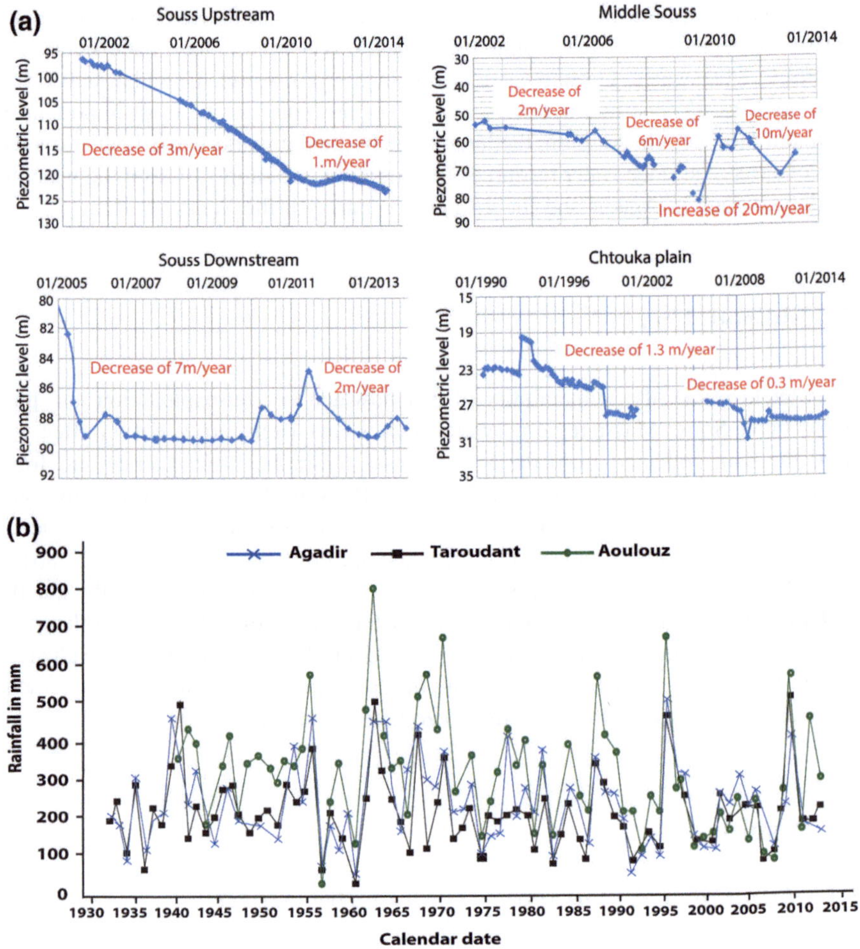

Fig. 2 Piezometric level of the aquifers and precipitation in Souss-Massa region: **a** Piezometric level variation at different points within the Souss-Massa aquifers (ABHSM 2015). **b** Annual variation of precipitation in three main meteorological stations (Agadir, Taroudant and Aoulouz) in the Souss-Massa Basin (Hsaissoune et al. 2016)

variations. The irrigation return flow (IRF) supplied by Youssef Ben Tachfine dam maintains the water table level in this area.

It should be pointed out that more than 25,000 boreholes have been drilled in the Souss-Massa basin. Most of them correspond to irrigation wells. The groundwater remains largely overexploited due to the succession of drought years and the extending of irrigated areas. This has affected the quantity and the quality of the water resources which became vulnerable to pollution.

Table 1 Simplified grid for the assessment of the overall quality of groundwater (SEEE 2014)

Quality parameters	Electrical conductivity (μs/cm)	Chloride Cl^- (mg/l)	Nitrate NO_3^- (mg/l)	Ammonia Nitrogen NH_4^+ (mg/l)	Oxidizable matter MO (mg/l)	Fecal coliform (UFC/100 ml)
Excellent	<400	<200	<5	≤ 0.1	<3	≤ 20
Good	400–1300	200–300	5–25	0.1–0.5	3–5	20–2000
Medium	1300–2700	300–750	25–50	0.5–2	5–8	2000–20000
Bad	2700–3000	750–1000	50–100	2–8	>8	>20000
Very bad	>3000	>1000	>100	>8	–	–

Table 2 Results of the physic-chemical and bacteriological analyses in the aquifers (ABHSM 2015)

Aquifer	Electrical conductivity (μs/cm)	Cl^- (mg/l)	NO_3^- (mg/l)	NH_4^+ (mg/l)	Oxidizable Matter (mg/l)	Fecal coliform (UFC/100 ml)
Souss	1086	161.02	8.4	0.2355	1.81	21
Chtouka	1589	288.61	30.5	0.1522	2.05	14
Tiznit	1007	108.02	28.2	0.0272	1.83	0

4.2 *Quality of Groundwater*

The Moroccan national system for the assessment of groundwater quality considers 6 indicative parameters of the physic-chemical, organic, nitrogen and bacterial pollution. The Table 1 shows the different parameters that determine water quality and rank them according to their quality from very poor to excellent.

Table 2 summarizes the average value of the physic-chemical and bacteriological analyses results of the water samples carried out in the three aquifers of the Souss-Massa region. These analyses were achieved by the Souss-Massa Hydraulic Basin Agency (ABHSM) from February 2014 to January 2015.

The results in the study area indicate that nitrate contents in groundwater vary widely in the space (Table 2). The values range from 0.015 to 124.95 mg/l with an average of 22.37 mg/l (ABHSM 2015). The spatial distribution of nitrate concentrations shows that areas highly polluted by nitrates are located in the Chtouka Ait Baha, Massa and Tiznit. Other areas appear to be spared from nitrate pollution. Nitrate levels in this sectors remain in compliance with the drinking water standard of 50 mg/l. Anthropogenic activities, and in particular agriculture, are the major factors that accentuate the process of water pollution by nitrates. Indeed, groundwater, are currently threatened by the abusive use of nitrogen fertilizers (Tagma et al. 2009).

In the Souss-Massa basins, the average ammonia nitrogen content in groundwater is 0.138 mg/l with a maximum value of 1.65 mg/l in Souss downstream and a minimum value of 0.001 mg/l recorded in the Souss upstream. The distribution of groundwater quality by ammoniacal nitrogen class in the Souss-Massa aquifers shows that in general Souss and Tiznit aquifers are in good quality on the contrary to Chtouka aquifer (ABHSM 2015; Malki et al. 2016).

The hydrochemical data shows a spatial variation of chloride. The results of the analyses show that anomalies (value > 1000 mg/l) occur in the Chtouka plain. Indeed, the Souss and Tiznit aquifers contain waters of good to medium chemical quality where the concentration of chloride does not exceed 700 mg/l. The results of measure for electrical conductivity show that along the shore the values are high except at in the dune sector located to the South of the Oued Souss. Thus we distinguish: (i) an average value of 1086 μS/cm in the Souss aquifer; (ii) an average value of 1589 μS/cm in the Chtouka aquifer; (iii) an average value of 1007 μS/cm in the Tiznit aquifer (ABHSM 2015). These salinity anomalies have multiple origins such as: (i) marine intrusion due to frequent and intensive pumping during drought periods which cause the drop of the groundwater level and a possible rise of the saline bevel to the lands; (ii) the geological formations of aquifers which are rich in evaporite and schist (Bouchaou et al. 2008).

Regarding oxidizable materials and fecal coliforms, analytical results show that in general groundwater is good to excellent quality. The overall quality of groundwater in the Souss-Massa region seems that the Souss aquifer is characterized by good quality unlike the Chtouka and Tiznit aquifer (Fig. 3).

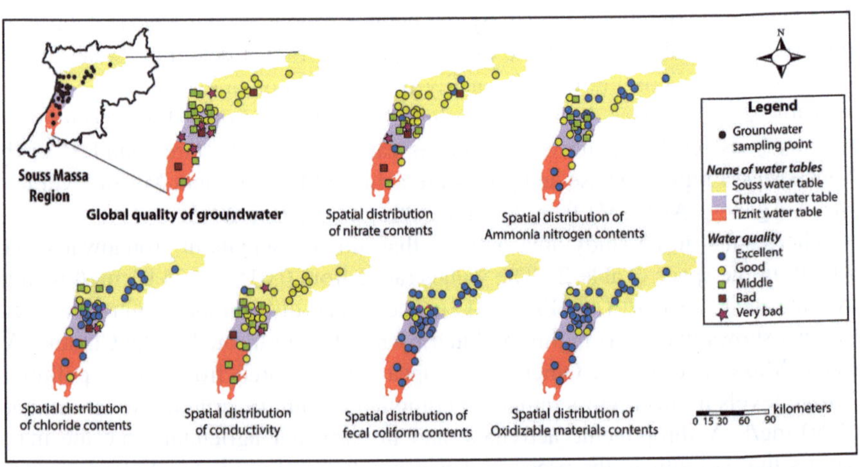

Fig. 3 Global water quality of Souss-Massa aquifers according to the sampling campaign results by the ABHSM from February 2014 to January 2015

5 Discussion

The Souss-Massa region is an important agricultural and touristic pole in the country. As a result, the Souss-Massa region is now confronted with multiple constraints such as overexploitation of aquifers, high population growth and water resources pollution. Accordingly, it is necessary a good management of water resources. The region is characterized by intensive agriculture, which is mainly based on irrigation and consumes more than 90% of water resources. The continuous extraction of groundwater decreases the groundwater level, and the use of deep wells and drilling is mandatory. The water resources become limited, therefore more expensive to extract and can rise to 0.13 Euro/m^3. By way of comparison, the cost of pumping is higher than that of wastewater treatment (ABHSM 2006).

Agriculture remains the country's largest source of employment. Intensive use of groundwater for irrigation will have a huge impact on farmland. The decline in agricultural profitability and the quality of products are among the consequences of this deterioration of water resources. According to recent studies, a decrease of 8% in the added value of the agricultural sector in the Souss-Massa region will be due to this water stress (Elame and Doukkali 2012). On this basis, farmers' incomes will decline, especially small farmers who are most dominant and unable to bear the cost of lowering water levels. In addition, rural unemployment will increase and this will encourage migration. This leads to the abandonment of agricultural land, as in the case of Sebt el Guerdane, Taroudant, Ouled Teima and Chtouka. Groundwater also plays an essential role in maintaining ecosystems in arid regions. In the Chtouka Ait Baha province, all ecosystems (agroecosystems, wetlands, forests, etc.), depend on groundwater for their continuity. These ecosystems have a great impact on the socio-economic development of the region and provide valuable services to humans and other species. These systems are of great value because they support high biodiversity, produce diverse agricultural foods and provide habitat for several rare and endangered animal species. This water stress experienced by the region will have adverse effects on groundwater ecosystems (Hirich et al. 2016).

The combined use of groundwater and surface water should be devoted to the supply drinking water, and agriculture should make greater use of treated wastewater. The development of the wastewater treatment field, whether physical, chemical or biological pathways, allow to have water which can be reused in agriculture and it is in accordance with irrigation standards. Our next research will focus on the development of an integrated set of technological tools and management systems to improve wastewater treatment and reuse of treated wastewater for irrigation, in order to reduce the water stress in the Souss-Massa region. Various wastewater treatment technologies are developed and adapted to the local context. Among these processes: (i) Lagoon piped with nitrification/denitrification and disinfection capacity: these lagoons are characterized by alternating aerobic and anoxic zones, making them suitable for the nitrification/denitrification process and biological removal of BOD (Biological oxygen demand) and phosphorus. (ii) Innovative nitrifying percolator filters with high surface area: in the case of small and medium-sized communities, nitrifying percolator filters can be an attractive alternative to activate sludge units, due to the low energy consumption required for waste water aeration. We will develop filters characterized by innovative high surface area to reduce retention time by biofilm of substantial thickness, and to improve the nitrification/denitrification performance. (iii) Artificial marshes with bacteria promoting plant growth: artificial marshes have the potential to eliminate nitrogen, phosphorus, heavy metals and residual BOD due to the combined effect of plants and the microbial community of the rhizosphere. Bacteria that promote plant growth can play a key role in artificial swamps by actively acting with plants in the process of degradation. (iv) Disinfection of wastewater by catalytic beds activated by solar UV radiation: this technique is a low cost, low environmental impact technology and can be an effective alternative to conventional disinfection systems. The effluent stored in a tank at night, will be treated during the day. (v) Integrated flotation/flocculation process: flotation is an attractive alternative to sedimentation due to reduced treatment volumes and increased treatment efficiency. The proposed process leads to higher purification efficiencies not only in terms of suspended solids but also in terms of BOD. This technique is characterized by a very low hydraulic retention time and low recycling rate. All these technologies are designed for different types of wastewater (urban, industrial, etc.) and will be tested on a pilot scale in the laboratory to check their effectiveness before applying them on a large scale.

6 Conclusion

Water resources in Souss-Massa region are overexploited. Also, the area is exposed to frequent droughts and scarcity of precipitation. This produces significant water stress in the region, which requires good governance of water resources. Without forgetting that, human factors specifically the development in the sectors of industry, tourism and agriculture which is a major consumer of water, play an important role in the degradation of these resources. Rain fed crops will be most affected by drought and arid climate. Irrigated crops will also be affected due to the reduction in the quantity of mobilized water resources. Consequently, all development activities will be constrained by the scarcity of water resources if no institutional strategy and appropriate management are applied. Integrated solutions for the treatment reuse and sustainable management of water resources in agriculture must be developed. This has become necessary in order to reduce the vulnerability of water resources and promoting sustainable agriculture, food security and economic growth.

Acknowledgements We would like to thank of ABHSM for their cooperation in providing all the necessary data. We are grateful to the project entitled MADFORWATER for its financial support of this research.

References

ABHSM (2006) Agence du Bassin Hydraulique du Souss-Massa. Etude de révision du plan directeur d'aménagement intégré des ressources en eau (PDAIRE) des bassins du Souss-Massa. Volume 10: Demande en eau agricole, Juin 2006

ABHSM (2015) Agence du Bassin Hydraulique du Souss-Massa. Etat de la qualité des ressources en eau dans les bassins hydrauliques de Souss-Massa et Drâa, Janvier 2015

Bouchaou L, Michelot JL, Vengosh A, Hsissou Y, Qurtobi M, Gaye CB, Bullen TD, Zuppi GM (2008) Application of multiple isotopic and geochemical tracers for investigation of recharge, salinization, and residence time of water in the Souss-Massa aquifer, Southwest of Morocco. J Hydrol 352:267–287

Bouchaou L, Tagma T, Boutaleb S, Hssaisoune M, El Morjani Z (2012) Climate change and its impacts on groundwater resources in Morocco: the case of the Souss-Massa basin. Climate change effects on groundwater resources: a global synthesis of findings and recommendations. Int Assoc Hydrogeol 27:129–144

Elame F, Doukkali R (2012) Water valuation in agriculture in the Souss-Massa Basin (Morocco). In: Ragab R, Rodriguez-Clemente R (eds) Choukr-Allah R. Integrated water resources management in the Mediterranean region. Springer, Netherlands, pp 109–122

Haddouch M, Elame F, Abahous H, Choukr-Allah R (2016) Socio-economics and governance of water resources in the Souss-Massa River Basin. In: Choukr-Allah R, Ragab R, Bouchaou L, Barceló D (eds) The Souss-Massa River Basin, Morocco. Springer, Switzerland, pp 335–350

Hirich A, Choukr-Allah R, Nrhira A (2016) Groundwater-dependent ecosystems in the Souss-Massa River region: an economic valuation of ecosystem services. In: Choukr-Allah R, Ragab R, Bouchaou L, Barceló D (eds) The Souss-Massa River Basin, Morocco. Springer, Switzerland, pp 163–196

Hsaissoune M, Boutaleb S, Benssaou M, Bouchaou L (2016) Physical geography, geology, and water resource availability of the Souss-Massa River Basin. In: Choukr-Allah R, Ragab R, Bouchaou L, Barceló D (eds) The Souss-Massa River Basin, Morocco. Springer, Switzerland, pp 27–56

Malki M, Choukr-Allah R, Bouchaou L, Ait Brahim Y, Hirich A, Reichert B (2016) Evolution of groundwater quality in intensive agricultural zone: case Of Chtouka-Massa Aquifer, Morocco. Arabian J Geosci 9:566

SEEE (2014) Fiche sur le nouveau système d'évaluation de la qualité des eaux, Maroc. Secrétariat d'état auprès du ministère de l'Energie des Mines, l'Eau et de l'Environnement

Seif-Ennasr M, Zaaboul R, Hirich A, Caroletti GN, Bouchaou L, El Morjani Z, Beraaouz H, Rachael A, Choukr-Allah R (2016) Climate change and adaptive water management measures in Chtouka Aït Baha region (Morocco). Sci Total Environ 573:862–875

Tagma T, Hsissou Y, Bouchaou L, Bouragba L, Boutaleb S (2009) Groundwater nitrate pollution in Souss-Massa basin (Southwest Morocco). Afr J Environ Sci Techn 9:301–309

WWDR (2016) Water and jobs. The United Nations World Water Development Report. Annual report

Hydrogeological Model of Mijas Mountain Aquifers Under Different Climate Conditions (Malaga, Spain)

J. Martín-Arias, P. Martínez-Santos and B. Andreo

1 Introduction

The global change is a worldwide phenomenon that concerns society as a whole, as the effects on population and the environment are not totally known. In the Iberian Peninsula, a reduction of rainfall (from 10 to 40 mm per trimester) and an increase of the average temperature value (6 °C at the end of 21st Century) have been predicted (Ruiz Sinoga et al. 2010; Gutiérrez et al. 2006). Global change effects must be studied to know what can occur in groundwater reservoirs worldwide, due to their importance in agricultural and urban uses (Escribano Francés et al. 2017). This is particularly pertinent in semiarid regions, where groundwater could be an important water source.

Hydrogeological models are useful tools to calculate changes in water storage and water quality in the face of global change (Panagopoulos 2012). The purpose of this communication is to simulate groundwater levels in Mijas mountain aquifers, with an equivalent porous media model using Processing Modflow software. Future scenarios were developed, depending on the variation of recharge and pumping well rate conditions, in order to assess the changes in water levels and to solve the adverse effects in all the study area.

J. Martín-Arias (✉) · B. Andreo
Departamento de Geología y Centro de Hidrogeología de la Universidad de Málaga. Facultad de Ciencias, Universidad de Málaga, Campus de Teatinos s/n, 29071 Málaga, Spain
e-mail: j.martin@uma.es

B. Andreo
e-mail: andreo@uma.es

P. Martínez-Santos
Departamento de Geodinámica, Facultad de Ciencias Geológicas. Universidad Complutense de Madrid, Ciudad Universitaria, C/José Antonio Novaís 2, 28040 Madrid, Spain
e-mail: pemartin@ucm.es

M. L. Calvache et al. (eds.), *Groundwater and Global Change in the Western Mediterranean Area*, Environmental Earth Sciences,
https://doi.org/10.1007/978-3-319-69356-9_23

2 Study Area

The Mijas mountain (80 km^2) is located in the province of Malaga, southern Spain (Fig. 1). It comprises 6 municipalities: Alhaurín el Grande, Alhaurín de la Torre, Torremolinos, Benalmádena, Mijas and Churriana neighbourhood (Málaga).

The study area presents a typically Mediterranean climate, with temperate wet winters and warm dry summers. The average annual rainfall is 642 mm in 1970/71–2013/14 period, with a precipitation gradient from the west (720 mm/year in Mijas village) to the east (581 mm/year in Torremolinos city). The average annual temperature is 18.6 °C.

The mountain range is included within the Blanca geological unit in Triassic Alpujarride Complex of Betic Cordillera. The geological bedrock is formed by metapelites. These are overlain by dolomitic marbles and calcareous marbles, where karstification may develop (Andreo et al. 1998). Since karstification is relatively limited, groundwater flow through the carbonate units largely takes place in a diffuse manner. The geological structure is formed by ESE-WNW folds, where the metapelites anticline cores and faults have divided the studied area into four aquifer systems (Andreo 1997).

Recharge takes place through the direct infiltration of rainfall, while discharge occurred in a natural way by the several springs around the range (Fig. 1). After 1975, discharge through pumping wells began to replace spring flows. Springs are currently inactive as a consequence of intensive pumping. All municipalities use groundwater to supply urban uses (pumping wells are the only source of water supply for some of the neighboring urban nuclei), but the amount of water pumped

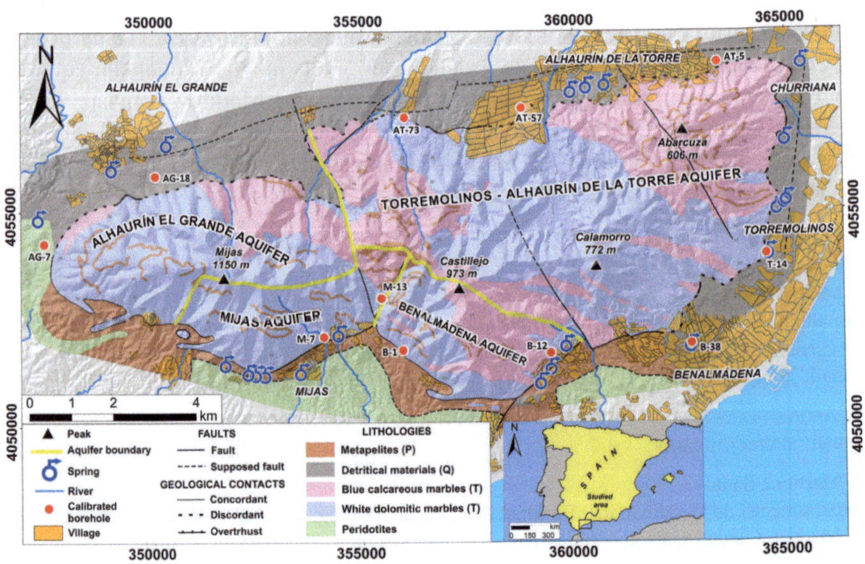

Fig. 1 Location and geological map of the study area

in each aquifer is different. Alhaurín el Grande aquifer presents the smallest pumping rate. This translates into a lesser perturbation of the piezometric surface. The Mijas, Benalmádena and Torremolinos-Alhaurín de la Torre aquifers are subject to intensive pumping.

Urban supply is the main water user across all four systems (29 hm^3 in 2015). Agricultural uses are comparatively small (1.5 hm^3 in 2015) and restricted to the northern and western sides of the Mijas mountain range.

3 Methodology

A conceptual flow model for the Mijas mountain range was developed based on previous geological and hydrogeological information. Rainfall and temperature data were collected from weather stations around the study area. Average recharge data has been calculated for each aquifer by Thornthwaite, Hargreaves, Kessler and Blaney-Criddle methods. Pumping rates were supplied by water supply companies, farmers and private groundwater users.

Processing Modflow 8.042 was used for modelling purposes. This software has been successfully applied to karstic areas in the past (Panagopoulos 2012; Martínez-Santos and Andreu 2010), often through assuming equivalent porous media models (Scanlon et al. 2003). Calibration was carried out based on water table logs for the 1979–2015 period. Representative piezometers were selected based on the length of historical records, as well as on the location of the boreholes.

Five future scenarios have been created for a 12-year period until 2027, when the 2nd management cycle of the Water Framework Directive will be finished. Data of future pumping wells rate has been taken from Andalusia Statistic Institute (IECA 2013) and 5 rainfall events were introduced in the model with previous data series, taking account the normalized period value of different 12-year period.

4 Results and Discussion

Mijas mountain model has been calibrated at 11 boreholes around the study area (Fig. 1). Figure 2 presents the steady and transient-state calibration of the model. The coefficients of determination were 0.9707 and 0.9643, respectively.

Future pumping rates have been established based on population forecasts for the study area. According to the Andalusian Statistic Institute (IECA 2013) population can be expected to grow by 12.2% in the area by the year 2027. On the other hand, recharge was established based on 1970/71–2013/14 rainfall data. The average value of precipitation, for all possible 12 year periods, have been normalized to be compared. Rainfall normalized values are in a range of −0.48 (the driest period) to 0.47 (the wettest period). The final periods selected have been

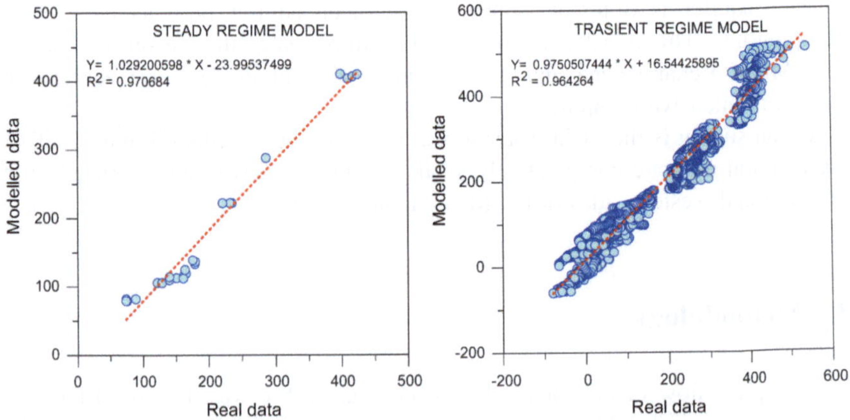

Fig. 2 Coefficient of determination of steady-state and transient regime model

Table 1 Rainfall information (mm/year) in all the entire study area used in 12-year period future scenarios models

Scenario	Hidrological period	Average rainfall	Normalised value
Wet	1987/88–1998/99	734	0.395
Intermediate-wet	2000/01–2011/12	730	0.286
Intermediate	1980/81–1991/92	635	−0.006
Intermediate-dry	1977/78–1989/90	554	−0.287
Dry	1970/71–1981/82	529	−0.383

presented in Table 1. All scenarios would have, at least, one dry and wet hydrological year.

The results of five future scenarios for four boreholes across all four groundwater systems are shown in Fig. 3.

Modelled piezometric levels in intermediate-dry scenario would have the lowest elevation in Torremolinos—Alhaurín de la Torre and Benalmádena aquifers. Distribution of dry and wet hydrological years would reflect this situation. The first half of the simulation period, rainfall data in dry scenario would be higher than intermediate-dry scenario, and a lower decrease of modelled water level, in the driest scenario, would occur at the beginning of the simulation period.

Different water level evolutions have been observed with similar annual rainfall data in all aquifer systems. 2025/26–2026/27 in AG-18 (Table 2), and 2023/24–2024/25 in M-7 (Table 3) water level evolutions have been compared.

Distributed rainfall data in some hydrological year would produce a higher recharge and a minor piezometric decrease (2026/27 in AG-18 and 2023/24 in M-7). Nevertheless, when precipitation may be concentrated in few months (2025/26 in AG-18 and 2024/25 in M-7 boreholes), the recharge would not be so

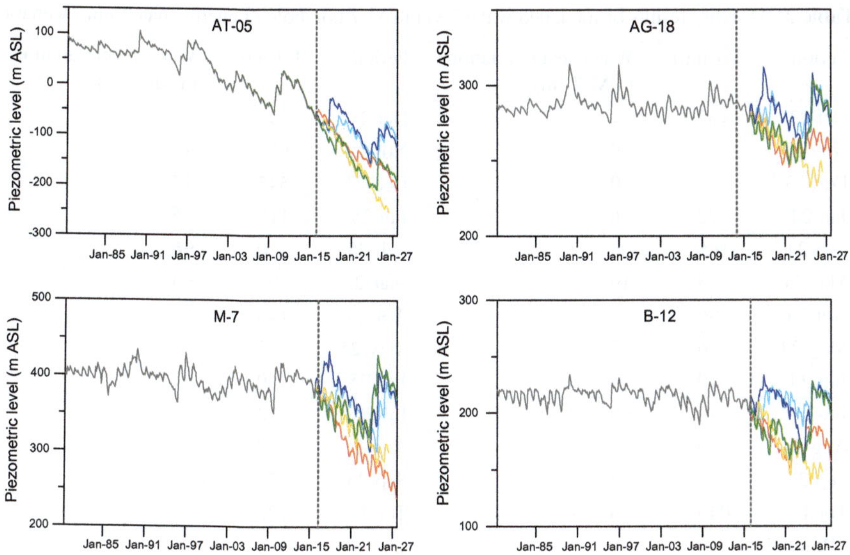

Fig. 3 Results of future scenarios model in AT-5 (Torremolinos—Alhaurín de la Torre aquifer), AG-18 (Alhaurín el Grande aquifer), M-7 (Mijas aquifer) and B-12 (Benalmádena aquifer) boreholes. Calibrated and future scenarios period are divided by the vertical grey dashed line

Table 2 Monthly results of modelled water level in AG-18 borehole in intermediate future scenario

Period	P (mm)	Water level variation in AG-18 (m)	Period	P (mm)	Water level variation in AG-18 (m)
Oct-25	39	−2	Oct-26	169	1
Nov-25	51	−1	Nov-26	30	0
Dec-25	84	−1	Dec-26	64	0
Jan-26	26	−2	Jan-27	71	0
Feb-26	179	1	Feb-27	147	2
Mar-26	153	1	Mar-27	25	0
Apr-26	58	−2	Apr-27	29	−2
May-26	4	−4	May-27	0	−4
Jun-26	9	−2	Jun-27	54	−2
Jul-26	0	−1	Jul-27	3	−1
Aug-26	0	−4	Aug-27	0	−5
Sep-26	58	2	Sep-27	18	2
Total	661	−14	Total	610	−8

P Rainfall

Table 3 Monthly results of modelled water level in M-7 borehole in intermediate future scenario

Period	P (mm)	Water level variation in M-7 (m)	Period	P (mm)	Water level variation in M-7 (m)
Oct-23	152	4	Oct-24	91	0
Nov-23	315	19	Nov-24	693	40
Dec-23	2	0	Dec-24	545	14
Jan-24	172	6	Jan-25	118	−5
Feb-24	165	8	Feb-25	0	−6
Mar-24	8	19	Mar-25	98	−3
Apr-24	96	2	Apr-25	149	−3
May-24	36	7	May-25	7	−3
Jun-24	10	5	Jun-25	0	−2
Jul-24	0	−6	Jul-25	2	−4
Aug-24	0	−6	Aug-25	0	−3
Sep-24	85	−3	Sep-25	20	2
Total	1040	56	Total	1722	26

P Rainfall

Table 4 Monthly results of modelled water level in AG-18 in dry scenario

Period	P (mm)	Water level variation in AG-18 (m)	Period	P (mm)	Water level variation in AG-18 (m)
Oct-16	107	1	Oct-21	87	1
Nov-16	65	1	Nov-21	0	1
Dec-16	73	1	Dec-21	265	6
Jan-17	87	1	Jan-22	168	5
Feb-17	24	0	Feb-22	36	2
Mar-17	141	2	Mar-22	4	2
Apr-17	15	−1	Apr-22	0	−1
May-17	12	−4	May-22	0	−3
Jun-17	2	−2	Jun-22	7	−1
Jul-17	0	−1	Jul-22	23	−1
Aug-17	0	−4	Aug-22	0	−4
Sep-17	25	1	Sep-22	0	1
Total	551	−5	Total	589	9

P Rainfall

effective, and the rise of piezometric levels would not be as higher as distributed rainfall.

On the other hand, too much distributed rainfall data during one hydrological year could also cause a low effective recharge. 2016/17 and 2021/22 piezometric evolutions, on AG-18 borehole, have been compared in Table 4. Precipitation would be distributed over almost all year in 2016/17, and its volume could not be enough to cause any rise of the water table. However, in 2021/22, 88% of rainfall would be concentrated in 3 months (October and December 2021, and January 2022) and the recharge could be more effective.

Finally, all piezometric level would be lower in 2027 year in relation to 2015 year (Fig. 3), even in the wettest scenario, caused by the high pumping wells rate imposed.

5 Conclusions

The hydrogeological model of Mijas mountain aquifers provides a valuable contribution to the hydrogeological understanding of the study area. It is also useful in terms of establishing those parameters which need to be better determined at the field scale.

The outcomes of this study suggest that recharge in the study area depends not only on the volume of precipitation, but also on its distribution over time. Two hydrological years with similar year rainfall data may result in different piezometric evolutions at the same representative point. For instance, if most of the rainfall were concentrated in few months, this would cause the water table to rise sharply in those months. However, pumping would eventually offset and exceed the gain. With such a scenario, the piezometric level, at the end of the hydrological year, would be lower than at the beginning. In contrast, a more evenly distributed rainfall may result in the water table remaining at roughly the same level at the beginning and the end of a given year.

From a water resources perspective, the expected population growth is likely to raise sustainability issues in the mid term. Thus, measures to cater for climate-related eventualities should be developed. Within this context, scenarios are advocated as valuable means to think about the future and devise potential solutions.

Acknowledgements This communication has been prepared under Projects 8.06/5.44.4100 of the University of Málaga and CGL2015-65858R. The authors would also like to thank to water supply companies of Alhaurín de la Torre, Torremolinos and Benalmádena, farmers and groundwater users of Mijas mountain range area, for their cooperation.

References

Andreo B (1997) Hidrogeología de acuíferos carbonatados en las sierras Blanca y Mijas. Cordillera Bética. Sur de España. Dissertation. University of Málaga

Andreo B, Carrasco F, Cuenca J, Téllez A, Vadillo I (1998) El karst en los mármoles alpujárrides de las sierras Blancas y Mijas (provincia de Málaga). In Durán JJ, López Martínez J (eds) Karst en Andalucía, Instituto Tecnológico Geominero de España, Madrid, pp 131–143

Escribano Francés G, Quevauviller P, San Martín González E, Vargas Amelin E (2017) Climate change policy and water resources in the EU and Spain. A closer look into the water framework directive. Environ Sci Policy 69:1–12

Gutiérrez JM, Pons MR (2006) Modelización numérica del cambio climático: bases científicas, incertidumbres y proyecciones para la Península Ibérica. Revista de Cuaternario y Geomorfología 20:15–28

Instituto de Estadística y Cartografía de Andalucía—IECA (2013) Proyección de la población de Andalucía, provincias y ámbitos subregionales 2013–2070. Conserjería de economía y conocimiento. Junta de Andalucía

Martínez-Santos P, Andreu JM (2010) Lumped and distributed approaches to model natural recharge in semiarid karst aquifers. J Hydrol 388:389–398

Panagopoulos G (2012) Application of MODFLOW for simulating groundwater flow in the Trifilia karst aquifer, Greece. Environ Earth Sci 67:1877–1889

Ruiz Sinoga JD, García Marín R, Martínez Murillo JF, Gabarrón Galeote MA (2010) Precipitation dynamics in southern Spain: trends and cycles. R Meteorol Soc

Scanlon BR, Mace RE, Barrett ME, Smith B (2003) Can we simulate regional groundwater flow in a karst system using equivalent porous media models? Case study, Barton Springs Edwards aquifer, USA. J Hydrol 276:137–158

Global Change and Groundwater in Catalonia: Contributions from the Three "Climate Change Reports" (CADS-IEC 2005, 2010, 2016)

J. Mas-Pla

1 Introduction

Since the earliest reports of the Intergovernmental Panel for Climate Change (e.g., IPCC 2014), mitigation and adaptation to the new climate scenarios has been an obligation for all administrations. As regards groundwater, adaptation to climate change has entered most of the hydrological management agendas of the water authority agencies, and the relevance on how it will determine water supply in the next decades is seriously considered. Climate change is expected to significantly modify the global hydrological cycle, and it will largely affect humans mainly through its impacts on water resources, including groundwater. Direct impacts of climate change on natural processes (groundwater discharge, recharge storage and quality) may be even worsened by the human response to these impacts, such as increased groundwater abstraction or just a non-action response to climate threats. The strategic relevance of water resources, whether surface water or groundwater, for human supply, food security and ecological preservation is nowadays out of discussion (Clifton et al. 2010; Green et al. 2011; Taylor et al. 2013).

Because of its importance, the Advising Council for the Sustainable Development in Catalonia (*Consell Assessor per al Desenvolupament Sostenible de Catalunya*, CADS) and the Institute for Catalan Studies (*Institut d'Estudis Catalans*, IEC) promote periodically reports on the climate change issue. These scientific reports analyse, from a multidisciplinary perspective the indicators of climate change, its causes and its possible impacts in Catalonia. They also focus on

J. Mas-Pla (✉)
Institut Català de Recerca de l'Aigua (ICRA), C/Emili Grahit, 101, 17003 Girona, Spain
e-mail: jmas@icra.cat

J. Mas-Pla
Grup de Geologia Aplicada i Ambiental (GAiA-Geocamb), Deptartament de Ciències Ambientals, Universitat de Girona, Girona, Spain

M. L. Calvache et al. (eds.), *Groundwater and Global Change in the Western Mediterranean Area*, Environmental Earth Sciences,
https://doi.org/10.1007/978-3-319-69356-9_24

the appropriate actions to use existing skills and knowledge to tackle with anthropogenic climate change.

Each one of the three CADS reports, published in 2005, 2010, and 2016, have considered water resources as a critical issue, and have devoted them a specific chapter. These present a comprehensive description of the climate variations influences on the hydrological cycle, and groundwater stands as an important factor since it is a key supply for water demand in Catalonia. Indeed, about a 40% of the total supply in the inner basins is supplied by groundwater, and it provides a 35 and 50% of the urban and agricultural demands, respectively.

Therefore, the CADS reports on climate change have provided an overlook and an analysis of the changes that hydrogeological systems in Catalonia will undergo because of the new climatic scenarios. After the third issue has been published, this conference on "Groundwater and Global Change in the Western Mediterranean" stands as a convenient opportunity to summarize the contributions of the CADS reports on groundwater response to climate change and its hydrological, environmental and human implications.

It is worth mentioning, for the sake of completeness, the report published by the Catalan Water Agency (ACA) on water resources and climate change, which provides a full insight into the hydrological facts and knowledge thus far (ACA 2009).

2 Temperature and Precipitation Predictions

As part of the Western Mediterranean area, climatic models have continuously forecasted water scarcity. Such scarcity will be caused by a decrease of rainfall rate and rise of temperatures that will, in turn, increase evapotranspiration. From a broad perspective, both consequences will directly affect surface runoff and groundwater recharge.

Each report has updated climate tendencies observed in Catalan observatories and have issued climatic predictions according to the most recent global climatic models. The latest available data by Calbó et al. (2016) calculate the seasonal and annual temperature and precipitation projection for 2021 and 2050 using distinct climatic databases (MERCAT, ESTCENA, EuroCORDEX, CMIP5, and DCPP), as well as distinct downscaling methods (dynamic and statistic), for the different geographical areas in Catalonia: Pyrenees, inland areas, and coastal areas. Downscaling resolution varied according the method and the area of application, and it ranged between 0.04 and 0.20°. Such scaling approaches are necessary, especially as regards rainfall, in areas such as Catalonia where the complex orography and land-sea contrast are largely misrepresented by global models.

The climate projections given by these authors reveal a robust temperature rising trend over Catalonia for the coming decades over all geographical/climatic areas of Catalonia. Using the median from all downscaled datasets, the surface temperature increase can be of +0.8 °C in the current decade (2021), and even reach +1.4 °C by mid-century (2050), as compared to the reference (period 1971–2000). The rising trends may be even more powerful for the Pyrenean region, particularly in summer. As for rainfall, the climate projections show a decreasing trend, but with an uncertain slope. Indeed, variation of rainfall rate in the current decade is barely significant. By 2050, on the contrary, decrease in rainfall is clear, with a median in distribution of estimated values around −10% in spring, summer and autumn. Average annual decrease in Catalonia is 6.8%, being larger in the coastal areas (−8.3%) and smaller in the Pyrenees (−5.3%). These projections are based on moderate emission scenarios of radiative forcing (i.e., A1B, RCP4.5), and consequently, other estimated values associated with future climate change might be slightly higher than those herein introduced.

Regarding extreme events, Llasat et al. (2016) describe an increase of heavy rainfall storms and floods, mainly during the summer, and they conclude that droughts will increase their frequency and persistency, summer periods will be longer and, eventually, with less water resources available to face the demand.

3 Groundwater as a Key Resource Under Scarcity Conditions

The first report describes the whole effect of climate change on the water cycle, considering groundwater as a resilient resource in front of the climatic consequences upon the water balance (Mas-Pla 2005). Groundwater expected behavior under the predicted hydrologic scenarios was illustrated using three distinct examples:

1. The response of alluvial aquifers, in particular the fluvio-deltaic formations, to rainfall recharge oscillations under continuous withdrawal rates. Using a piezometer on the Daró alluvial aquifer, water table oscillations illustrated the quick decline of stored water resources during the 1996–2002 drought as an example of temporal aquifer overexploitation when wet years are sparse, as predicted by climate models.
2. The response of mountain areas to a growing vegetation cover: the effect of afforestation on infiltration rates because of the increase of evapotranspiration demand and, as a result, the progressive drying up of natural springs and drop of baseflow contribution to creek and mountain stream runoff, according to observations in the Montseny range.
3. The case of coastal aquifers, which concentrate intense groundwater exploitation areas along all the Catalan coast (Mas-Pla et al. 2014), where decreasing rainfall recharge and stream infiltration will compete with heavy withdrawal rates without potential supply alternatives. Declining potentiometric levels will induce seawater intrusion, which will be concurrently aggravated by the effects of sea level rise.

The influence of climate change on the stream-aquifer interaction, in particular on the groundwater contribution to runoff as baseflow, was also stressed in the report as an environmental impact that has its solution on appropriate aquifer system management.

Finally, adaptation to climate change for distinct uses was treated as a balance between potential available resources and human (and ecological) demand. Among the most sustainable options to satisfy water demand, groundwater exploitation of regional systems, which are expected to be more resilient to climatic effects at least in the short-middle run, stands as an alternative; altogether with management actions as water saving strategies, conjunctive use of surface water and groundwater, and treated wastewater reclamation.

4 Groundwater Vulnerability to Climate Change

Assuming that adaptation strategies must be designed and anticipated at a regional level, these must be based on the knowledge of the hydrological system vulnerability to climate change; that is, vulnerability as the intrinsic capacity of a given system to cope, adapt or recover from the effects of adverse stresses given by a hazard. In this paper, 'hazard' refers to changes in water availability under global change, which may come from human as well as from climate pressures (Smit and Wandel 2006). Consequently, an analysis of an aquifer or groundwater body vulnerability requires a sound knowledge of its hydrogeological characteristics and status, and the hazards that may occur. In this sense, the second report evaluated and mapped groundwater vulnerability in Catalonia based on the analysis of human and climatic pressures (Mas-Pla 2010). After a convenient evaluation of both pressures, a final index classified each groundwater body as more or less vulnerable to climate changes under present use and exploitation regimes (Fig. 1).

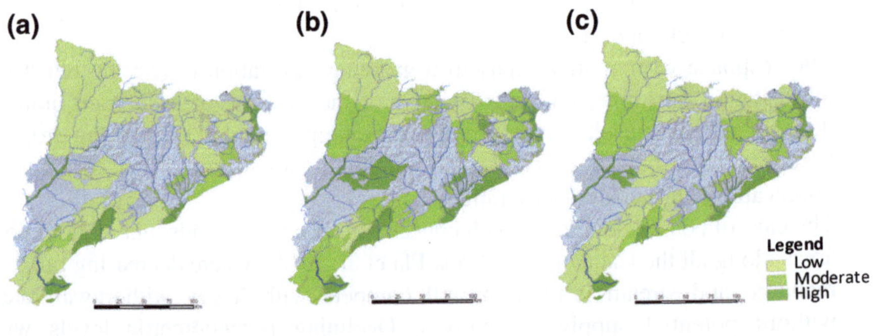

Fig. 1 Vulnerability maps of the groundwater bodies, according the water framework directive, related to **a** the decrease of stream discharge, **b** the decrease of direct rainfall recharge, and **c** groundwater quality issues. After Mas-Pla (2010)

As regards groundwater, several vulnerability settings were evaluated:

1. *Aquifer vulnerability due to the decrease of runoff and, therefore, of stream recharge* (Fig. 1a). Water scarcity derived from decreasing rainfall rates and increasing evapotranspiration will reduce runoff and, consequently, limit aquifer recharge. Catalonia's groundwater bodies were analysed according to the dependence of their water balances on surface water inputs. For instead, alluvial systems are considered as the most vulnerable, especially where stream water is usually captured by pumping wells. Moreover, aquifers in irrigated areas that depend on surface water diversions will also show a high vulnerability since a drop of stream discharge will need to be compensated by groundwater exploitation. Coastal aquifers, specifically in the lower reaches, where a delicate equilibrium exists among surface water, groundwater levels and seawater intrusion, will be the most vulnerable; and, among them, the paradigmatic case of the Ebro River delta.
2. *Aquifer vulnerability associated to a decrease of direct recharge through infiltration* (Fig. 1b). Rainfall recharge is expected to decrease merely because of the drop of rainfall. Additionally, evapotranspiration will still reduce infiltration volumes in forested areas as well as in arable lands. Outcrop soil characteristics and aquifer lithologies determine the infiltration rate, so predicted rainfall decrease and geological features can be combined to map the vulnerability of groundwater bodies to direct recharge losses.
3. *Aquifer vulnerability linked to groundwater quality issues* (Fig. 1c). In this case, human pressures, as pollutant activities obviously unlinked to climate events, are the main factor, jointly with the soil intrinsic properties, to determine aquifer vulnerability. Falling aquifer recharge from rainfall and streams will limit dilution, and contribute to the loss of groundwater quality. Nevertheless, locations where polluted surface water affects groundwater quality were identified as causes for increasing aquifer vulnerability.

5 Estimating Water Resources Availability Under Climate Change

Hydrological climate change effects are basically modifications of the water balance in a given basin. These modifications involve a variation of the magnitude of the distinct parameters (mainly, rainfall, evapotranspiration, runoff and infiltration) and the relative distribution among them. Groundwater flow terms, as in-out flows across the basin vertical boundaries (e.g., Menció et al. 2010), will also change due to modifications of the potentiometric head distribution; yet they will be negligible in the short-mid run and, consequently, they are not be herein considered.

Therefore, an estimation of the water balance under future scenarios is viable as climatic (temperature and rainfall) predictions for 2021 and 2050 are already known given by for three distinct geographical areas; namely, Pyrenees, inland and coastal areas, (Calbó et al. 2016).

Mas-Pla et al. (2016) conducted such estimations assuming that the amount of annual available resources (R; runoff and groundwater infiltration) can be simply expressed as the difference between precipitation (P) and evapotranspiration (ET). Available resources for 2021 and 2050 were then estimated according to a water budget that uses the predicted values of precipitation and temperature. Actual evapotranspiration estimates are based on Zhang et al. (2001) approach, which relates this component with the potential evapotranspiration, based on future temperature values, and the land-use of the basin. Main hydrological basins were subdivided in smaller areas, according to lower-order streams, and data from 168 meteorological station (Servei Meteorològic de Catalunya) were used to estimate present and future water budgets.

The mean ratio between available resources and precipitation (R/P) for the entire basins in Catalonia is 0.268 for the present (2015), and predicted ratios are 0.235 and 0.218 for 2021 and 2050, respectively. Ratio changes are found in distinct geographical areas, being as 0.434, 0.413 and 0.380 for 2015, 2021 and 2050 at the Pyrenees, while for the same years, present and predicted ratios are of 0.194, 0.167 and 0.161 for the inland basins, and 0.241, 0.196 and 0.174 for the coastal areas (Fig. 2). Detailed results for each sub-basin are listed in the third CADS report.

In other words, a mean reduction of the available resources of 11 and 18% is expected for 2021 and 2050 in Catalonia. Estimations also point out that Pyrenees basin headwaters will also reduce this ratio R/P in a 4.5% mean value or 2050, being larger in those sub-basins of the Ebro River (5% at 2050) than for the Ter (3.5%) and Llobregat (1.2%) rivers. These results are relevant as the upper reaches fed the reservoirs that provides water to the metropolitan areas of Barcelona and Girona-Costa Brava, and the irrigation zones along the Segre River valley.

Nevertheless, estimation of the total available resources does not provide information on the variation of groundwater resources for each sub-basin. Further modelling should be performed with the shortcoming of the absence of field data for calibrating the model outputs. Qualitative, local analysis could be conducted to discuss which fraction of the total available resources eventually infiltrates, yet no conclusive figures exist. Estimations by Ortuño et al. (2009) and Alcalá and Custodio (2014) are the closest references.

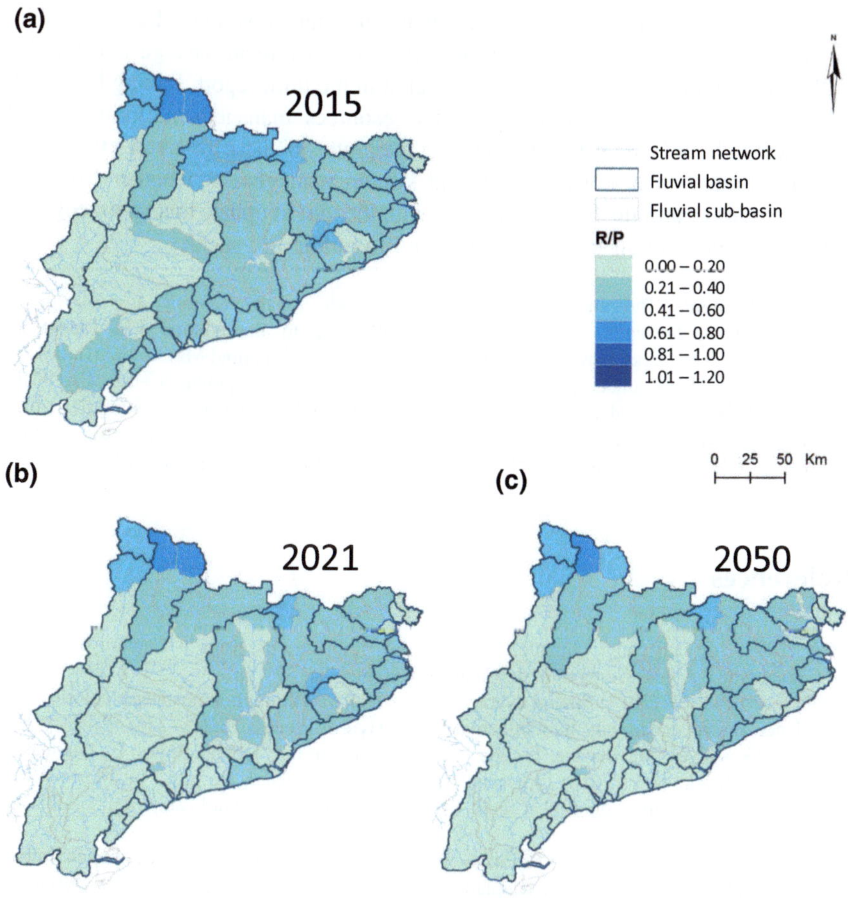

Fig. 2 Regional distribution of the R/P ratio, as the quotient between available resources and precipitation, for each sub-basin including the inner basins and those of the affluents of Ebro basin within (or partially within) Catalonia at **a** present (2015), and at the future scenarios of **b** 2021, and **c** 2050. After Mas-Pla et al. (2016)

6 Conclusions

Adaptation to climate change requires a comprehensive analysis of the natural systems response to the new environmental scenarios at a regional scale. The CADS reports provide such examination, being water resources, and groundwater in particular, a major issue of concern. Each report, as summarized in this text, provides a contribution based on the current knowledge with the aim of outlining the key issues that could be useful for water resources management.

In this sense, all three contributions provide a reflection on the final relevance of climate change upon water resources. Original data and maps offer a realistic assessment at regional and local levels that will later on support decision-making processes. Nevertheless, it must be recognized that climate change research has proliferated in the last years, as already stated in the third report, linking hydrology with environmental protection and water resources management. Future CADS documents should focus to sum up and integrate all these research efforts concerning climate change, hydrology, and system management whether from academic groups or administration agencies, with the aim to update this knowledge and make it available to all involved agents and to public in general.

Acknowledgements The author expresses his gratitude to the *Consell Assessor per al Desenvolupament Sostenible de Catalunya (CADS)*, and to the editors: J.E. Llebot and J. Martín-Vide. This paper is written under the University of Girona fund MPCUdG2016/061. All figures are reproduced under permission from Creative Commons public license. CADS-IEC reports can be downloaded, including Spanish and English executive summaries, from, http://canviclimatic.gencat.cat/en/politiques/politiques_catalanes/ladaptacio_al_canvi_climatic/Informes_sobre_canvi_climatic_Catalunya/index.html (last accessed on May 2017).

References

ACA (2009) Aigua i canvi climatic. Diagnosi dels impactes previstos a Catalunya. Agència Catalana de l'Aigua, Generalitat de Catalunya. Barcelona, 329 pp

Alcalá F, Custodio E (2014) Spatial average aquifer recharge through atmospheric chloride mass balance and its uncertainty in continental Spain. Hydrol Process 28:218–236

Calbó J, Gonçalves M, Barrera Escoda A, García-Serrano J, Doblas-Reyes F, Guemas V, Cunillera J, Altava V (2016) Projeccions climàtiques i escenaris de futur. In: Martín-Vide J (ed) Tercer Informe sobre el Canvi Climàtic a Catalunya. Publicació del CADS-IEC, pp 113–133

Clifton C, Evans R, Hayes S, Hirji R, Puz G, Pizarro C (2010) Water and climate change: impacts on groundwater resources and adaptation options. Water Sector Board of the Sustainable Development Network of the World Bank Group, Water Working Notes 25

Green TR, Taniguchi M, Kooi H, Gurdak JJ, Allen DM, Hiscock KM, Treidel H, Aureli A (2011) Beneath the surface of global change: Impacts of climate change on groundwater. J Hydrol 405:532–560

IPCC (2014) Climate Change 2014: synthesis report. Contribution of Working Groups I, II and III to the Fifth Assessment Report of the Intergovernmental Panel on Climate Change, 151 pp

Llasat C, Corominas J, García Sellés C, Quintana Seguí P, Turco M (2016) Riscos d'origen climatic. In: Martín-Vide J (ed) Tercer Informe sobre el Canvi Climàtic a Catalunya. Publicació del CADS-IEC, pp 137–160

Mas-Pla J (2005) Recursos hídrics, dinámica hidrológica i canvi climàtic. In: Llebot JE (ed) Informe sobre el canvi climàtic a Catalunya. Publicació del Consell Assessor per al Desenvolupament Sostenible i l'Institut d'Estudis Catalans, pp 485–516

Mas-Pla J (2010) Vulnerabilitat territorial dels recursos hidrològics al canvi climàtic. In: Llebot JE (ed.) Segon Informe sobre el Canvi Climàtic a Catalunya. Publicació del CADS-IEC, pp 309–342

Mas-Pla J, Ghiglieri G, Uras G (2014) Seawater intrusion and coastal water resources management. Examples from Catalonia and Sardinia. Contrib Sci 10:171–184

Mas-Pla J, Batalla RJ, Cabello A, Gallart F, Llorens P, Pascual D, Pla E, Pouget L, Sànchez A, Termes M, Vergonyós L (2016) Recursos hidrològics. In: Martín-Vide J (ed.) Tercer Informe sobre el Canvi Climàtic a Catalunya. Publicació del CADS-IEC, pp 161–187

Menció A, Folch A, Mas-Pla J (2010) Analyzing hydrological sustainability through water balance. Environ Manage 45:1175–1190
Ortuño F, Jódar J, Carrera J (2009) Canvi climàtic i recàrrega d'aqüífers a Catalunya. In: Agència Catalana de l'Aigua (ed) Aigua i Canvi Climàtic. Diagnosi dels impactes previstos a Catalunya, pp 143–152
Smit B, Wandel J (2006) Adaptation, adaptive capacity and vulnerability. Glob Environ Change 16:282–292
Taylor RG, Scanlon B, Döll P [… and 23 authors more] (2013) Ground water and climate change. Nature Climate Change 3:322–329
Zhang L, Dawes WR, Walker GR (2001) Response of mean annual evapotranspiration to vegetation changes at catchment scale. Water Resources Research 37:701–708

Qualitative Evaluation of Climate Change Effects on Nitrate Occurrence at Several Aquifers in the Catalonia Inner Basin

J. Mas-Pla, A. Menció and L. Portell

1 Introduction

Groundwater nitrate pollution is a ubiquitous worldwide problem that has been in the managers, as well in the researchers, agenda for decades. Multiple efforts have been devoted to improve agricultural practices, so nitrogen leaching due to fertilization is minimized, as well as to track its movement through the unsaturated zone and within the aquifer. Biogeochemical processes have also been researched to identify in situ denitrification processes which can be subsequently used to induce nitrate removal from groundwater. Despite all these efforts and knowledge, present nitrate concentrations in groundwater reflect the impact of decades of nitrogen inputs. Dilution stands as one of the major natural processes that diminish nitrate concentration in aquifers. Mixing between polluted water resources and less-polluted inputs reduces this environmental pressure. Dilution, at the end, relies on the input of non-polluted water fluxes.

In many regions on Earth, climate change will represent a shift towards drier conditions (IPCC 2014). Consequently a modification of their water balance and a diminution of the aquifer dilution capacity will occur. Therefore, an increase of nitrate concentration can be predicted. The goal of this contribution is to explore the effects of climate change upon the occurrence of nitrate in groundwater at a regional scale and in the long run. To conduct such objective, an approach founded on the basin water balance analyzes is used. In this sense, we consider the variation that each component of the hydrological cycle will experience under new climatic

J. Mas-Pla (✉) · L. Portell
Institut Català de Recerca de l'Aigua (ICRA), c/ Emili Grahit, 101, 17003 Girona, Spain
e-mail: jmas@icra.cat

A. Menció
Grup de Geologia Aplicada i Ambiental (GAiA-Geocamb), Deptartament de Ciències Ambientals, Campus de Montilivi, Universitat de Girona, 17003 Girona, Spain

M. L. Calvache et al. (eds.), *Groundwater and Global Change in the Western Mediterranean Area*, Environmental Earth Sciences,
https://doi.org/10.1007/978-3-319-69356-9_25

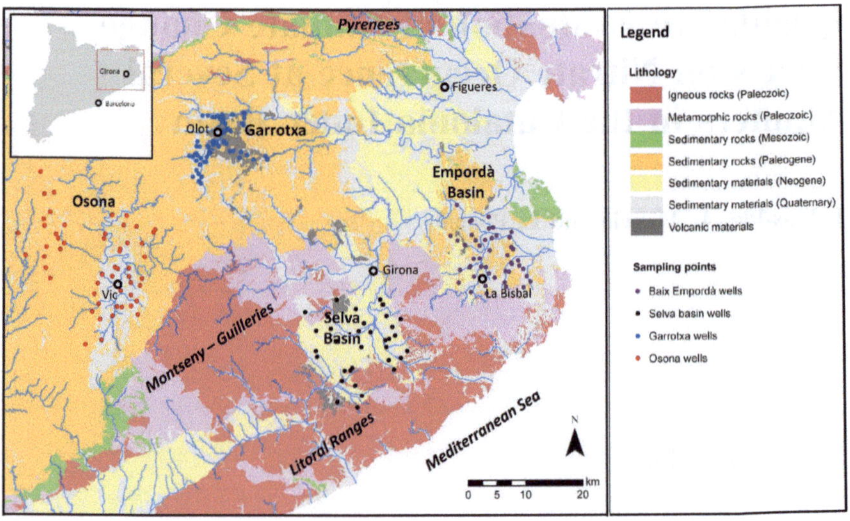

Fig. 1 Geographical locations and geological setting of the studied areas

scenarios and similar water uses, and how these changes will affect groundwater nitrate content.

During the last fifteen years, the extent of nitrate pollution and the occurrence of denitrification processes from a hydrogeological and hydrogeochemical perspectives has been studied by several projects in distinct aquifers of Catalonia, all of them declared as vulnerable to nitrate pollution as a result of EU directives (Fig. 1).

Our approach refers to the paradigmatic study on the impact of climate change on future nitrate concentrations in groundwater of the UK by Stuart et al. (2011). They focus on the idea that climatic changes would likely affect groundwater recharge mechanisms and, the way they will affect groundwater receptors.

2 Methodology

This study consists on the interpretation of already existing data from nitrate polluted aquifers under the future climatic scenarios (i.e., horizons 2021 and 2050), and the subsequent modifications of the water budget for each hydrogeological system. Data from each aquifer have been reported in the following references: according to published results in the Selva basin (Folch et al. 2011; Puig et al. 2013), Empordà basin (Puig et al. 2017), Osona region (Otero et al. 2009; Menció et al. 2011; Boy-Roura et al. 2013b), and Garrotxa (Bach 2015; unpub. data). Details in land use and pollution levels in these aquifers are summarized by Menció et al. (2016).

Climatic conditions for the above-mentioned horizons were calculated after a downscaling process from global circulation models by Calbó et al. (2016). This reference provides the averaged temperature increment and rainfall variations for different geographical/climatic areas in Catalonia. Water budgets for 2021 and 2050 in the hydrographic basins of Catalonia, with special interest of those corresponding to the studied aquifers, were estimated by Mas-Pla et al. (2016) using the climatic projections by Calbó et al. (2016) and the percentage of land-use as determined by Zhang et al. (2001) approach. To differentiate between run-off and infiltration terms, we use a map of average recharge based on the chloride mass balance by Alcalá and Custodio (2014).

We herein profit from these references and use their results to discuss the impact of climate change on nitrate pollution for distinct hydrogeological settings using a water balance approach (e.g., Fitts 2012, pp. 13–16; Menció et al. 2010).

3 Results

3.1 Hydrology and Water Quality in the Selected Aquifers

Data from the studied regions are summarized for each aquifer (Table 1); in particular: (1) aquifer type (water-table, leaky, or confined), (2) origin of recharge: direct rainfall (representing soil infiltration and, therefore, leaching of soil nutrients and applied fertilizers), interaction with streams (considering discharge regime), and occurrence of recharge from mountain fronts or the basement according to the geological setting; and (3) mean and median nitrate content, percentage of samples with $[NO_3] > 50$ mg/L, and (4) mean, median of $\delta^{15}N_{NO_3}$ values, and percentage of denitrification based on the multi-isotopic approach. They provide a brief indication of their present state and of potential alternatives under future climatic scenarios.

3.2 Hydrology and Water Quality in the Selected Aquifers

Present climate projections derived from climatic/atmospheric models provide estimates for temperature (T) and precipitation (P) variations for the next decades. In most cases, downscaling processes are necessary to refine such predictions, and to offer more appropriate data to hydrologists so future water budgets can be estimated with the predicted parameters. Calbó et al. (2016) calculated the seasonal and annual temperature and precipitation projection for 2021 and 2050 using distinct climatic databases, as well as different downscaling methods, for each geographical area in Catalonia: Pyrenees, inland areas and coastal areas. Major results are given in Table 2. Scaling is necessary, especially as regards rainfall, as in areas such as Catalonia, the complex orography and land-sea contrast are largely misrepresented by global models.

Table 1 Summary of the main hydrogeological and chemical status of the aquifers considered in this study. Data after references cited in the text

	La selva	Osona	Empordà	Garrotxa
Aquifer lithology: Bedrock, sedimentary infilling, alluvial	S-A	B-A	S-A	B-A
Recharge origin: local, regional	L-R	L-R	L-R	L
Aquifer type: unconfined/multilayered/leaky	U-M	M	U-M	U-M
$[NO_3]$ (mg/L): mean/median/% > 50 mg/L	65/59/51%	165/92/72%	100/67/57%	35/29/22%
$\delta^{15}N_{NO_3}$(‰): mean/median/% > 15‰[a]	10.8/10.4/7%	14.0/13.4/28%	12.9/12.2/22%	9.9/9.4/6%
Land-use: forested/agricultural	A	A	A	F - A
Groundwater exploitation: seasonal, non-seasonal	NS (livestock) S (agriculture)	NS (livestock) S (agriculture)	NS (urban)—S (agriculture)	S (agriculture)
Alternative resources:	Water reclamation	Ter river (mid-basin)—water reclam.	Ter river (lower-basin)—water reclamation	Fluvià river
Other pressures, hazards:	Local as, F high levels	(none)	Seawater intrusion—over-exploitation	(none)

[a]Values of $\delta^{15}N_{NO_3}$ > 15‰ are assumed to be indicative of denitrification processes

Table 2 Annual median temperature and rainfall variations for the 2021 and 2050 horizons in Catalonia with respect to 1971–2000. Data in parenthesis correspond to the 5th and 95th percentile. Data after Calbó et al. (2016)

	Climatic area	Pyrenees	Inland areas	Coastal areas	Catalonia
2021	ΔT (°C)	0.8 (0.5/1.1)	0.7 (0.5/1.0)	0.7 (0.5/1.0)	0.8 (0.5/1.0)
	Rainfall (%)	−0.2 (−7.8/8.0)	0.7 (−14.1/8.0)	−2.4 (−20.7/6.0)	−2.4 (−13.4/5.8)
2050	ΔT (°C)	1.6 (0.9/2.2)	1.4 (0.9/2.1)	1.4 (0.9/2.0)	1.4 (0.9/2.0)
	Rainfall (%)	−5.3 (−16.1/−1.2)	−6.5 (−23.7/1.4)	−8.5 (−27.1/2.3)	−6.8 (−22.0/−0.7)

Table 3 Summary of available resources, as R/P, according to published data

		LA SELVA	OSONA	EMPORDÀ	GARROTXA
R/P—2015[a]		0.292	0.287	0.238	0.380
R/P—2021[a]		0.278	0.265	0.229	0.357
R/P—2050[a]		0.254	0.258	0.206	0.338
Average groundwater recharge, R(gw)/P[b]		0.120	0.200	0.150	0.220
2021	R(sw)/P	0.158	0.065	0.079	0.137
2050	R(sw)/P	0.134	0.058	0.056	0.118

sw surface water
gw groundwater
[a]Mas-Pla et al. (2016)
[b]Alcalá and Custodio (2014)

3.3 Water Budget Estimations for Future Climate Scenarios

Table 3 shows the mean quotient between available resources and the precipitation (R/P; where R stands for "available resources" and P for "precipitation") in the hydrological basins of the aquifers considered in this study using predictions for 2021 and 2050. The hydrologic balance considers predicted rainfall to estimate precipitation inputs, and predicted temperature and present land-use cover at a sub-basin scale to estimate the actual evapotranspiration. These R/P values are just representative of the direct recharge within the basin. Other recharge fluxes: stream inflow to the aquifer and regional flow systems, are not evaluated.

Nevertheless, a main difficulty at a basin level consists in separating between surface water and groundwater resources for the predicted values. In this sense, the average groundwater recharge values for the selected aquifers estimated by Alcalá and Custodio (2014) are also included in Table 3, and as they depend of soil conditions (which are assumed unchanged), groundwater recharge ratio (R(gw)/P) is assumed constant for 2021 and 2050.

4 Discussion

Based on the above data, climate change effects on groundwater quality are discussed based on the water balance approach and the learnings from the field work in the studied aquifers. We assume that for the next decades, major changes on land and groundwater uses will not occur unless any large development project is planned, which is the case in most these regional aquifers; yet, planned irrigation systems optimization and a potential reduction of nitrogen inputs would always contribute to reduce the fertilization impact.

4.1 Hydrological Concerns

As pointed out from climate predictions, water scarcity will primarily affect surface runoff and direct groundwater recharge. Surface runoff is relevant when discussing nitrate in alluvial aquifers, especially for loosing stream reaches, where runoff is the main input of low-nitrate content input to groundwater. In this sense, what does now constitute a recharge pole in the alluvial fluvio-deltaic aquifers of the Baix Ter and Baix Fluvià systems, it will reduce its contribution in the near future. If the consequences of diminishing discharge in the lower reaches of a river are to be examined, a set of outcomes can be enumerated. Explicitly, this implies several hydrological effects: (1) less aquifer recharge from the losing reaches, (2) less instream flow, (3) a diminution of the aquifer water table; and, (4) a reduction of the flow discharging back to the river in the gaining stream reaches. Direct groundwater recharge from rainfall will determine the rate and concentration to which nitrogen soil will percolate to the aquifer.

As regards to groundwater quality, a minor dilution associated with less stream and rainfall recharge will introduce nitrate to the soil, and later to the aquifer, at larger concentrations. In these circumstances, the overall hydraulic gradient will also be flattened due to a decrease of the inland hydraulic head at the boundaries, whether the stream itself or the surrounding hydrogeological, and an increase of the sea level. This will affect the alluvial aquifer residence time and, therefore, determine a longer stay of nitrate mass in the aquifer. Reduced outflows towards the stream or wetlands will reduce the rate of nitrogen assimilation that takes place in riparian areas. Denitrification processes will depend on the subsurface geochemical environment. Usual heterotrophic denitrification linked to the occurrence of organic matter and lack of oxygen, as observed in the Baix Ter area, can certainly be modified by soil production and oxygen solubility changes related to higher temperatures. Indeed, dry soil moisture lowers soil enzyme activity (Sardans et al. 2008), and increasing summer droughts will reduce mineralization and N and C fluxes whereas increasing summer precipitation could enhance losses (Borken and Matzner 2009). Nevertheless, nitrogen isotopes for the two alluvial aquifers (Baix Ter and Baix Fluvià) indicate that, under present conditions, only one out of four samples show some degree of denitrification. This suggests that nitrate mass may accumulate under future dominant hydrological conditions. Larger denitrification occurrence in Osona (nearly in a 30% of the samples) occur in marl deposits where pyrite oxidation enhances nitrate removal, and they are not associated with alluvial aquifers.

Alluvial aquifers in la Selva, Osona and Garrotxa hydrogeological systems are of less entity both in volume of groundwater storage and input/outflow magnitudes. Such smaller alluvial deposits will be even more depending on the stream discharge originated in the range headwaters.

From a management perspective, two actions can be adopted to minimize the effects of climate change. First, allocating wells in places and in aquifer levels more resilient to recharge variations. Secondly, reducing groundwater exploitation rates by increasing agricultural efficiency, urban water saving actions and diminishing losses in the distribution network, and ultimately, reclaim treated urban wastewater as an effective alternative to compensate water scarcity. Heavily urbanized areas with intense agricultural water demand are the first allocation places for treated wastewater, at least for specific uses as the total treated water volumes will be by far smaller than the agricultural demand. Presently, and more relevant under the future climatic scenarios, treated wastewater applied as irrigation will sum up its nitrogen content with that from applied manure, resulting in an excess load that will affect the final soil and groundwater nitrate concentration.

4.2 Environmental Concerns

In the effort to protect streams and wetlands as valued groundwater dependent ecosystems, water balance variations must be taken into consideration. Because of a diminishing recharge, nitrate mass will tend to accumulate in aquifers due, basically, to larger residence times caused by less inflow and minor outflow. Groundwater nitrate flowing to rivers in their gaining reaches or to wetlands may increase its content in surface water bodies. Nitrogen, as a nutrient, is highly assimilated by plants in riparian areas and pond boundaries, preventing eutrophication. Pumping rates will also interfere in hydrogeological dynamics. Capture zones affecting stream reaches and wetlands will progressively dry them up. Recovering natural flow fields after the irrigation season will feed the groundwater dependent stream reaches and wetlands with high-nitrate groundwater recharge. This is the reason why, especially in wetlands and ponds, preserving the natural habitat in their shores, where nitrogen assimilation/immobilization processes occur, is so essential.

4.3 Economic, Social and Political Concerns

Beyond hydrological and environmental concerns, climate change issues associated with nitrate pollution have strong economic and social sides. It is obvious that groundwater high nitrate concentrations are related to cattle rising (mainly pig) activities, being crucial for the economy of rural areas. Given the importance of this economic sector, it is not plausible that manure production will decrease in a medium run and, in consequence, arable lands will be fertilized at its maximum capacity to get rid of manures and slurries. As in any pollution case, input control is a key factor in controlling the spread and evolution of the pollutant. Alternative manure treatment processes should then be encouraged from the administration. In the context of climate change, as mentioned, lower recharge rates in the agricultural areas located on alluvial aquifers will hamper dilution. Hence, the management problem becomes twofold: water scarcity and pollution.

Solutions will need social and political support. From the social perspective, population, and politicians, need to understand the fact that under the new climatic scenarios the cost of good quality water resources will increase. From the socio-political side, education focused to recognize the value of a scarce resource should be focused not only to citizens, but also to farmers and, especially, first-sector investors whose profits are presently based on water resource exploitation strategies that would be unappropriated in the next decades. The livestock sector has to face it. From the political responsibility, a commitment is needed to impulse directives and measures focused on water resources planning that recognizes the urgency for actions. The water balance approach indicates that the "nitrate problem" may become even worse if the no-action attitude prevails. Active political involvement is paramount to force action and to adjust government as well as private responsibilities on preserving groundwater quality resources.

5 Conclusions

Under present nitrogen inputs in soils, groundwater nitrate concentration in the studied area will likely tend to increase/persist due to climate effects on the hydrologic cycle, land-use changes and human water demand. Recharge will be limited, and with it, its dilution capacity to decrease nitrate concentrations. Therefore, scientific-based opinions must be explained to stakeholders and decision makers so groundwater quality issues are included in their agendas together with water supply. Since agriculture is a primordial economical as well as territorial-integrating activity, and fertilization will necessarily go on, the nitrate-problem becomes twofold:

1. The implementation of better agricultural practices, yet their outcome is uncertain as seen from increasing local nitrate concentration tendencies despite the application of EU directives, mainly where livestock is the main economic activity.

2. New water supply alternatives must be sought and planned with time, so they can be appropriately programmed and executed, and their costs included in today's operational budgets so to avoid considering them as externalities when they will be urgently needed.

Acknowledgements This study is part of the REMEDIATION project (Spanish National Program on Water Resources; CGL2014-57215-C4-2-R), and also by the University of Girona fund MPCUdG2016/061.

References

Alcalá FJ, Custodio E (2014) Spatial average aquifer recharge through atmospheric chloride mass balance and its uncertainty in continental Spain. Hydrol Process 28:218–236

Borken W, Matzner E (2009) Reappraisal of drying and wetting effects on C and N mineralization and fluxes in soils. Glob Change Biol 15:808–824

Boy-Roura M, Nolan BT, Menció A, Mas-Pla J (2013) Regression model for aquifer vulnerability assessment of nitrate pollution in the Osona region (NE Spain). J Hydrol 505:150–162

Calbó J, Gonçalves M, Barrera Escoda A, García-Serrano J, Doblas-Reyes F, Guemas V, Cunillera J, Altava V (2016) Projeccions climàtiques i escenaris de futur. In: Martín Vide J (ed) Tercer Informe sobre el Canvi Climàtic a Catalunya. Pub. of Consell Assessor per al Desenvolupament Sostenible, pp 113–133

Fitts C (2012) Groundwater Science, 2nd edn. Academic Press, London

Folch A, Menció A, Puig R, Soler A, Mas-Pla J (2011) Groundwater development effects on different scale hydrogeological systems using head, hydrochemical and isotopic data and implications for water resources management: the Selva basin. J Hydrol 403:83–102

IPCC (2014) Climate change 2014: synthesis report. Contribution of Working Groups I, II and III to the fifth assessment report of the Intergovernmental Panel on Climate Change, 151 p

Mas-Pla J, Batalla RJ, Cabello A, Gallart F, Llorens P, Pascual D, Pla E, Pouget L, Sànchez A, Termes M, Verbonyós L (2016) Recursos hidrològics. In: Martín Vide J (ed) Tercer Informe sobre el Canvi Climàtic a Catalunya. Pub. of Consell Assessor per al Desenvolupament Sostenible, pp 161–187

Menció A, Folch A, Mas-Pla J (2010) Analyzing hydrological sustainability through water balance. Environ Manag 45:1175–1190

Menció A, Mas-Pla J, Otero N, Soler A (2011) Nitrate as a tracer of groundwater flow in a fractured multilayered aquifer. Hydrol Sci J 56(1):108–122

Menció A, Mas-Pla J, Soler A, Otero N, Regàs O, Boy-Roura M, Puig R, Bach J, Domènech C, Folch A, Zamorano M, Brusi D (2016) Nitrate pollution of groundwater; all right …, but nothing else? Sci Total Environ 539C:241–251

Otero N, Torrentó C, Soler A, Menció A, Mas-Pla J (2009) Monitoring groundwater nitrate attenuation in a regional system coupling hydrogeology with multi-isotopic methods: the case of Plana de Vic (Osona). Agr Ecosyst Environ 133:103–113

Puig R, Folch A, Menció A, Soler A, Mas-Pla J (2013) Multi-isotopic study (15N, 34S, 18O, 13C) to identify nitrate evolution in a large-scale groundwater system. Appl Geochem 32:129–141

Puig R, Soler A, Widory D, Mas-Pla J, Otero N, Domènech C (2017) Characterizing sources and natural attenuation of nitrate contamination in the Baix Ter aquifer system (Spain) using a multi-isotope approach. Sci Total Environ 580:518–532

Sardans J, Peñuelas J, Estiarte M (2008) Changes in soil enzymes related to C and N cycle and in soil C and N content under prolonged warming and drought. Appl Soil Ecol 39:223–235
Stuart ME, Goody DC, Bloomfield JP, Williams AT (2011) A review of the potential impact of climate change on future nitrate concentrations in groundwater of the UK. Sci Total Environ 409:2859–2873
Zhang L, Dawes WR, Walker GR (2001) Response of mean annual evapotranspiration to vegetation changes at catchment scale. Water Resour Res 37:701–708

Investigating the Impact of Climate Change on Groundwater Recharge Using a High Precision Meteo Lysimeter in a Dune Belt of the Doñana National Park

L. Molano-Leno, C. Kohfahl, D. J. Martínez Suárez, F. Ruiz Bermudo, A. N. Martínez Sánchez de la Nieta, S. Engelhardt, K. Vanderlinden and J. V. Giráldez

1 Introduction

Dune belts are fundamental for groundwater recharge in coastal aquifers and constitute therefore a key location for the quantitative and qualitative monitoring of water resources in ecological habitats. Beyond the large number of methods for recharge estimations, at the moment weighing lysimeters yield the most precise measures for recharge, evapotranspiration and precipitation (Peters et al. 2014). Nonetheless, precise weighing lysimeters have been mostly installed for agricultural purpose in crop areas and therefore only limited knowledge exists about recharge dynamics and its dependence on meteorological parameters in dune belts (Schrader et al. 2013).

The Spanish Geological and Mining Survey (IGME), started recently a research project to monitor the natural recharge in the dune belts of the Doñana Natural Reserve. A high precision weighing meteo-lysimeter with lower boundary control

L. Molano-Leno · C. Kohfahl (✉) · D. J. Martínez Suárez · F. Ruiz Bermudo ·
A. N. M. S. de la Nieta
Instituto Geológico y Minero de España, Pza. de España—Torre Norte,
41013 Sevilla, Spain
e-mail: kohfahl@zedat.fu-berlin.de

L. Molano-Leno
e-mail: l.molano@igme.es

S. Engelhardt
UMS AG, 81379 Munich, Germany

K. Vanderlinden
IFAPA, Centro Las Torres-Tomejil, Alcalá Del Río, 41200 Seville, Spain

J. V. Giráldez
Departamento de Agronomía, Universidad de Córdoba, 14071 Córdoba, Spain

M. L. Calvache et al. (eds.), *Groundwater and Global Change in the Western Mediterranean Area*, Environmental Earth Sciences,
https://doi.org/10.1007/978-3-319-69356-9_26

was installed in September 2015 for continuous monitoring of recharge and other soil and meteorological parameters.

Recharge in the dune belt is essential for the conservation of the groundwater dependent wetlands of the Doñana National Park, which is threatened by intensive agricultural irrigation and water supply for tourism. In general, groundwater recharge is a complex process and depends on a large number of factors such as morphology, vegetation, meteorological and soil parameters. Climate change may impact groundwater recharge due to increasing temperatures and changing seasonal patterns of precipitation but also due to change in vegetation. The main objective of this study is to explore the impact of different meteorological conditions on groundwater recharge in dunes belts within semiarid climate, and to derive its dependence on regional climate trends predicted by climate models.

Sediment analysis from previous projects have shown that the Doñana dune belt material is made up of medium coarse sands of a well-sorted grain size distribution (Kohfahl et al. 2014). The aeolian dune sands are composed of quartz and feldspars. Some secondary minerals, such as carbonates, clay, sulphides or Fe-oxides, might have formed afterwards in situ due to subsequent weathering processes.

The climate of the area is sub-humid Mediterranean with Atlantic influence, dry summers and humid winters. The average rainfall, which occurs between October and March, is between 500 and 600 mm, with a great interannual variability, between 250 and 1,100 mm (Custodio et al. 2009). The average annual air

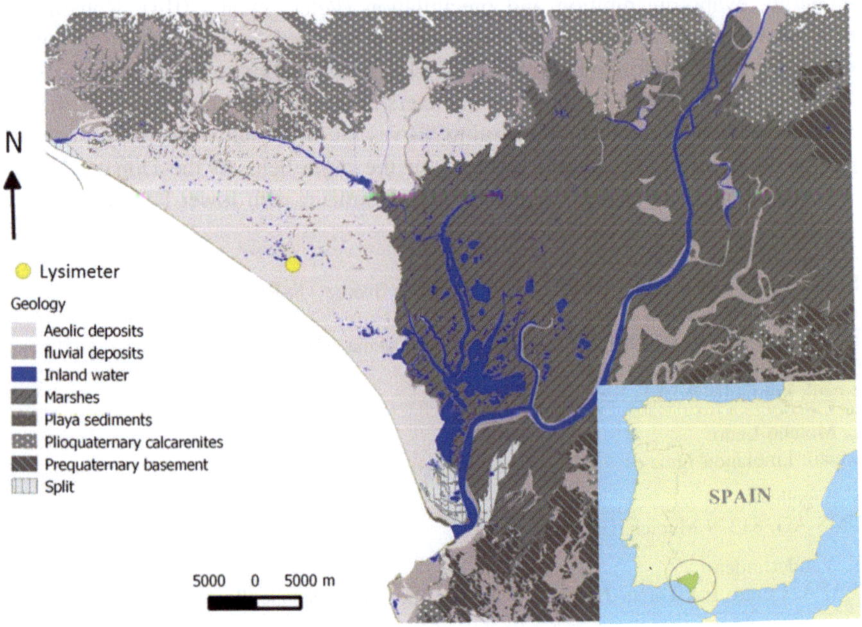

Fig. 1 Field site and Geology (modified after National Geological Map (MAGNA), map sheet 1033)

temperature is about 17 °C near the coast and 18 °C in the center of the Park. There are around 3,000 h of sunshine every year (Manzano et al. 2009) (Fig. 1).

2 Materials and Methods

The site is equipped with a UMS (UMS AG, Munich, Germany) weighing lysimeter (a cylinder of 1 m^2 surface area, 1.50 m height and a weighing resolution of 10 g), six CS650 soil moisture sensors (Campbell Scientific, Logan, UT) installed beside the lysimeter at 0.30, 0.60, 1.20, 1.60, 2.20, and 3 m depth, and 2 automatic meteorological stations (Vantage PRO2 Davis, California, USA; UMS AG, Munich, Germany). The lower boundary condition at the bottom of the lysimeter is controlled using a tensiometer. A peristaltic pump maintains the bottom of the lysimeter at the same potential as measured by the field tensiometer installed outside the lysimeter. Rain and drained water from the lysimeter is collected and sampled for analysis of chemical species including stable isotopes. Table 1 shows the measurements that are continuously performed. Physico-chemical soil properties such as density, grain size, mineralogy and metals were also analysed at different depths.

To filter measurement noise, the raw data are corrected to accurately calculate precipitation (P) and recharge (R) from lysimeter data. The time-series of lysimeter weight and drainage water quantity can be affected by a wide variety of singular disturbances, which add to the measurement noise. Examples are the withdrawal of water from the drainage sampling vessel, sudden changes in weight when vegetation is removed, maintenance work, technical staff stepping on the lysimeter

Table 1 Measured parameters and intervals

Measured parameter	Time interval (minutes)
Soil mass lysimeter	1
Water mass drained from lysimeter	1
Soil water tension	10
Soil moisture	10
Wind direction	10
Wind velocity	10
Net radiation	10
Precipitation	10
Air humidity	10
Air and soil thermal profile	10
Soil bulk density	Once
Grain size distribution	Once
Mineralogy	Once
Metals content	Once

surface, and so on. Detection and correction of such singular events was performed by suitable filters and manually in some cases.

Intrinsic noise in lysimeter data was reduced by smoothing through the AWAT filter (Peters et al. 2014). This method has been recently applied in other studies (Hoffmann et al. 2016). For the application of the AWAT algorithm, the parameters maximum window width and maximum threshold were set to 31 min and 0.24 mm, respectively. Several short gaps in the lysimeter data time series occurred due to power failures.

According to the soil water balance, P distributes into increase of soil water storage (Δw_{lys}), R, and evapotranspiration (ET) as shown in Eq. 1.

$$\mathrm{P} = \mathrm{R} + \mathrm{ET} + \Delta w_{lys} \tag{1}$$

P can then be calculated according to Schrader et al. (2013) as in Eqs. 2 and 3.

$$\Delta W = \Delta w_{lys} + \Delta w_{drain} \tag{2}$$

$$P = \begin{cases} \Delta W, & \Delta W > 0 \\ 0, & \Delta W \leq 0 \end{cases} \tag{3}$$

where Δw_{lys} [kg] is the mass change of the lysimeter during each time interval which corresponds to the water storage change, Δw_{drain} [kg] is the mass change in the drainage sampling vessel, P [kg] is the sum of precipitation recorded by the lysimeter.

ET was also calculated but results are not presented here due to errors produced by noise elimination unsolved at the moment.

3 First Results and Discussion

Data from cumulative Δw_{drain} and Δw_{lys} are represented in Fig. 2a without previous treatment. Sudden drops in the weight of the drainage vessel correspond to the withdrawal of water. This kind of errors are automatically, or sometimes manually, corrected.

The results show some differences between the field and lysimeter tensiometer data (Fig. 2b). In September 2016 the tensiometer located in field was replaced reducing significantly the previous differences. The graphs 2c, d and e represent temperature, soil moisture, and salinity measured from the six CS650 soil moisture sensors at different depths. This data will be used for recharge estimation with Hydrus 1D benefitting from the possibility to compare it with direct measurements.

Figure 3 shows air and soil temperature at different heights and depths. Wind speed, wind direction, net radiation, and relative humidity, measured at 10 min intervals. Also these data are required as input parameters to set up a model with HYDRUS 1D for calculating recharge. Temperature will be used to assess energy

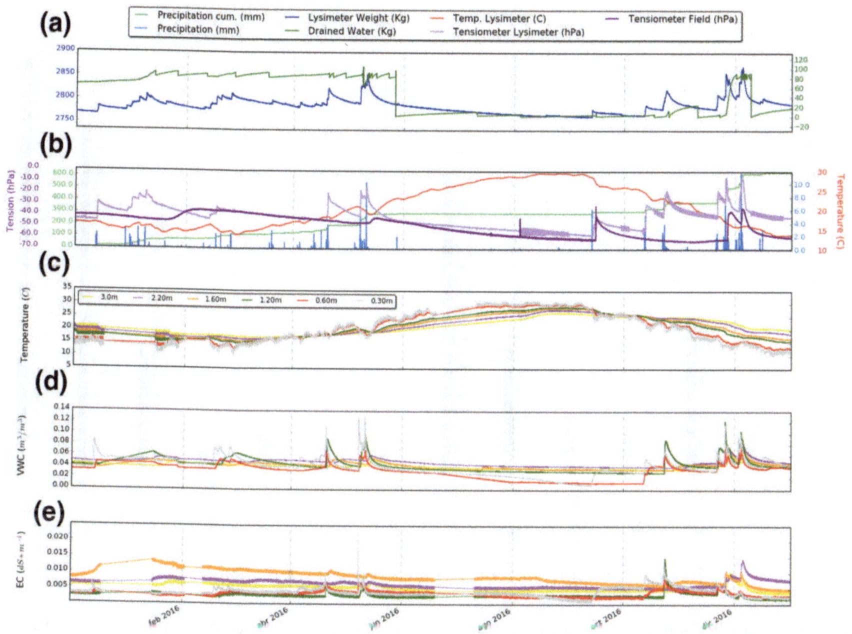

Fig. 2 **a** Weight of lysimeter and drainage vessel (kg); **b** P measured by tipping bucket rain gauge (mm) and its cumulative P (mm), suction measured by tensiometer in field and in lysimeter (hPa), temperature from lysimeter at 1.40 m depth (°C); **c** Temperature profile beside the lysimeter measured by CS650 soil moisture sensors (°C); **d** Volume Water Content (VWC) measured by CS650 soil moisture sensors (m^3/m^3); **e** Bulk electrical conductivity (EC) measured by CS650 soil moisture sensors (dS/m)

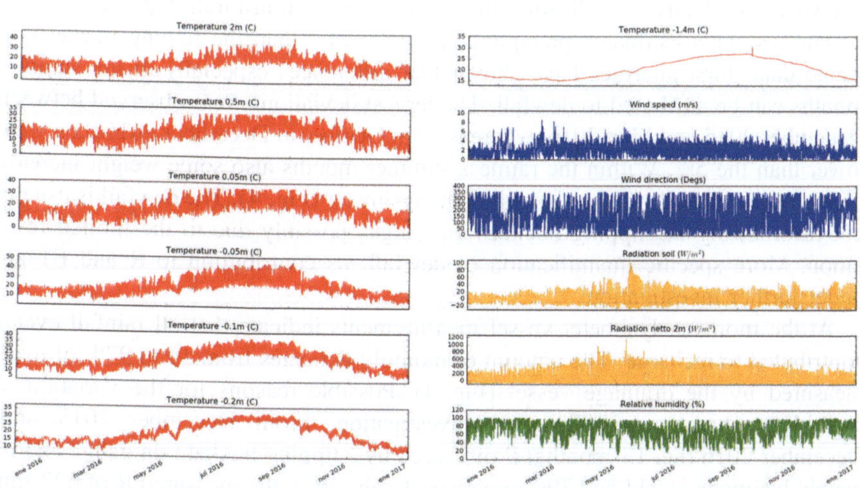

Fig. 3 Temperature at the indicated depths, wind speed and direction, ground and net radiation, and air relative humidity measured every 10 min. Values given in m refer to distance from soil surface

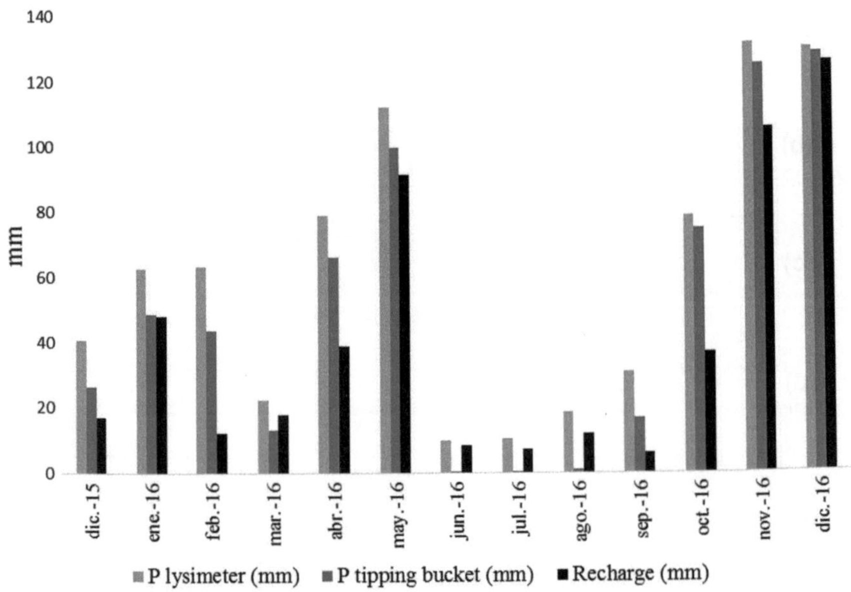

Fig. 4 Monthly recharge and rainfall, both calculated using lysimeter data (P lysimeter), rainfall measured with the tipping bucket rain gage (P tipping bucket), and recharge (R), measured as weight increase due to drained water of the lysimeter into the sample vessel, from December 2015 to December 2016

transport, which is included in the model HYDRUS (Radcliffe and Šimůnek 2010). Electrical bulk conductivity measured with CS650 sensors will be converted to porewater conductivity, indicating the extent of evapotranspiration processes.

The lysimeter estimated precipitation was greater than the tipping bucket rain gauge data. First analyses have shown that differences especially in the summer months can be attributed to dewfall. The largest deviations were observed between December 2015 and May 2016, whereas from October to December 2016 they were lower than the 5%. Within the rainless summer months also some weight increase of 10–18 mm/month was measured by the lysimeter (Fig. 4). The dewfall is usually not recorded by the tipping bucket rain gauges possibly due to the limited resolution. More specific quantification of dewfall, its contribution to R and ET are under further investigation.

At the moment, lysimeter-vessel measurements indicate that all rainfall events contributed to recharge. The amount of monthly R ranges from 19 to 97% of the P measured by the drainage vessel (Fig. 4). Possible reasons for the variation of monthly recharge are still under investigation. From December 2015 until December 2016 (i.e. 13 months) P measured by a tipping bucket rain gauge and by weight lysimeter is 644 and 790 mm respectively. With the measured R of 527 mm this gives a 13 months recharge amount of 82 and 67%.

To better understand the uncertainty of measurements a soil water model is currently being set up and calibrated by lysimeter and soil moisture sensor-acquired data to better quantify (i) the effect of meteorological parameters on R rates, (ii) to simulate the effect of uncertainty in precipitation measurements on simulated R, (iii) to verify the dew effect and (iv) to quantify the effect of the lower boundary condition on measured hydrological components by the lysimeter.

4 Conclusions

The collected data in the first hydrological year have shown differences in P measurements between the tipping bucket rain gauge and the lysimeter with around 5% higher values measured by the lysimeter. A positive increase in ΔW has been detected during summer, in spite of missing rainfall, which is attributed to dewfall. Furthermore, differences of the relative amount of monthly R have been observed and are under evaluation. A HYDRUS 1D model is currently set up to evaluate all these processes. With the ongoing studies we will try to answer the questions that have been exposed in this research.

Acknowledgements The infrastructure has been co-financed by European Research Funds (SE Scientific Infrastructures and Techniques and Equipment 2013, IGME13-1E-2113). This research has been funded by the CLIGRO Project (MICINN, CGL2016-77473-C3-1-R) part of the Spanish National Plan for Scientific and Technical Research and Innovation. We are grateful for the support and collaboration of the Biological Station of Doñana, the Biological Reserve of Doñana and the administration of the Doñana National Park. This work is also part of the activities subsidized within the National System of Youth Guarantee (MINECO activity with reference PEJ-2014-85121) co-financed under the Youth Employment Operational Program, with financial resources from the Youth Employment Initiative (YEI) and the European Social Fund (ESF).

References

Custodio E, Manzano M, Montes C (2009) Las aguas subterráneas en Doñana: Aspectos ecológicos y sociales

Hoffmann M, Schwartengräber R, Wessolek G, Peters A (2016) Comparison of simple rain gauge measurements with precision lysimeter data. Atmos Res 174–175:120–123

Kohfahl C, Fernández de los Santos N, Mediavilla Laso C, Martínez Sánchez de la Nieta A, Ruiz Bermudo F (2014) Natural dynamics and control of soil gas oxygen in wetlands: fibre-optic oxygen measurements at two field sites in the Doñana National Park (Spain). Wetlands 34 (6):1197–1206

Manzano M, Custodio E, Higueras H, Puig R, Soler A (2009) Influencia de la gestión del acuífero sobre los humedales del manto eólico de Doñana. Boletín Geológico y Minero 120(3):377–392. ISSN: 0366-0176

Peters A, Nehls T, Schonsky H, Wessolek G (2014) Separating precipitation and evapotranspiration from noise—a new filter routine for high-resolution lysimeter data. Hydrol Earth Syst Sci 18(3):1189–1198

Radcliffe DE, Šimůnek J (2010) Soil physics with hydrus: modeling and applications. CRC Press, Boca Raton, Fl, USA

Schrader F, Durner W, Fank J, Gebler S, Pütz T, Hannes M, Wollschläger U (2013) Estimating precipitation and actual evapotranspiration from precision lysimeter measurements. Procedia Environ Sci 19:543–552

Chemical Tracers and Stable Isotopes Mixing Models for Groundwater Quality and Recharge Study in the Moroccan High Atlas Mountains

B. N'da, L. Bouchaou, B. Reichert, L. Hanich, S. Danni, Y. Ait Brahim, A. Chehbouni, E. Beraaouz and J. L. Michelot

1 Introduction

Morocco, due to its semi-arid to arid climate, has limited potential for water resources; however, it faces considerable challenges due to a sharp increase in water demand, and a decrease in water resources availability. Morocco is facing significant population growth, accelerated urbanization, and intensive use of water resources for irrigation, drinking water and tourism. The distribution of water throughout the Moroccan territory is marked by a strong inequality in time and space. Groundwater, which accounts for nearly 20% of the resource potential, has been widely used, resulting in a water table drop of about 2 mm/year on average (Stoffel et al. 2002).

In this context of vulnerability of water resources, the High Atlas represents a rampart: due to the orographic effect, it allows precipitation of the air masses coming from the Atlantic coast. Like the other mountain ranges in the world, the High Atlas represents an important reservoir of fresh water, hence its nickname as a water tower in Morocco.

B. N'da (✉) · L. Bouchaou · S. Danni · Y. Ait Brahim · E. Beraaouz
Laboratory of Applied Geology and Geo-Environment, Ibn Zohr University, BP. 8106, Cité Dakhla, 80060 Agadir, Morocco
e-mail: baudnda2013@gmail.com

B. Reichert
Steinmann-Institute, Geology, Bonn University, Nussallee, 53115 Bonn, Germany

L. Hanich
Laboratory of Geo-Resources, Faculty of Science and Technology, Cadi Ayyad University, BP. 549, Marrakech, Morocco

A. Chehbouni
CESBIO, 15 rue Abou Derr, BP. 8967, 10000 Rabat Agdal, Morocco

J. L. Michelot
UMR IDES 8148 CNRS—Université Pari-Sud, 11-Bât 504, 91405 Orsay Cedex, France

M. L. Calvache et al. (eds.), *Groundwater and Global Change in the Western Mediterranean Area*, Environmental Earth Sciences,
https://doi.org/10.1007/978-3-319-69356-9_27

Generally in the arid to semi-arid zones, as in the case of the Moroccan High Atlas, the most common practice is the storage of runoff water from the mountains in containment reservoirs for irrigation that increases the water deficit. In addition, future climate scenarios predict a 15% decrease in precipitation, with an increase in the duration of the dry period throughout Morocco by 2050 (Driouech et al. 2010). These announced climate changes could increase the water problem in the region, which is already under high water stress. This region, home to about 20% of the population of Morocco, could suffer from a major water crisis in the years to come, linked to anthropogenic effects and climate change.

So, the implementation of an integrated water resource management system in the Moroccan High Atlas region is crucial. However, the implementation of an integrated management system implies a good knowledge of the different sources of water supply as well as their respective contributions. Precipitation in the High Atlas Mountains falls as rain and snow, with snow dominance at high altitudes, but the snow cover is not permanent and varies from one winter to the next.

Previous studies have shown that runoff from high Atlas precipitation (rain and snow) is the main recharge of aquifers beneath the adjacent plains Haouz, Souss and Ouazarzate (Bouragba et al. 2011). Using chemical and isotopic tracers, we have carried out a study on the respective contribution of precipitation (rain and snow) to the groundwater recharge, and their impact on the water quality of the region.

The main objective of the study is to determine, from the three sub-watersheds (Oukaimeden, lake Ifni and M'goun), the contribution of precipitation (snow and rain) in the High Atlas Mountain to the recharge aquifers beneath of adjacent plains (Haouz, Souss and Ouazarzate) and the impact of snowmelt on the water quality.

2 Study Sites and Methods

2.1 Study Sites

The study area is located in the High Atlas Mountains between longitude 7° 30′ and 9° 00′ West and latitude 30° 40′ and 31° 30′ North. The High Atlas is a long mountain range characterized by high altitudes with several summits over 3500 m, some of which exceed 4000 m, including jbel M'Goun (4071 m) and jbel Toubkal (4165 m) which is the highest peak in Morocco. The climate in the study area is semi-arid to arid, with annual rainfall ranging from 200 mm/year on adjacent plains to 600 at 900 mm/year in high mountains (altitude >700 m asl).

The investigations were carried out in three sub-watersheds in the High Atlas, namely the Lake Ifni site in the Souss upstream, the Oukaimeden site in the upstream Tensift and the M'goun site in the Draa watershed (Fig. 1). On the

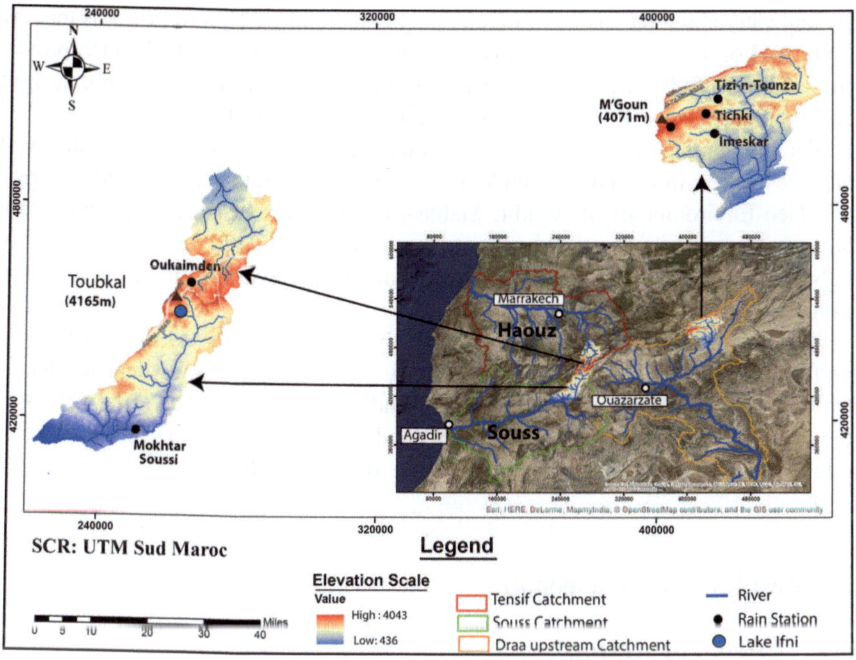

Fig. 1 Location map of the three study sites in the Moroccan High Atlas Mountains

oukaimeden site, granitic, volcanic and metamorphic formations dominate in high altitude (>2000) while the downstream of the basin is dominated by Triassic sedimentary sandstones. The Ifni sites is characterized mainly by crystalline Paleozoic rocks and the M'goun site by Jurassic carbonate and dolomitic rocks with karstic aquifers.

2.2 *Sampling and Analyses Methods*

Several sampling campaigns were carried out on the three sites concerning precipitation, surface water and groundwater. Two types of analysis are carried out each time, chemical and isotopic analyzes.

At the Oukaimden sites, a monthly monitoring focuses snowmelt collected with the passive capillary sampling (PCS) design which preserves the isotopic composition of infiltrating melt water, reducing the effect of isotopic fractionation by snow sublimation (N'da et al. 2016), surface and groundwater sampling each year between November and May from 2011 to 2014. On the Lake Ifni site, sampling was carried out on 18 sampling points concerning snow, rain, surface water and groundwater along the main river under the high altitudes (2248 m) towards the low (700 m). For the M'Goun site, the data used were acquired in the IMPETUS project from 2000 to 2004. Campaigns concerned precipitation (rain and snow), surface and groundwater (wells and sources) at different altitudes (Fig. 3).

For the samples taken at the site of oukaimeden and Lake Ifni, the chemical analyzes were carried out in LAGAGE laboratory (Laboratory of Applied Geology and Geo-Environment) of Agadir. Stable isotope were performed by cavity ring down spectrometry using a laser spectrometer (Picarro L2120) at Duke University (USA) prior to use in the Applied Geology and Geo-Environment Laboratory, Ibn Zohr University, Agadir, Morocco.

The chemical and isotopic analyzes of the samples of the M'Goun site were performed at the GSF Institute of Hydrology of Neuherberg. Analytical precisions for $\delta^{18}O$ and $\delta^{2}H$ were estimated as ±0.1 and ±1.5%, respectively. The values of the isotopic results are presented in the standard notation delta per mille (%) and referenced to Vienna Standard Mean Ocean Water (VSMOW).

3 Results and Discussion

3.1 Chemical Results

The water taken from the various sites (Oukaimden, Lake Ifni, Mgoun) is generally lowly mineralized. Nevertheless, the waters taken from the M'goun site are more mineralized (with TDS ranging from 148 to 2436 mg/L), than the waters analyzed in the Ifni Lake watershed (29.4–420 mg/L) and than the Oukaimeden site (3.7–294 mg/L). The Piper diagrams show that the waters sampled display basically calcium and magnesium bicarbonate types in the Oukaimeden and Ifniwatersheds, while in the M'goun watershed present different chemical types (SO_4–Ca, Mg; Cl–Na and HCO_3–Ca) (Fig. 2).

The majority of the samples follows the carbonate dissolution line at the Oukaimeden and Ifni sites (Fig. 3a, b), whereas in the Mgoun site, the majority of the samples follows the evaporite dissolution line (Fig. 3c). The waters sampled display basically calcium and magnesium bicarbonate types in the Oukaimeden and Ifni watersheds, which mainly reflects the predominance of limestones and dolomitic rocks dissolution in the High Atlas, while in the M'goun watershed, different chemical facies, largely related to evaporite rocks dissolution that are exposed in the basin (Fig. 4).

The results obtained show a decrease in the water salinity as well as the chemical elements that control this mineralization from high altitudes of the different watersheds upstream, versus low altitudes downstream of the basins. The waters

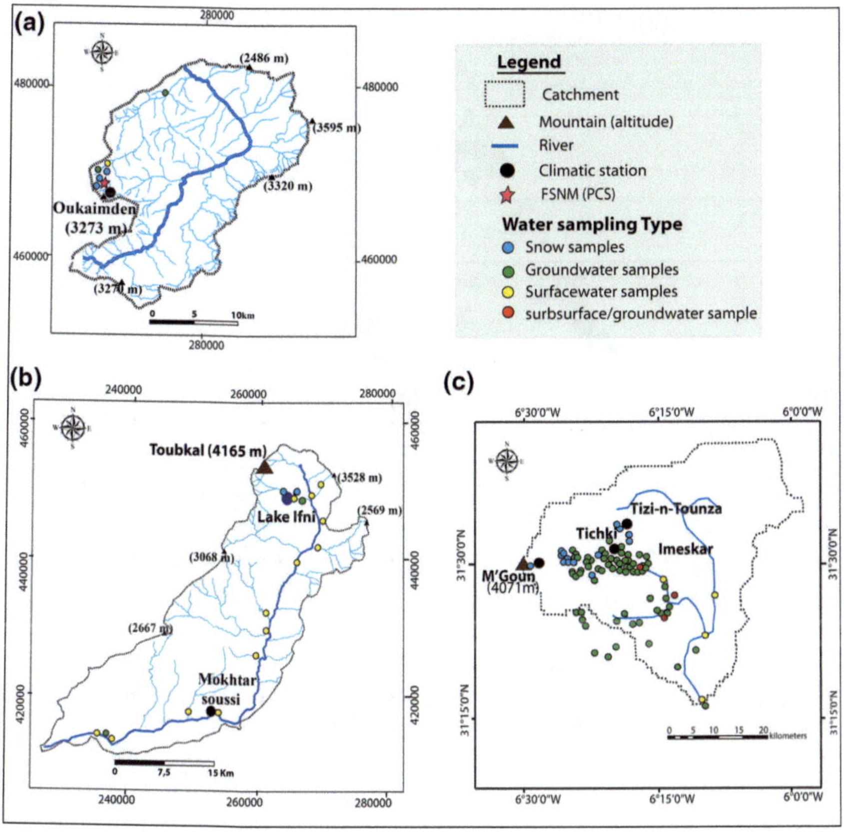

Fig. 2 Sampling points in study area: **a** Oukaimeden site with the passive capillary sampler (PCS) installed during 2013 and 2014, **b** Lake Ifni site, **c** M'goun site

upstream of the basins at the high altitudes of the High Atlas have low mineralization, and have generally a fairly good quality compared to the waters downstream of the different basins and in the adjacent plains. The waters from the High Atlas, mainly composed of snow melt water tend to dilute neighboring watersheds waters.

3.2 *Isotopic Results*

The isotopic signals of the samples in the High Atlas are characterized by a large variation in space and time. The heavy isotope contents obtained at the various sites vary between −16.86 and 1.05‰ in oxygen 18, and between −127.80 and 15.80‰

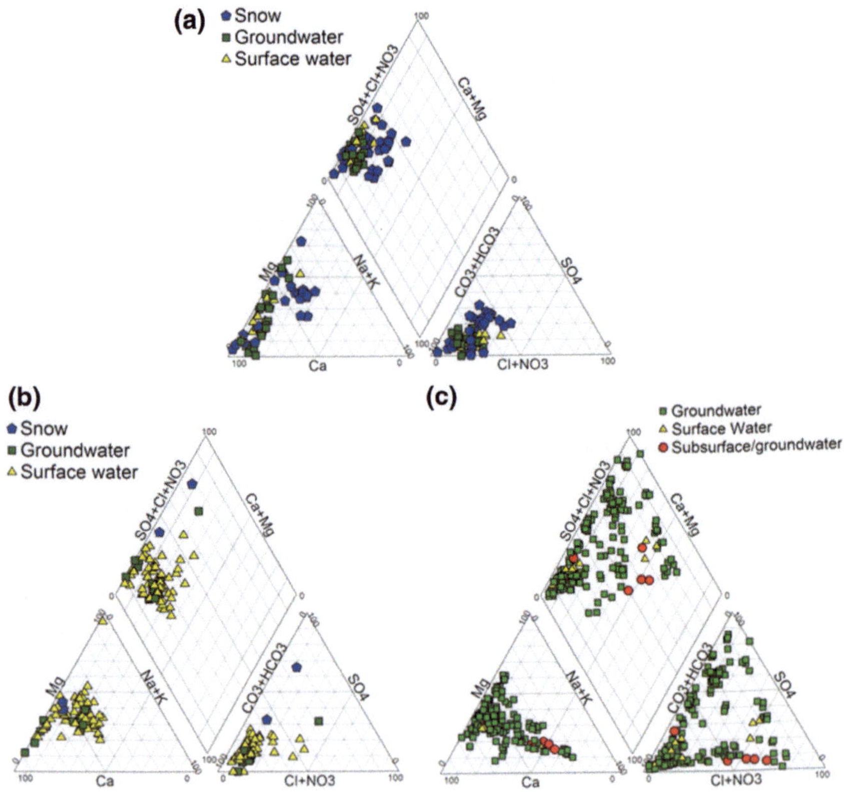

Fig. 3 Piper diagram showing water types within the three studied areas: **a** Oukaimeden site, **b** Lake ifni site, **c** M'goun site

in deuterium. Variation of heavy isotope levels (oxygen 18 and deuterium) in Morocco; is spatially due to the latitude and altitude effects, temporally due to the seasonal effect, and the mass effect. The majority of the samples are close to the Global Metoric Water Line (GMWL: $\delta^2H = 8\delta^{18}O + 10‰$, after Craig 1961) and the Local Meteoric Water Line for the High Atlas Mountains ($\delta^2H = 8\delta^{18}O + 13.5‰$, after Raibi et al. 2006), except some rain, surface and groundwater sample indicating a enrichment in oxygen-18 linked to evaporate process. The evaporation can affect the relative abundances of oxygen-18 and deuterium, particularly in semi-arid areas (Mook 1982).

Most samples fall along the meteoric lines which imply an oceanic origin of these waters. The isotopic values of majority of surface and groundwater samples are strongly influenced by snow, which means that the contribution of water from snow is greater than that of rainfall (Fig. 5).

To confirm the probable origin of surface water and groundwater in the studied areas, the results were compared with the isotopic gradient of $\delta^{18}O = -0.27\%$ per 100 m asl, defined for the High Atlas (Bouchaou et al. 1995) (Fig. 6). Results

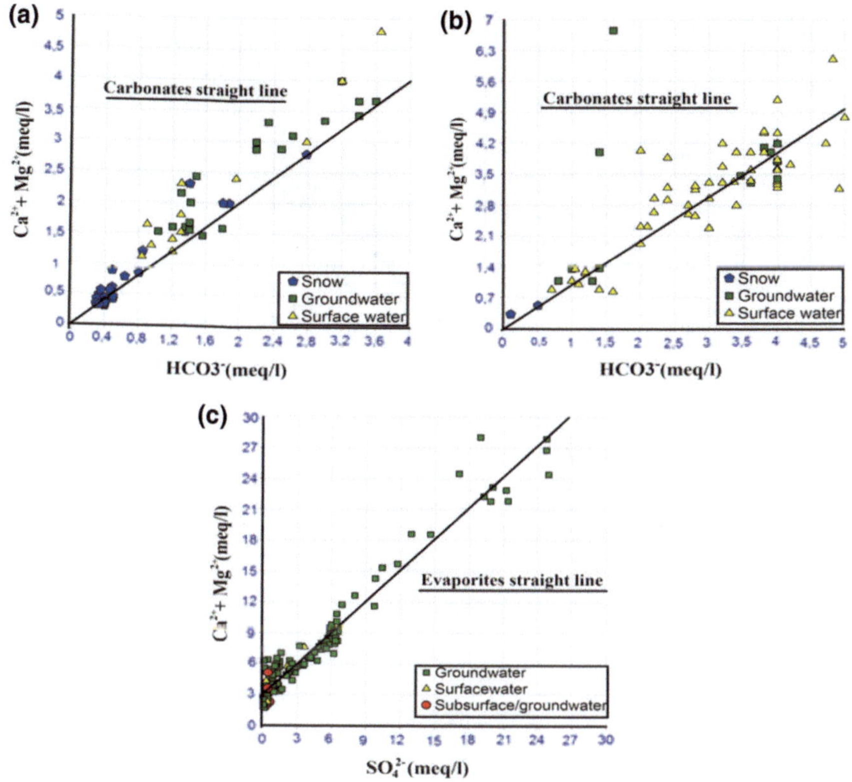

Fig. 4 Relationship between (Ca^{2+}+Mg^{2+}), (HCO^{3-}) and ($SO_4{}^{2-}$) contents: **a** Oukaimeden site, **b** Lake ifni site, **c** M'goun site

indicate that water originates from altitudes between 2000 m and 3500 m asl, corresponding to the High Atlas relief.

Based on mean isotopic values obtained in the upstream Souss basin (Bouragba et al. 2011), Haouz (Boukhari et al. 2015) and Ouarzazate aquifer (Cappy 2006), rough estimations of the snowmelt contribution to groundwater in the study sites and adjacent plain aquifers, were performed using the following isotope balance equation:

$$\delta^{18}O_{GW} = (1 - X)\delta^{18}O_{SN} + X\delta^{18}O_R$$

with GW (Groundwater); SN (Snow); R (Rain).

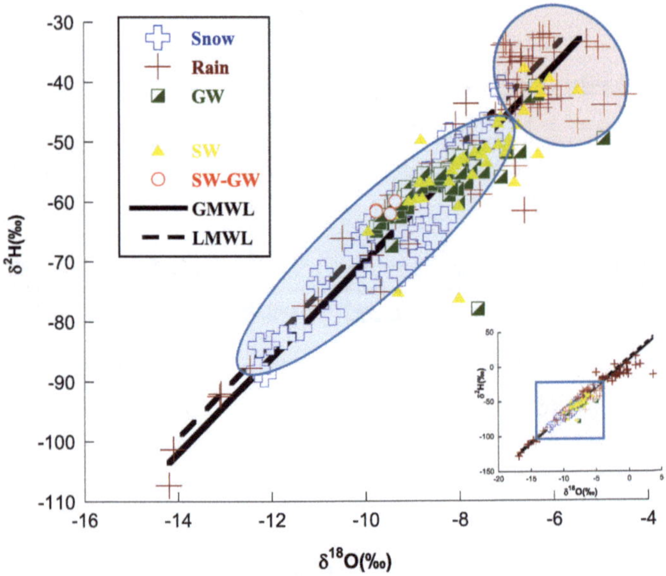

Fig. 5 Relationship between δ^{18}O and δ^2H of snow, rain, groundwater (GW), surface water (SW) and subsurface-groundwater (SW-GW) in the three study sites

Fig. 6 Estimation of the mean recharge altitude of groundwater (GW) in the three study sites and adjacent plains (Souss, Haouz and Ouarzazate) using the regional isotopic gradient line for the High Atlas (Bouchaou et al. 1995). The sample for the Souss, Haouz and Ouarzate aquifers represent average isotopic contents in the upstream of aquifers

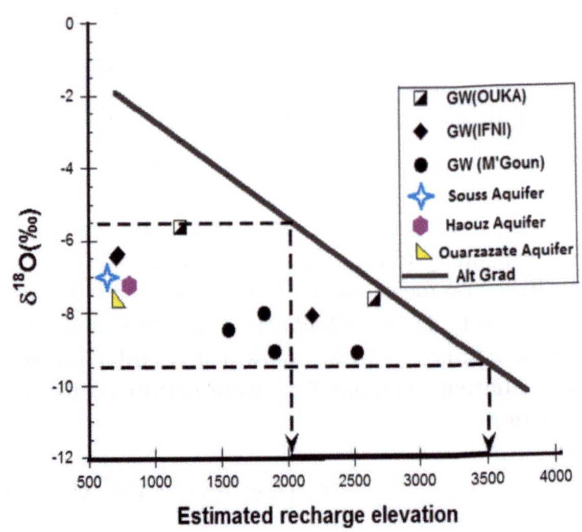

Estimate precipitation (rain and snow) respective contribution indicates that snow melt water contributes to the groundwater recharge between 57 and 71% in Haouz plain, between 42 and 52% in Souss plain and up to 75% in Ouazarzate plain, while rainfall contributes between 25 and 52% on the three adjacent plains aquifers (Table 1).

Table 1 Estimation of snowmelt contribution to groundwater in the Atlas Mountains and adjacent plain aquifers of Haouz, Souss and Ouazarzate

	Average ^{18}O snowmelt value	Average ^{18}O rain value	Average ^{18}O GW value	Estimate snowmelt contribution (%)	Estimate rain contribution (%)
Oukaimden	−7.98	−6	−7.42	71	29
Haouz plain	−7.98	−6	−7.12	57	43
Ifni	−7.98	−6.27	−7.29	52	48
Souss plain	−7.98	−6.27	−7	42	58
Ouazarzate plain	−7.98	−6.27	−7.69	75	25

4 Conclusion

The chemical characterization of the waters in the High Atlas reveals that these waters are generally less mineralized and induce a dilution of the salinity in the neighbouring plains waters (Haouz, Souss and Ouazarzate). The isotopic data confirm that the waters of the region originate from the high altitudes of the High Atlas (between 2000 and 3500 m) and indicate that the contribution of snowmelt to the recharge is greater than rain contribution. Using the stable isotopes mixing model balance the contribution of precipitation (snow and rain) was estimated, and varies between 42 and 75% for snow and between 25 and 52% for rain. Global warming has an impact on the snow cover, its duration and accentuates the snow sublimation (López-Moreno et al. 2009). Given the importance of the contribution of snowmelt to the recharge of stream flow and groundwater, the climate change could impact on the water resources in the area and on the Southern oasis ecosystems survival.

References

Bouchaou L, Michelot JL, Chauve P, Mania J, Mudry J (1995) Apports des isotopes stables à l'étude des modalités d'alimentation des aquifères du Tadla (Maroc) sous climat semi-aride. Comptes rendus de l'Académie des sciences. Série 2. Sciences de la terre et des planètes 320 (2):95–101

Boukhari K, Fakir Y, Stigter TY, Hajhouji Y, Boulet G (2015) Origin of recharge and salinity and their role on management issues of a large alluvial aquifer system in the semi-arid Haouz plain, Morocco. Environ Earth Sci 73(10):6195–6212

Bouragba L, Mudry J, Bouchaou L, Hsissou Y, Krimissa M, Tagma T, Michelot JL (2011) Isotopes and groundwater management strategies under semi-arid area: case of the Souss upstream basin (Morocco). Appl Radiat Isot 69(7):1084–1093

Cappy S (2006) Hydrogeological characterization of the Upper Draa Catchment: Morocco. Ph.D. dissertation, University Bonn, 216 pp

Craig H (1961) Isotopic variations in meteoric waters. Science 133(3465):1702–1703

Driouech F, Déqué M, Sánchez-Gómez E (2010) Weather regimes—Moroccan precipitation link in a regional climate change simulation. Global Planet Change 72(1):1–10

López-Moreno JI, Goyette S, Beniston M (2009) Impact of climate change on snowpack in the Pyrenees: horizontal spatial variability and vertical gradients. J Hydrol 374(3):384–396

Mook WG (1982) The oxygen-18 content of rivers. Mitt Geol-Paläont Inst Univ Hamburg, SCOPE/UNEP Sonderband 52:565–570

N'da B, Bouchaou L, Reichert B, Hanich L, Ait Ibrahim Y, Chehbouni G, Beraaouz EH, Michelot JL (2016) Isotopic signatures for the assessment of snow water resources in the Moroccan High Atlas Mountains: contribution to surface and groundwater recharge. Environ Earth Sci 75:755, 1–11. doi:10.1007/s12665-016-5566-9

Raibi F, Benkaddour A, Hanich L, Chehbouni A, Chtioui M (2006) Variation de la composition des isotopes stables des precipitations en climat semi-aride (cas du bassin-versant de Tensift Maroc). GIRE3D, Marrakech, Maroc

Stoffel M, Monbaron M, Maselli D (2002) Montagne et plaines: adversaires ou partenaires: Exemple du Haut Atlas, Maroc. Atlas 2002/Département de géosciences géographie/Univ. de Fribourg (Suisse)

Preliminary Study of the Impact of Guadalhorce River Mouth Channeling (Málaga, Spain) on Groundwater and Related Wetlands

J. M. Nieto-López, B. Andreo, J. A. Barberá-Fornell,
J. M. Ramírez-González and M. Rendón-Martos

1 Introduction

Prevention of floods in riverine areas is a great concern for the humankind. This concern is increasingly higher in densely populated urban settlements and their infrastructures for residential, industry or entertainment uses. Several anthropogenic actions have been conducted worldwide to minimize the flood risks, such as the modification of the original river courses or their channelling, among others.

Some examples are the construction of dams (Limonero dam—Málaga, Spain), the dredging of river mouths (Martín Sánchez River—Panamá) or their channeling (Turia River—Valencia Spain).

Historically, many floods events have occurred in the Guadalhorce River mouth (Málaga, Southern Spain), very often with catastrophic results: material damages in adjacent urban and industrial areas, but also personal ones. After the historic floods in 1989, when several people died, regional authorities decided to channel the Guadalhorce River, and split it in two branches, at its ending stretch, leaving isolated an area between these two river arms. The artificial river channel is 7 km length, 350 m wide and 2 m depth, also at the mouth (−2 m a.s.l.). These works took place between 1997 and 2003, originating changes in land use that provoked the disappearing of several hectares of reed farming.

J. M. Nieto-López (✉) · B. Andreo · J. A. Barberá-Fornell
Department of Geology and Centre of Hydrogeology, University of Málaga (CEHIUMA), 29071 Málaga, Spain
e-mail: nietolopezjm@uma.es

J. M. Ramírez-González
Andalusian Regional Agency of Environment and Water, 29001 Málaga, Spain

M. Rendón-Martos
Andalusian Regional Office of Environment and Land Management, 29520 Fuente de Piedra, Spain

M. L. Calvache et al. (eds.), *Groundwater and Global Change in the Western Mediterranean Area*, Environmental Earth Sciences,
https://doi.org/10.1007/978-3-319-69356-9_28

A wetland complex formed by eight lagoons was originated in the land surface between both river branches. Its appearing was conditioned by the digging works (extraction of sands and gravels) for the construction of buildings in Málaga city, between 1977 and 1982. This area is environmentally protected since 1989.

Servicios Omicron (1995) studied the environmental impacts of the channeling works on the Guadalhorce River, particularly on the potential damages affecting surface water dependent habitats, as well as on local groundwater dynamics. Other authors carried out hydrogeological investigations in the Guadalhorce River Valley (IGME 1983) and its lower catchment (GHUMA-EMASA 1996).

The main objective of this research is to characterize the potential environment impacts of the river channeling on the surface waters and groundwater dynamics in the Bajo Guadalhorce coastal area.

2 Hydrogeological Setting

The study area is located at the southwest to the city of Málaga (Fig. 1), in the Bajo Guadalhorce coastal area. The underlying Quaternary (unconfined) aquifer is composed of alluvial sediments (gravels, sands, silts and clays) that crop out in 115 km^2 land surface. It is the most recent deposit of the Málaga sedimentary basin, whose infilling is constituted by slightly deformed materials (Sanz de Galdeano and

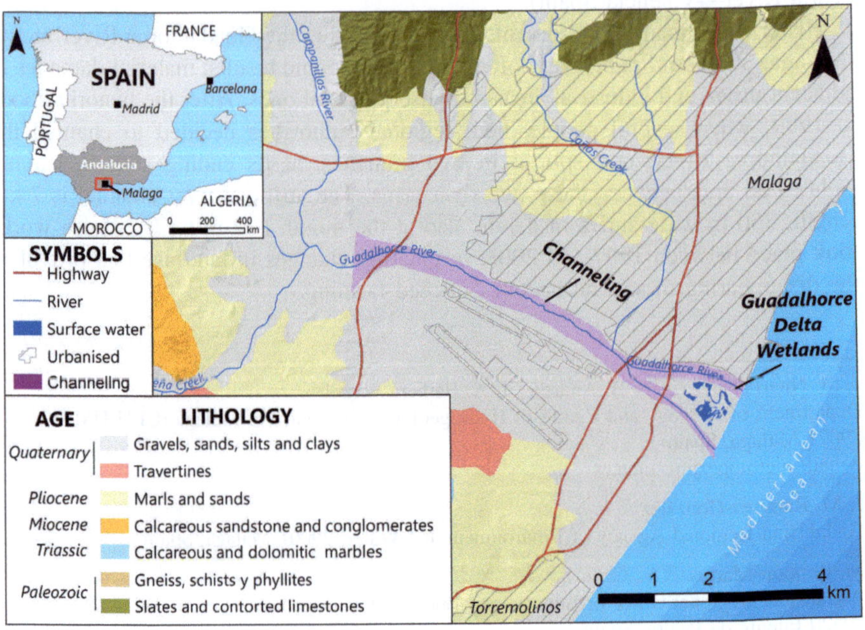

Fig. 1 Geology of the study area, channelling of Guadalhorce River and coastal wetlands

López Garrido 1991). The older lithologies are Upper Miocene calcareous sandstones and conglomerates with Pliocene conglomerates (confined aquifer), marls and sand layers overlying them. The latter group of lithologies is <20 m thick, acting as semi-confined aquifer (IGME 1983; Linares et al. 1996). The Quaternary alluvial aquifer (30–50 m thick) is found stratigraphically over the former sediments (IGME 1983; Linares et al. 1996).

The Bajo Guadalhorce River Valley is a traditional farming area near Málaga city and its surroundings settlements. Groundwater from the Quaternary and Pliocene aquifers have been traditionally exploited (Linares et al. 1988) and very often negative water table (WT) have been observed during the summer season at the coastline, which contributed to the seawater intrusion in the aquifers. However, the channeling works (1997–2003) in the river favored the change in land use, reducing the irrigated areas and, consequently, the groundwater pumping. Simultaneously, the groundwater abstraction for water supply to Málaga city ceased.

3 Methodology

Groundwater levels have been measured in the Bajo Guadalhorce aquifers to determine the potential influence of the Guadalhorce River channeling works on it. The piezometric record comprises measurements carried out by the Spanish Geological Survey—IGME—(1977–2001), by the Andalusian Regional Agency of Environment and Water (2006–present) and by the Centre of Hydrogeology of the University of Málaga—CEHIUMA—(1996, 2013–present).

Since 1995, water level data of Grande and Eucaliptal wetlands are available and they have been used to analyze if the Guadalhorce River channeling has influenced the dynamic of the wetland complex in the delta area. Regular water level readings and hydrochemical sampling have been carried out by the Andalusian Regional Office of Environment and Land Management (1997–1998), by the Andalusian Regional Agency of Environment and Water (2002–2013) and by the Centre of Hydrogeology of the University of Málaga (1995, 2013–present).

4 Results and Discussion

Distinctive groundwater dynamics in the Quaternary and Pliocene aquifers have been observed before and after the Guadalhorce River channeling. The variations of the water table in the Quaternary aquifer (Fig. 2) are lower than 2 m. Meanwhile, greater changes (>6 m) were recorded in the Pliocene aquifer, as consequence of its confinement, but also to the groundwater pumping. During and after the channeling works, narrow head variations occurred in both Pliocene and Quaternary aquifer. This could be explained by the hydraulic effect of the river channelling and/or the

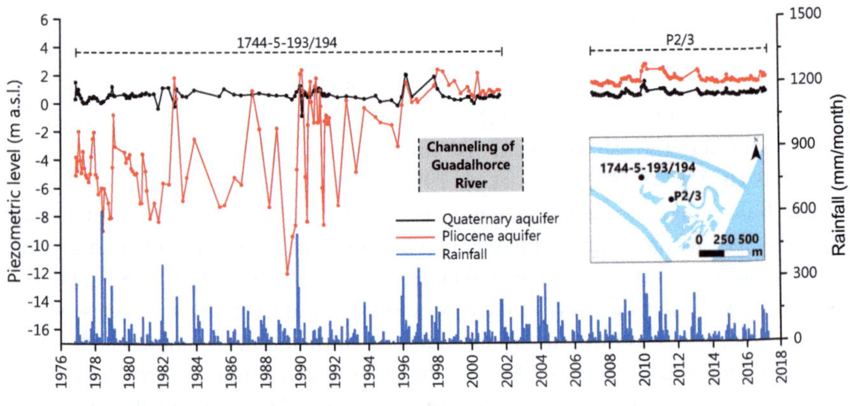

Fig. 2 Water table of various representative points of the coastal sector of the Bajo Guadalhorce aquifers. Points 1744-5-193/194 and P2/3 represent pairs of piezometers that catch groundwater from the Quaternary (1744-5-194 and P3) and the Pliocene (1744-5-193 and P2) aquifers

drastic reduction of pumping because of the decrease of the farming activities, but also by the stop of the groundwater exploitation for water supply of the Málaga city.

Figure 3 shows the water table in the Quaternary aquifer during 1996 (GHUMA-EMASA 1996) and 2016, before and after the river channeling. A preferential groundwater flow path can be deduced in 1996 from the aquifer toward the Guadalhorce River at its ending stretch. Groundwater moves through the wetlands, so they would act as transitional hydraulic features where groundwater passes toward the Guadalhorce River and the Mediterranean Sea.

Actually, the water table distribution is relatively similar and the Guadalhorce River still behave as gaining in relation with the Quaternary aquifer. However, the 1 m a.s.l. potentiometric line is 2 km toward the northwest respect to 1996. This is due to the constructive design of the river channelling, in which the new branch (river bed on it is −2 m a.s.l.) favoured the creation of a small estuary in the river, modifying the natural discharge area of the aquifer. In addition, the water table in the estuary where the wetlands are located is the shallowest, with dome morphology.

Because of the land use change, wetlands became a singular (maybe the more relevant) environmental feature in the area. Figure 4 shows the time series of the available data of electrical conductivity (EC) of Grande and Eucaliptal wetland waters.

Both wetlands have water with EC values lower than 10 mS/cm before river channelling, when only a few measurements are available, whereas to the end of the works (2003) values are above 20 mS/cm. After a few years, these values have progressively increasing to reach maximum one of 102 mS/cm, in summer, and minimum ones of 40–50 mS/cm, during wet periods. Recorded data show annual cycles, as well as a year to year positive trend, which can be explained by the (new created) estuary dynamics at the northern river branch, inducing groundwater fluxes directly to it, instead flowing to wetlands from northeast.

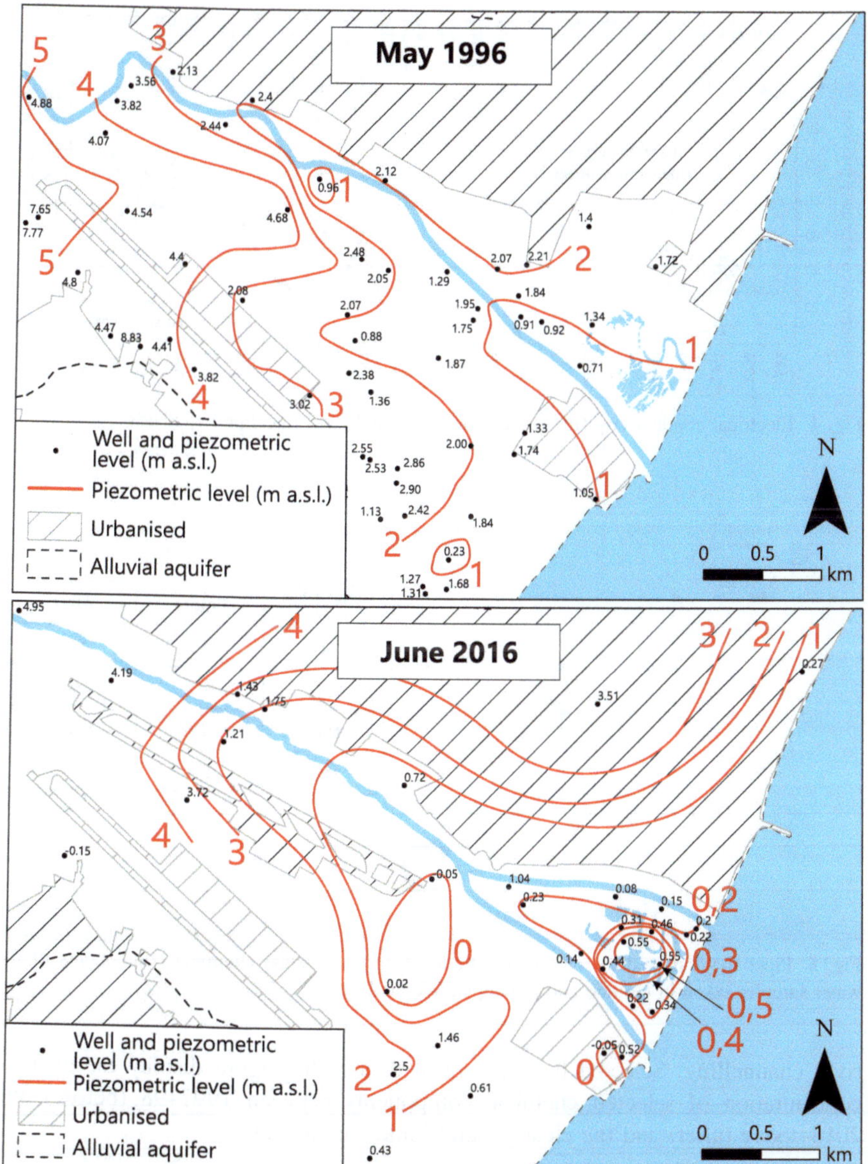

Fig. 3 Piezometric surfaces of the coastal sector of Bajo Guadalhorce Quaternary aquifer in May 1996 (up) and June 2016 (down)

Figure 5 shows the Piper diagram of the water samples taken in Grande and Eucaliptal wetlands between 1997 (Consejería de Medio Ambiente 1998) and today. Wetland water are characterized by sodium-chloride chemical facies. However, an increase in Na^+ and Cl^- contents is observed in wetland waters after

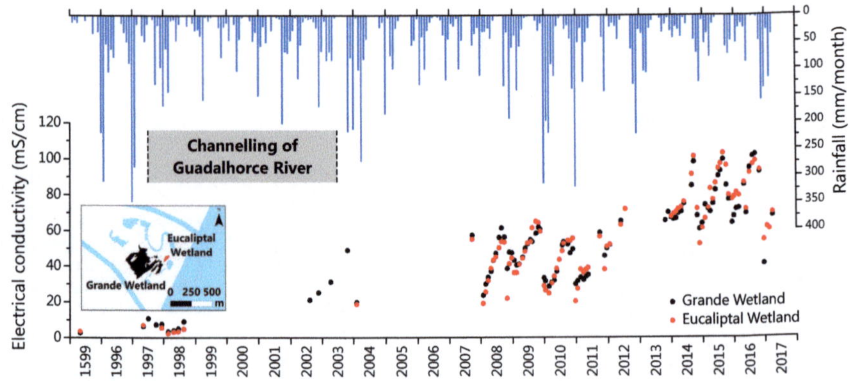

Fig. 4 Electrical conductivity (EC) time series of Grande and Eucaliptal wetlands

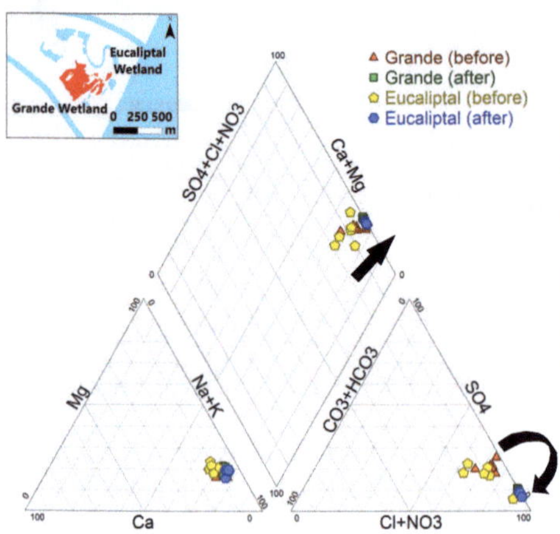

Concentration (mg/l)	Before	After
Cl^-	1830	45136
So_4^{2-}	615	6484
Na^+	1139	23621
Mg^+	147	3450

Fig. 5 Piper diagram and average concentration (mg/L) of some chemical components of the water samples taken in Grande and Eucaliptal Wetlands

river channelling. Selected hydrochemical data in the figure display the average concentration of selected chemical components between 1997–98 (before) and 2013–today (after) and the clear general ionic enrichment.

5 Conclusions

The infrastructures that prevent or minimize the flood risk in riverine areas can be hydraulically effective but very often cause severe environmental impacts on their associated aquatic ecosystems. The frequent floods in the Guadalhorce River mouth (Málaga, Southern Spain) made necessary to implement protection measurements, so a channelling and splitting of the river were constructed. In this work the evolution of the piezometric levels of two overlapped aquifers is analyzed, as well as the electrical conductivity and chemical constituents of the water of two wetlands (located between the two arms of the river).

The results show the hydraulic influence of the river channelling on groundwater and wetland dynamics. This is due to two main factors: the substantial change of the land use (from irrigated to non-irrigated land surfaces), which means groundwater pumping reduction, and the creation of a new river branch, which lets groundwater flow toward the new estuary created in the river and the Mediterranean Sea. Before the channelling works, water exploitation in this area was significant and the groundwater of the Quaternary aquifer flowed into the river in its ending stretch, acting the wetlands as transitional hydraulic features between the aquifer and the river and the sea. After the channelling, most of the pumping of the area ceased and the new branch of the river provoked that only the meteoric waters feed to the wetlands, being actually in a piezometric dome that acts as a recharge component to the aquifer rather than a transmissive feature in terms of groundwater flow. Therefore, the most probable causes of the salinization of the wetland waters is the interaction with the Mediterranean Sea and the salt concentration because of the evaporation on surface.

The artificial channelling of rivers has a great influence on the environment, both biodiversity and surface/groundwater dependent ecosystems. It is necessary, therefore, to conduct specific hydrogeological studies to accurately assess the potential environmental impacts of civil engineering works.

Acknowledgements This work has been possible thanks to the collaboration of the Andalusian Regional Agency of Environment and Water and the Andalusian Regional Office of Environment and Land Management. It is a contribution of the Research Group RNM-308 of the Andalusian Government to the project "Hydrological and environmental restoration of wetlands in the delta of Guadalhorce River (Málaga, Spain) reusing treated wastewater", founded by the Coca-Cola Foundation (Atlanta, US).

References

Consejería de Medio Ambiente (1998) Caracterización hidroquímica y cartográfica de las zonas húmedas de Cádiz, Málaga y Almería. Volúmenes I, II y III

GHUMA-EMASA (1996) Estudio hidrogeológico del Bajo Guadalhorce para la captación de agua salobre para la planta desaladora de Málaga. Technical report

IGME (1983) Investigación hidrogeológica de las cuencas del sur de España (sector occidental). Informe técnico nº 5, Sistema Acuífero nº 37 (Detrítico de Málaga). Instituto Geológico y Minero de España. Technical report

Linares L, López-Geta JA, Parra y Alfaro JL (1988) Acuífero detrítico del Bajo Guadalhorce, TIAC'88, Tecnología de la intrusión en acuíferos costeros, Almuñécar (Granada), pp 287–315

Linares L, López Arechavala G, López-Geta JA, Rubio JC (1996) Definición geométrica de los acuíferos plio-cuaternarios del valle Bajo del Guadalhorce (Málaga). VI Simposio de Hidrogeología, Sevilla 19:435–444

Sanz de Galdeano C, López Garrido AC (1991) Tectonic evolution of the Málaga Basin (Betic Cordillera). Regional implications. Geodinamica Acta 5(3):173–186
Servicios Omicron SA (1995) Estudio de impacto ambiental del proyecto de adecuación del tramo bajo del río Guadalhorce. Unpublished report

Groundwater Responses to Climate Change in a Coastal Semi-arid Area from Morocco; Case of Essaouira Basin

S. Ouhamdouch, M. Bahir, P. M. Carreira and K. Zouari

1 Introduction

The climate change is already a reality in North Africa (NA) and it places additional constraints on its fragile ecosystems and natural resources that are limited (Mannava et al. 2013). In terms of current climate, NA has experienced climate extremes events that have had negative effects on water quality and have many forms of water pollution (Bates et al. 2008) (e.g. drought in Morocco during 1982–83 and 1994–95 and flood of 2009 in Tunisia).

Different studies on observed climate showed trend drier and warmer conditions in different regions of North Africa. In Morocco, the annual mean temperature and maximum temperature were 0.7 and 0.8 °C above normal, respectively (Blunden and Derek 2015). For precipitation, the coefficient of variation in Morocco range between 30 and 70% (Driouech et al. 2009). During a dry year, the deficit of rainfall exceed at national level 40% of climatological value (1981, 1994, 1998, 2001). National rainfall amount registered between 1961 and 2008 in Morocco show a negative trend (−5% decade^{-1}) (Driouech 2010). Like Morocco, the meteorological stations of Tunisia show negative trend in mean precipitation (from 18.3 mm decade^{-1} (northeastern) to 25.6 mm decade^{-1} (southeastern) for 1969–2009) (Driouech et al. 2013). This article aims to study the current state of water resources

S. Ouhamdouch (✉) · M. Bahir
Laboratoire de Géosciences et Environnement, Département de Géologie, Ecole Normale Supérieure, Université Cadi Ayyad, Marrakech, Maroc
e-mail: salah.ouhamdouch@edu.uca.ma

P. M. Carreira
Campus Tecnológico e Nuclear, Instituto Superior Técnico, Universidade Técnica de Lisboa, Lisboa, Portugal

K. Zouari
Laboratoire de Radio-Analyse et Environnement, Ecole Nationale d'Ingénieurs de Sfax, Université de Sfax, BP W3038, Sfax, Tunisie

M. L. Calvache et al. (eds.), *Groundwater and Global Change in the Western Mediterranean Area*, Environmental Earth Sciences,
https://doi.org/10.1007/978-3-319-69356-9_29

in the Essaouira basin, where climate change leads to a drastic decline of the hydraulic head and a deterioration in quality, by examining whether trends can be identified from observed data on precipitation and temperatures in the region and how they can be related to climate change.

2 Site Description

The Essaouira basin is located in southwestern Morocco on the Atlantic coast, between the Hadid anticlinal in the north and the Igouzoullene Wadi in the south (Fig. 1). The area is characterized by a semi-arid climate with highly variable rainfalls. The annual average does not exceed 300 mm and temperatures that can reach 45 °C. Hydrogeologically, the Essaouira basin is a set of hydrogeological systems more or less independent, which correspond to the various aquifer systems; the Plio-quaternary and Turonian aquifer in downstream part and the Cenomanian-Turonian aquifer in upstream part (Jalal 2001). Overexploitation, climate aridity and various sources of pollution in the area threaten both the groundwater quantity and quality. In order to improve the management of these precious resources, several studies have been carried out during the last years to reach a better understanding of the hydrological functioning of the aquifer system (Jalal 2001; Bahir et al. 2002, 2012) using different tools and approaches.

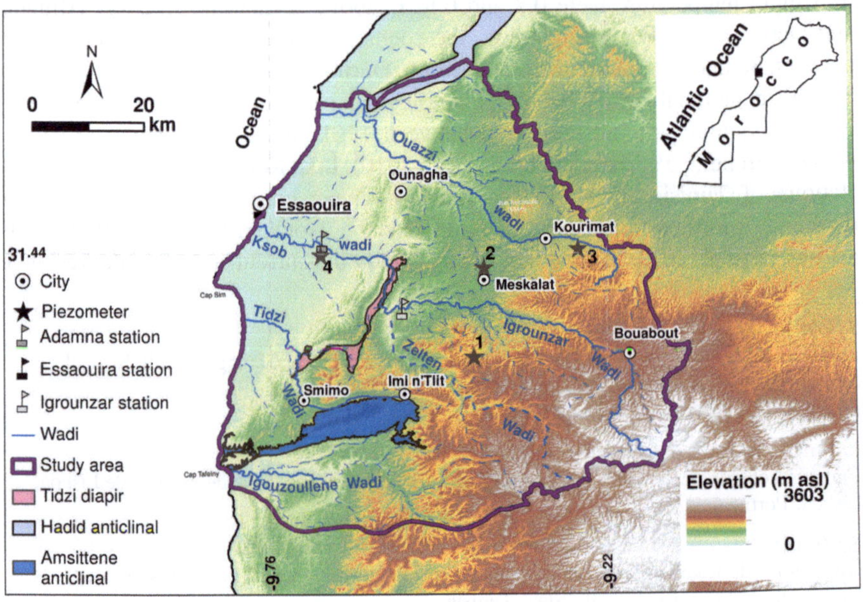

Fig. 1 Location of the study area

3 Methodology

This study is based on available meteorological time series data, temperature and precipitations, piezometry, hydrochemistry and isotopy which as could be used to predict climate change in Essaouira basin. A statistical method based on the use of the statistical test of homogeneity of Pettit (detection of ruptures) and that of trend of Mann-Kendall was applied using the software XLstat®. The Pettitt test (1979) is more particularly sensitive to a change of mean. It is based on the calculation of the variable $U_{t,T}$ defined by Eq. (1):

$$U_{t,T} = \sum_{i=1}^{T} \sum_{j=i+1}^{T} Dij \tag{1}$$

With $Dij = -1$ if $(xi - xj) > 0$, $Dij = 0$ if $(xi - xj) = 0$, $Dij = 1$ if $(xi - xj) > 0$.

As to the Mann-Kendall test (Mann 1945; Kendall 1975) is a non-parametric statistical test used to detect the presence of a linear trend (upward or downward) within a time series. Let the series Xi (x1, x2, … xn), this method defines the standard multivariable standard UMK as follows (2):

$$U_{MK} = \frac{S}{\sqrt{Var(s)}} \tag{2}$$

where $S = \sum_{i=1}^{n-1} \sum_{j=i+1}^{n} sgn(x_j - x_i)$, $Var(s) = \frac{n(n-1)(2n+5)}{18}$ and n is the number of observations. The direction of the trend is defined by the statistical coefficient of Mann-Kendall "U_{MK}". If U is positive, the trend is upward, but if U is negative, then the trend is downward. A GIS had been established to study the spatio-temporal variation of salinity.

4 Results and Discussion

4.1 Rainfall and Temperature Variation

In general, precipitation in Morocco decreases from the north to the south and from the west to the east. During the four decades, the precipitations show aleatory fluctuations (disorderly succession of dry and damp years), with a global negative trend. This decrease oscillates, depending on regions, between 3 and 30% (Babqiqi 2014). It is the results of natural variability and climate change effect. For the Essaouira Basin, the annual average rainfalls are around 300 mm and show significant year-to-year fluctuations (Fig. 2). The application of the Mann-Kendall test and that of Pettitt have shown a downward trend in precipitation with a deficit of

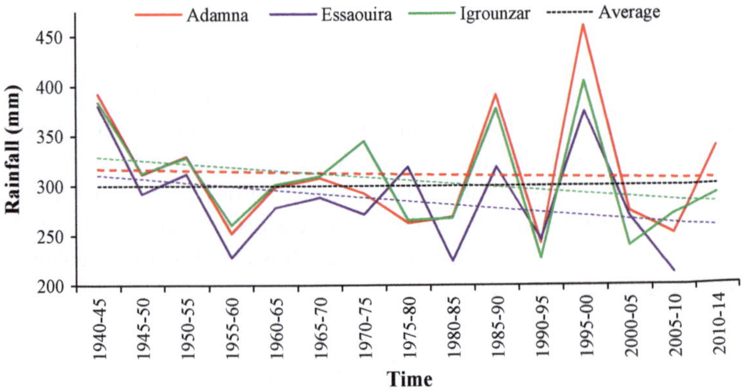

Fig. 2 Annual variation of precipitation in Adamna Essaouira and Igrounzar station from Essaouira basin for 1950–2014

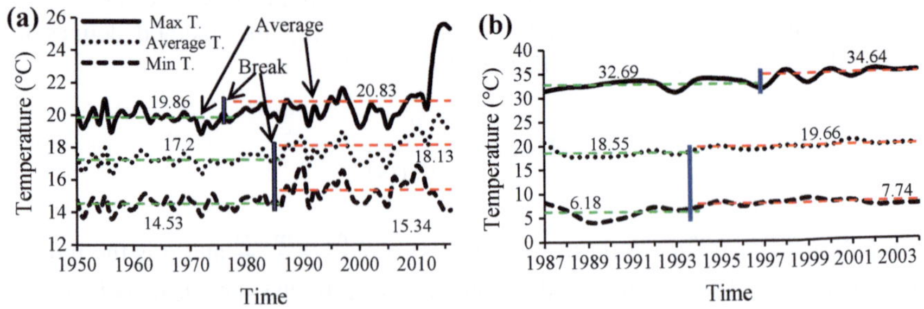

Fig. 3 Annual variation of temperature in: **a** Essaouira and **b** Igrounzar station from Essaouira basin for 1950–2016 and 1987–2004, respectively

about 12%. The calculated slope of Sen is −4.27, −3.43 and −3.72 for the Essaouira station, Adamna and Igrounzar, respectively.

At the national scale, temperatures show an upward trend ranging from 0.3 to 2.5 °C (Babqiqi 2014). Within the Essaouira basin, the temperatures oscillate around 18 °C at the Essaouira station and 19 °C at the Igrounzar station. The results of the statistical tests of Pettitt and Mann-Kendall applied on the time series of temperature at the Essaouira station and Igrounzar station, when considering a significance level of 5%, show an upward trend with a warming of the order of 0.9 °C for the Essaouira station and 1.1 °C for Igrounzar station (Fig. 3). This remains in perfect coherence with the upward trend in temperature observed at the global scale (IPCC 2013).

4.2 *Groundwater Level*

The groundwater level presents significant fluctuations between periods of high waters and those of low water. These fluctuations are mainly due to changes in rainfall. So the regime of the aquifer depends on precipitations. After a period of 10 years of monthly observation of groundwater level within 4 piezometers from Essaouira basin (Fig. 4), the groundwater level is in close relation with the precipitation, and shows a general downward trend. Given that the population within the study area only practices an agricultural activity of type "food crops", the downward trend of the water level can only be explained by the effect of climate change.

4.3 *Hydrochemistry and Stables Isotope*

The decreasing of groundwater level caused by limited recharged and anthropogenic activity has induced a degradation of water quality in Essaouira basin. In order to get an idea of the quality of the groundwater in the Basin of Essaouira, we based our investigation on the concentration of their salinity. This one ranges from 80 to 3500 mg/L (Fig. 5), and the majority of samples have content above 1350 mg/L. Generally, the water quality is average at near rivers, named locally "Wadi", (aquifer recharge areas) and poor to very poor elsewhere. More away from

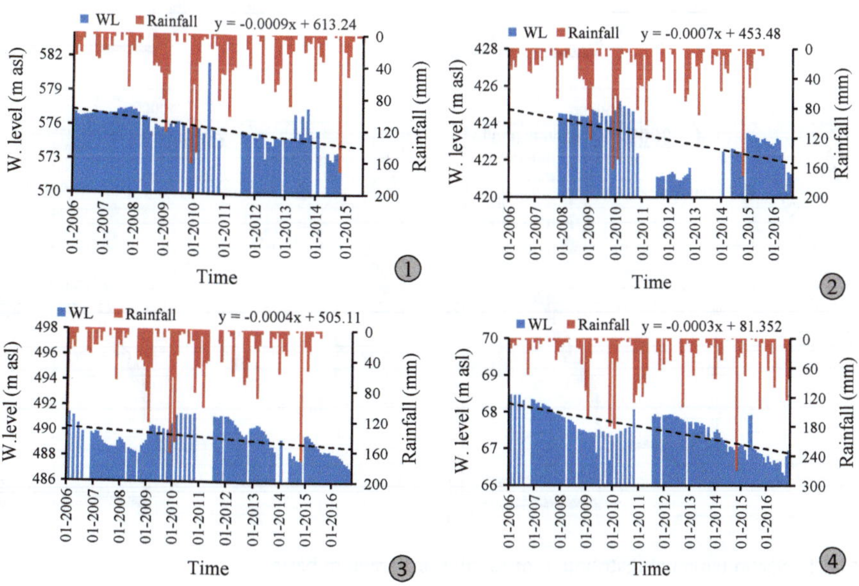

Fig. 4 Groundwater level evolution in Essaouira basin from 2006 to 2016 (see location on Fig. 1)

the Wadi, the recharge of groundwater is becoming low and therefore the dilution phenomenon is diminished. This promotes the increase in mineralization of groundwater and therefore the deterioration of their quality. However the average quality of water from Essaouira basin is poor.

The isotopic content of the water of Essaouira basin are between −6.21 and −3.77‰ and between −38.37 and −20.69‰ for ^{18}O and ^{2}H, respectively (Fig. 6). The correlation diagram ^{18}O–^{2}H; show that the majority of the sampled points are above the GMWL. This reflects a recharge by rapid infiltration without evaporation assured by oceanic precipitation. However, the aquifers recharge within Essaouira basin is closely related to the precipitation, confirming the piezometric results obtained and which makes this basin sensitive and vulnerable to any climatic variation and therefore climate change.

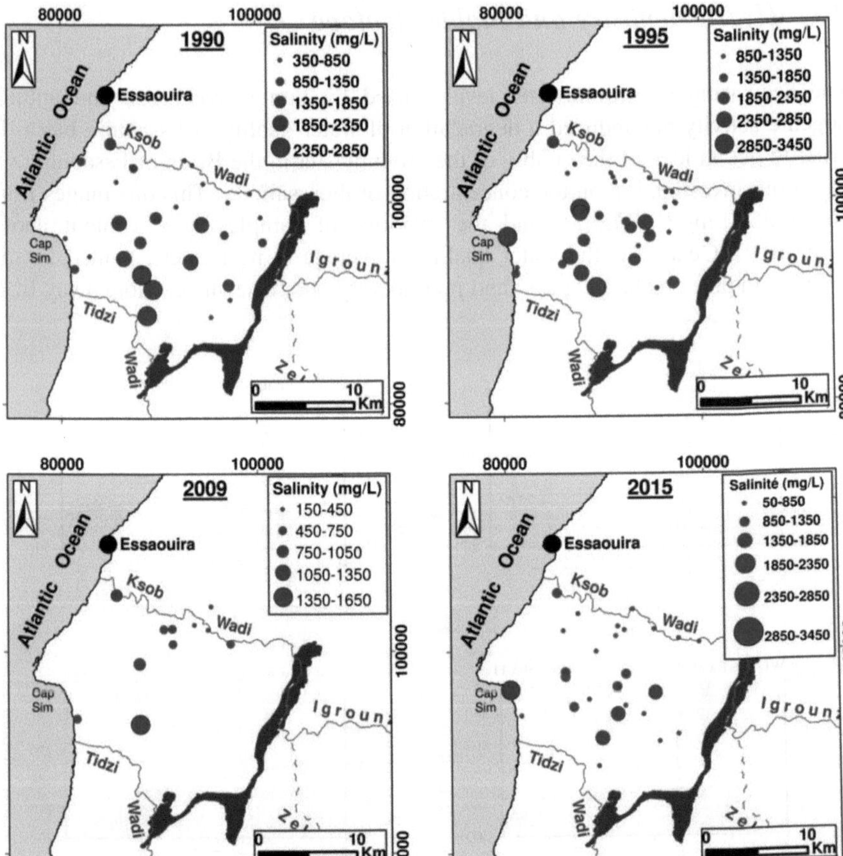

Fig. 5 Spatio-temporal distribution of salinity in Essaouira basin

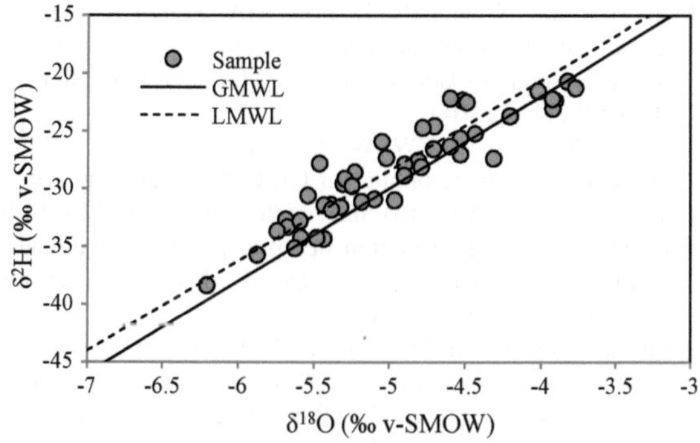

Fig. 6 Spatio-temporal distribution of salinity in Essaouira basin

5 Conclusion

In arid and semi-arid areas, rainfall remains the main source of groundwater recharge. The Essaouira Basin is no exception to this rule. According to the study of changing climatic parameters, precipitation and temperature, in the three meteorological stations of Essaouira basin (Essaouira, Adamna and Igrounzar), we observe an irregular temporal variation of rainfall with a general trend on the decline with a deficit of about 12%. On the other hand, we observe a significant trend towards the increase of temperature with a warming of the order of 0.9 °C for the Essaouira station and 1.1 °C for Igrounzar station. This trend of drying up of the regional climate, rainfall accompanied by a reduction confirms the observed warming globally. This drying up has like consequence: (i) a slight decline in the groundwater level, (ii) deterioration of water quality backing away from the wadi that represent the source of recharge of the aquifer during the rain. Isotopically, the aquifer recharge is assured by rapid infiltration without evaporation from oceanic precipitation. However, the complementary approaches hydroclimatic, potentiometric, hydrochemical and isotopic shows how climate change impacts on groundwater from the study area. These results should be taken into consideration for establishing a local strategy for the management of risks related to climate change.

References

Babqiqi A (2014) Changements Climatiques au Maroc: Etude du cas de la Région de Marrakech Tensift Al Haouz et implications sur l'agriculture à l'horizon 2030. Ph.D. thesis, Université cadi Ayyad, Marrakech, Maroc

Bahir M, Mennani A, Jalal M, Fakir Y (2002) Impact de la sécheresse sur les potentialités hydriques de la nappe alimentant en eau potable la ville d'Essaouira (Mogador, Maroc). Revue Sècheresse 13(1):13–19

Bahir M, Chkir N, Trabelsi R, Friha HA, Zouari K, Chamchati H (2012) Hydro-geochemical behaviour of two coastal aquifers under severe climatic and human constraints: comparative study between Essaouira basin in Morocco and Jeffara basin in Tunisia. Int J Hydrol Sci Technol 2(1):75–100

Bates BC, Kundzewicz ZW, Wu S, Palutikof JP (ed) (2008) Le changement climatique et l'eau, document technique publié par GIEC, Secrétariat du GIEC, Genève

Blunden J, Derek SA (2015) State of the climate in 2014. Bull Amer Meteor Soc 96(7):S1–S267

Driouech F (2010) Distribution des précipitations hivernales sur le Maroc dans le cadre d'un changement climatique: descente d'échelle et incertitudes. Ph.D. thesis, Université de Toulouse, Toulouse, France

Driouech F, Déqué M, Mokssit A (2009) Numerical simulation of the probability distribution function of precipitation over Morocco. Clim Dyn 32:1055–1063

Driouech F, Ben Rached S, El Hairech T (2013) Climate variability and change in North African Countries. In: Sivakumar MVK, Lal R, Selvaraju R, Hamdan I (eds) Climate change and food security in West Asia and North Africa. Springer, New York, London, pp 161–172

IPCC (2013) Climate change 2013, the physical science basis. In: Stocker TF, Qin D, Plattner G-K, Tignor M, Allen SK, Boschung J, Nauels A, Xia Y, Bex V, Midgley PM (eds) Contribution of working group I to the fifth assessment report of the intergovernmental panel on climate change. Cambridge University Press, Cambridge, United Kingdom and New York, NY, USA, 1535 p

Jalal M (2001) Potentialités hydrogéologiques du cénomano-turonien du basin synclinal de Meskala-Kourimat-Ida Ou Zemzem (Essaouira, Maroc). Ph.d. thesis, Université cadi Ayyad, Marrakech, Maroc

Kendall MG (1975) Multivariate nonparametric tests for trend in water quality. Water Resour Bull 24(3):505–512

Mann HB (1945) Nonparametric tests against trend. Econometrical 13:245–259

Mannava VK, Kumar S, Ruane AC, Camacho J (2013) Climate change in the West Asia and North Africa Region. In: Sivakumar MVK, Lal R, Selvaraju R, Hamdan I (eds) Climate change and food security in West Asia and North Africa. Springer, New York, London, pp 3–26

Pettitt AN (1979) A non-parametric approach to the change-point problem. Appl Stat 28(2): 126–135

Isotopic and Chemical Tracers for the Sustainable Management of Water Resources in Semi-arid Area: Case of Massa Catchment, Morocco

S. Oumarou Danni, L. Bouchaou, A. Elmouden, B. N'da, Y. Ait Brahim and M. Malki

1 Introduction

During recent decades, water quality continues to deteriorate as a result of climate change and human pressure, especially in arid zones (Fedrigoni et al. 2001). In Morocco, this phenomenon is much more accentuated owing to the overexploitation of water resources which are already limited .

Located in the south of Morocco, the Massa region (Fig. 1), which is an important agricultural area, has experienced an increasing deterioration of its water resources for more than a decade (ABHSMD 2008; Bouchaou et al. 2008; Tagma et al. 2009). Its climate is semi-arid to arid with an average precipitation of 150–200 mm/y.

The observed groundwater continuous depletion since the 1970's is similar to other aquifers and regions around the world where agriculture was intensively developed (Pulido-Bosch et al. 1991; Bosch et al.1992; Laftouhi et al. 2003; Ripoll et al. 2010; Famiglietti et al. 2011; Martos-Rosillo et al. 2013; Brahim et al. 2016a; Brahim et al. 2017). This overexploitation caused an important deficit in water resources, as it is the situation in Massa where the water deficit reached 58 mm3 (ABHSMD 2008). The situation could be aggravated due to climate change and continuous groundwater over-exploitation, if no groundwater safeguarding measures and strategies are implemented urgently. A continued groundwater depletion at this rate may will be unsustainable with potentially direct consequences for the

S. Oumarou Danni (✉) · L. Bouchaou · A. Elmouden · B. N'da · Y. Ait Brahim · M. Malki
Applied Geology and Geo-Environment Laboratory, Faculty of Sciences, Ibn Zohr University, BP 8106, Agadir, Morocco
e-mail: dannisaadou@gmail.com

Y. Ait Brahim
Institute of Global Environmental Change, Xi'an Jiaotong University, Xi'an, Shaanxi, People's Republic of China

M. L. Calvache et al. (eds.), *Groundwater and Global Change in the Western Mediterranean Area*, Environmental Earth Sciences,
https://doi.org/10.1007/978-3-319-69356-9_30

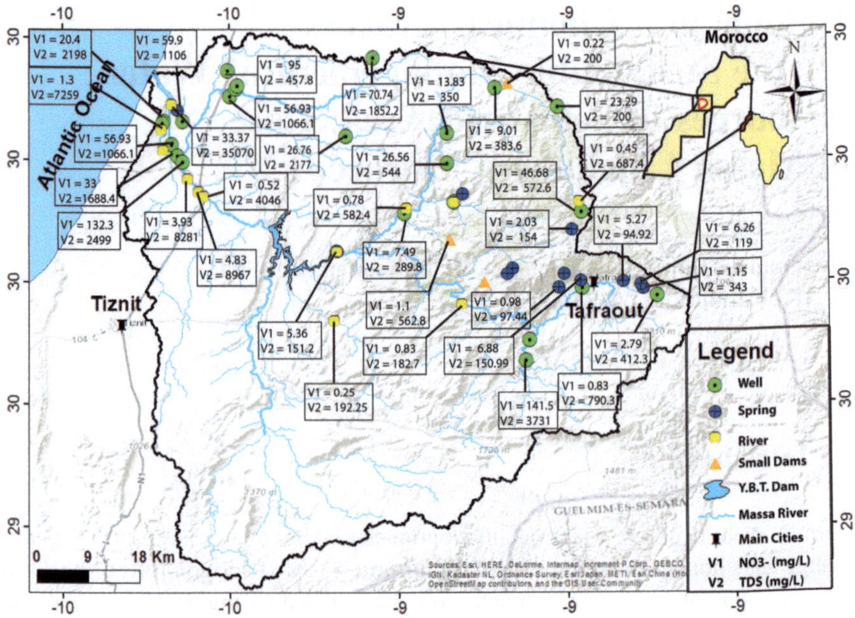

Fig. 1 Location of Massa catchment. The map shows different water samples collected within this study. V1 and V2 correspond to nitrate contents and total dissolved solids (TDS respectively)

economic and food security of the region. In these circumstances, studying the quality of water resources is of paramount importance.

In order to assess the quality of water resources in the Massa watershed, a multidisciplinary approach combining chemical and isotopic tracers is applied.

2 Geology and Hydrogeology Settings

The Massa plain is an alluvial depression located between the Anti-Atlas Mountains in the South East and the Atlantic Ocean in the North West. The age of the geological formations ranges from Paleozoic to Quaternary. The plain is composed of Plio-Quaternary sediments (sands, gravels and lacustrine limestone). The Anti-Atlas Mountains are characterized by carbonate and crystalline formations.

The soils of Massa plain are diversified; ranging from sandy-loam soils with low humus content and relatively high permeability to soils with high clay contents. The former soil types are characterized by a coarse texture with low content of clay minerals and occur mainly in the Chtouka–Massa. The later soil types are xeric, brown-to-black, and having clayey–sand texture with moderate amounts of clay minerals (Bouchaou et al. 2008)

The main aquifer of the area is the Chtouka aquifer, which is consisting essentially of Plio-Quaternary deposits, Miocene, and Eocene. They are represented by:

- Dune sandstones and limestone of 20–100 m thick containing intercalated with lake original limestone layer of 2–6 m thick and which can become more or less marl.
- A river-lake formation, made up of marl sand, marl, and limestone. Their thickness varies from 50 m to over 200 m in the southeast region of Agadir. Near Biougra, this formation tends to have coarser deposits (gravels and stones) with thickness that varies from 100 to 150 m (Hirich et al. 2016).

3 Methodology

With the aim of assessing the quality of Massa water resources, sampling campaigns during October 2016 was been carried out in 54 sites (Fig. 1). Samples were selected by conciliating different parts of the study area (the Anti-Atlas Mountains, along Massa River, the plain dominated by farms and the coastal area).

The samples were taken directly from wells, springs, dams and Massa River in double capped polyethylene bottles and sent to the Laboratory of Applied Geology and Geo-Environment at Ibn Zohr University, where chemical and stable isotope analyses were performed. Physico-chemical parameters (electrical conductivity (EC), pH, temperature, oxygen), were measured in situ. The concentrations of major ions were determined in the laboratory using volumetric titration method for Ca_2^+, Mg_2^+, Cl^- and HCO_3^-, absorption UV–Visible spectrophotometer for NO_3^- and SO_4^{2-}, and flame spectrometer for Na^+ and K^+.

Stable isotope measurements of $\delta^{18}O$ and δ^2H were performed by Cavity Ring down Spectrometry using a Picarro L2120. Analytical precisions for $\delta^{18}O$ and δ^2H were estimated as ±0.1 and ±1.5%, respectively. The values are expressed as relative differences (δ values) from the Vienna standard mean ocean water (V-SMOW) in per mill ‰.

4 Results

Water temperature and pH do not show large spatial variations. Temperatures fall within the range of 17.2–29 °C and pH ranges between 6.46 and 9.0. However, the mineralization of water is highly variable. The highest EC value of 50100 µS/cm is observed along the Massa River (near Massa village) and the lowest value of 65 µS/cm is in the Anti-Atlas Mountains.

The chemical analyses reported in the Piper diagram (Fig. 2) show three main water types: (i) Chloride sodium-potassium, (ii) bicarbonate calcium and magnesium, (iii) Chloride-sulphate and calcium-magnesium.

The results also show a large variability of NO_3^- with high contents in some sites. Its value range between 0.22 and 141.5. The highest concentrations of NO_3^-

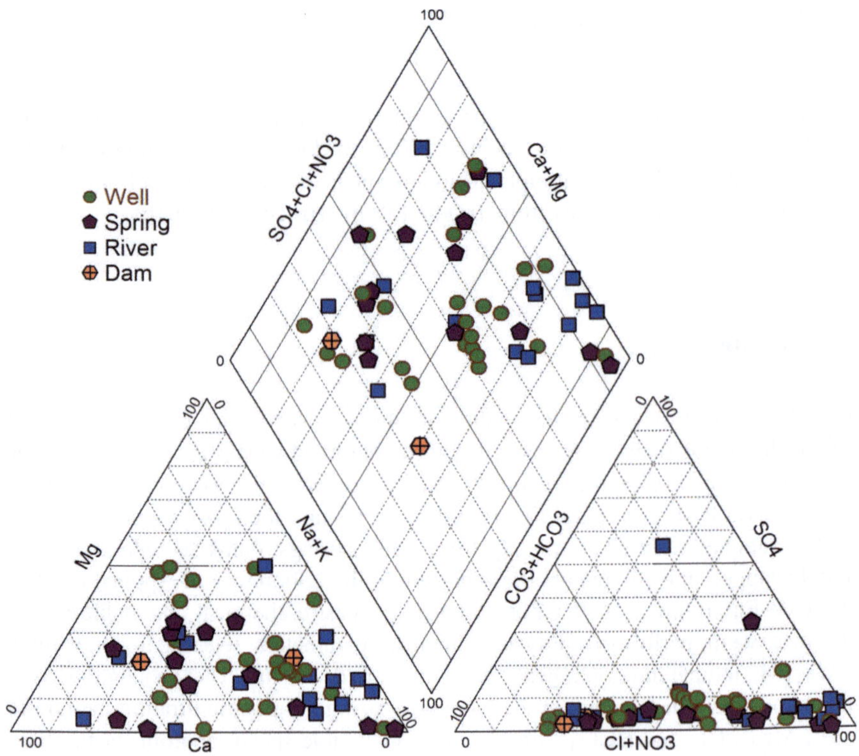

Fig. 2 Piper diagram, showing the major ion geochemistry of all water samples

are observed at the right bank of the Massa River, particularly in the agricultural plain area (Fig. 1).

Stable isotope values (alternatively: deviations) ($\delta^{18}O$, δ^2H) of the different samples collected are ranging from −7.73 to 3.75‰ for $\delta^{18}O$ and from −43.29 to 10.07‰ for δ^2H. The oxygen-18 versus deuterium relationship (Fig. 4b) shows that the samples are close to the global and local meteoric straight line, which reflects an oceanic origin of water (Fig. 2).

5 Discussion

The relation between chemical concentrations and total dissolved solids (TDS) can be a good indicator of the origin of groundwater mineralization. The plots for the water samples show a good correlation between major ions and TDS (Fig. 3). This confirms that groundwater salinity is mainly controlled by the dissolution of minerals that contain Cl^-, and SO_4.

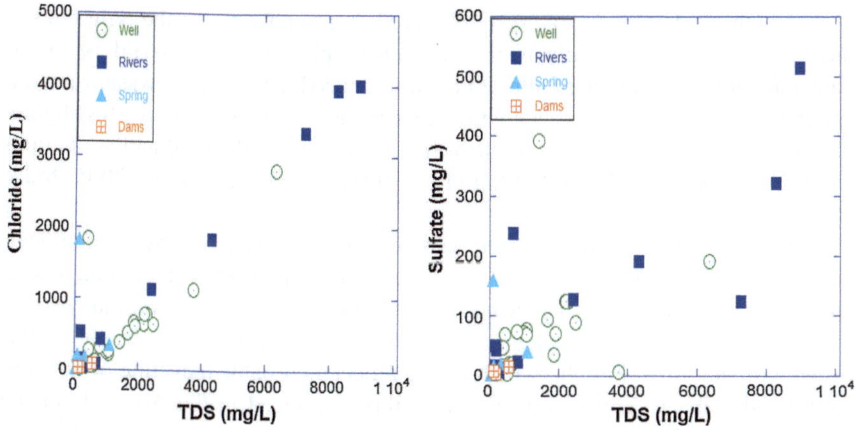

Fig. 3 Variations of chloride and sulfate concentrations (mg/L) versus total dissolved solids (mg/L) of all water samples

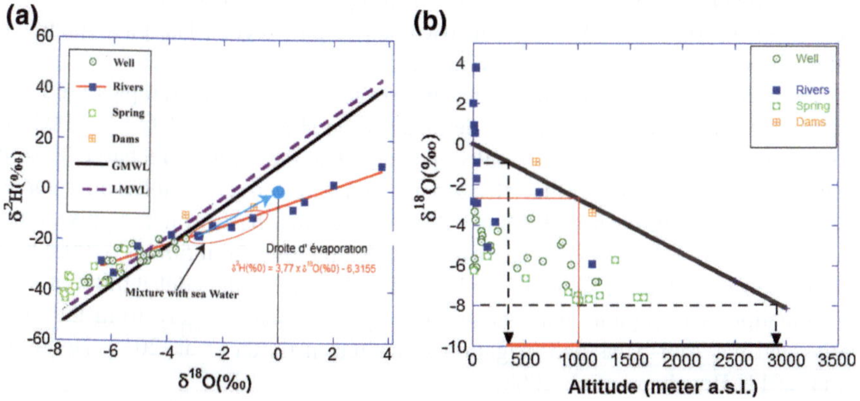

Fig. 4 **a** Relationship between $\delta^{18}O$ and $\delta^{2}H$ of all water samples in the Massa area as compared to the global meteoric water line (Craig 1961) and the local meteoric water line (Ouda et al. 2004); **b** Estimation of the recharge elevation using $d^{18}O$ versus elevation, and regional $\delta^{18}O$ Isotopic Gradient Line (Bouchaou et al. 1995)

The high values of EC observed along the Massa River (Fig. 1) could have a direct link with an intrusion of marine water (Fig. 4a). This phenomenon of marine intrusion had already been mentioned by several researchers who had carried out studies in the area (Hsissou et al. 1999; Krimissa et al. 2004; Bouchaou et al. 2008; Malki et al. 2016). However the few high values of EC observed far from the coastal zone could have a purely geochemical origin. As regards water chemistry, it directly reflects the geochemical composition of the geological formations of the area.

The high concentrations of NO_3^- observed in groundwater on the right bank of the Massa River, particularly in the agricultural plain (Fig. 1), would be due to contamination related to agricultural practices. Indeed the effects of wastewater and agriculture combination on rising NO_3^- concentration had already been highlighted in several areas where sources of nitrogen are wastewater disposal and agricultural practices (Gooddy et al. 2002; Tang et al. 2004; Wakida and Lerner 2005; Malki et al. 2016).

Surface water and groundwater sampled are characterized by a significant variation of $\delta^{18}O$. The majority of surface waters present a deuterium excess around +10, which implies an oceanic origin of these waters. However many spring (67%) and well (37%) samples have a deuterium excess higher than +10, (between +14 and +18) (Fig. 4a) which would reflect a probable recycling of moisture and altitude effects. This significant variability is noticed in the Atlas Mountains regions, where the stable isotope values of precipitation vary significantly over short distances due to altitude and microclimate effects (Brahim et al. 2016b and references therein). These processes indicate the impact of climatic variability on the stable isotopes signature.

The oxygen-18 versus deuterium relationship (Fig. 4a) indicates also a significant enrichment of river waters in $\delta^{18}O$ linked to evaporation process. This phenomenon is often observed in the arid and semi-arid regions (Simpson and Herczeg 1991).

To determine the altitude-recharge, the oxygen-18 values of samples are compared to the regional altitudinal line defined for the High Atlas: $\delta^{18}O = -0.0027‰$ (Bouchaou et al. 1995). The results indicate that the sample recharge area is between 500 and 3000 m (Fig. 4b). Apart from a few river samples all samples come from high altitudes (> 1000 m). This implies that basin water supply comes mainly from Anti Atlas Mountains with altitudes ranging from 500 to 3300. Other studies carried out in the same region have also shown that the contribution of local precipitation is negligible and that the recharge originates largely from the High Atlas and the Anti-Atlas and irrigation water return (N'da et al. 2016; Bouragba et al. 2011; Bouchaou et al. 2008).

The large signature variation reflects the difference between different system and the infiltration process.

6 Conclusion

Water resources in Southern Morocco, mainly Massa area, are experiencing a continuous qualitative and quantitative degradation. Water levels are showing a chronic deficit since 1970, mostly marked in agricultural areas, where water resources are subjected to strong human pressure and climate change. According to stable isotopes signature, surface water are impacted by evaporation due to aridity of climate in the area, while the groundwater show non-significant evaporation and the recharge area is located mainly in high evaluations corresponding the Anti atlas Mountains. The studied area indicate a wide variability of water quality; the high salinity is observed mainly in downstream of the YBT dam which can be explained by water-rock interaction, marine intrusion and anthropogenic impact. The impact of agricultural fertilisers is mainly highlighted in irrigated areas. Some local pollution from wastewater release was observed.

Acknowledgements This study is carried out within CRP project (F32007) with support of International Atomic Energy Agency (IAEA) and WASA project.

References

ABHSMD (2008) Plan directeur d'aménagement intégré des ressources en eau (PDAIRE) (Chapitre 3), p 120

Bosch AP, Martos FS, Vidal JLM, Navarrete F (1992) Groundwater problems in a semiarid area (Low Andarax River, Almeria, Spain). Environ Geol Water Sci 20(3):195–204

Bouchaou L, Chauve P, Mania J, Mudry J, Michelot J (1995) Apports des traçages artificiel et naturel à la connaissance des circulations et des zones de recharge en région semi-aride: cas de l'Atlas de Beni Mellal (Maroc). Revue des sciences de l'eau/Journal of Water Science 8(1): 43–55

Bouchaou L, Michelot JL, Vengosh A, Hsissou Y, Gaye CB, Bullen TD, Zuppi GM, Qurtobi M (2008) Application of multiple isotopic and geochemical tracers for investigation of recharge, salinization, and residence time of water in the Souss-Massa aquifer, southwest of Morocco. J Hydrol 352:267–287

Bouragba L, Mudry J, Bouchaou L, Hsissou Y, Krimissa M, Tagma T, Michelot JL (2011) Isotopes and groundwater management strategies under semi-arid area: case of the Souss upstream basin (Morocco). Appl Radiat Isot 69(7):1084–1093

Brahim YA, Seif-Ennasr M, Malki M, N'da B, Choukrallah R, Sifeddine A, Morjani ZEAE, Bouchaou HAL (2016a) Assessment of climate and land use changes: Impacts on groundwater resources in the Souss-Massa River basin. In: R. Choukr-Allah, R. Ragab, L. Bouchaou, D. Barcelo (eds) Chapter 71 in the Souss-Massa River Basin, Morocco, The Handbook of Environmental Chemistry, Vol 53. Springer Berlin, Heidelberg, pp 121–142

Brahim YA, Bouchaou L, Sifeddine A, Khodri M, Reichert B, Cruz FW (2016b) Elucidating the climate and topographic controls on stable isotope composition of meteoric waters in Morocco, using station-based and spatially-interpolated data. J Hydrol 543:305–315

Brahim YA, Saidi MEM, Kouraiss K, Sifeddine A, Bouchaou L (2017) Analysis of observed climate trends and high resolution scenarios for the 21st century in Morocco. Journal Mater Environ Sci 8(4):1375–1384

Craig H (1961) Isotopic variations in meteoric waters. Science 133:1702–1703

Famiglietti J, Lo M, Ho S, Bethune J, Anderson K, Syed T, Swenson S, De Linage C, Rodell M (2011) Satellites measure recent rates of groundwater depletion in California's Central Valley. Geophys Res Lett 3:38

Fedrigoni L, Krimissa M, Zouari K, Maliki A, Zuppi GM (2001) Origine de la minéralisation et comportement hydrogéochimique d'une nappe phréatique soumise à des contraintes naturelles et anthropiques sévères: exemple de la nappe de Djebeniana (Tunisie). C R Acad Sci Paris, Sci Terre Planètes Earth Planet Sci 332:665–671

Gooddy DC, Clay JW, Bottrell SH (2002) Redox-driven changes in porewater chemistry in the unsaturated zone of the chalk aquifer beneath unlined cattle slurry lagoons. Appl Geochem 17.7: 903–921

Hsissou Y, Mudry J, Mania J, Bouchaou L, Chauve P (1999) Utilisation du rapport Br/Cl pour déterminer l'origine de la salinité des eaux souterraines: exemple de la plaine du Souss (Maroc). Earth Planet Sci 328:381–386

Hirich A, Choukr-Allah R, Nrhira A (2016) Groundwater-Dependent Ecosystems in the Souss-Massa River Region: An Economic Valuation of Ecosystem Services. In: Choukr-Allah R, Ragab R, Bouchaou L, Barceló D (eds) The Souss-Massa River Basin, Morocco. The Handbook of Environmental Chemistry, vol 53, Springer, Cham

Krimissa S, Michelot JL, Bouchaou L, Mudry J, Hsissou Y (2004) Sur l'origine par altérationdu substratum schisteux de laminéralisation des eaux d'une nappe côtière sous climat semi-aride (ChtoukaMassa, Maroc). Compt Rendus Geosci 336:1363–1369

Laftouhi NE, Vanclooster M, Jalal M, Witam O, Aboufirassi M, Bahir M, Persoons É (2003) Groundwater NO_3^- pollution in the Essaouira Basin (Morocco). Compt Rendus Geosci 335 (3):307–317

Malki M, Bouchaou L, Hirich A, Ait Brahim Y, Choukr-Allah R (2016) Impact of agricultural practices on groundwater quality in intensive irrigated area of Chtouka-Massa, Morocco. Sci Total Env 574:760–770

Martos-Rosillo S, Rodríguez-Rodríguez M, Pedrera A, Cruz-SanJulián JJ, Rubio JC (2013) Groundwater recharge in semi-arid carbonate aquifers under intensive use: the Estepa range aquifers (Seville, southern Spain). Environ Earth Sci 70(6):2453–2468

N'da AB, Bouchaou L, Reichert B, Hanich L, Brahim YA, Chehbouni A, Michelot JL (2016) Isotopic signatures for the assessment of snow water resources in the Moroccan high Atlas mountains: contribution to surface and groundwater recharge. Environ Earth Sci 75(9):755

Ouda B, El Hamdaoui A, Ibn Majah M (2004) Isotopic composition of precipitation at three Moroccan stations influenced by Oceanic and Mediterranean air masses, IAEA, Vienna. TEC-DOC 1453:125–140

Pulido-Bosch A, Navarrete F, Molina L, Martinez-Vidal JL (1991) Quantity and quality of groundwater in the Campo De Dalias (Almeria, Se Spain). Water Sci Technol 24(11):87–96

Ripoll S, MacMillan T, Muñoz MJB, Velázquez E, López CM, Levidow L (2010) WP3: water scarcity and its virtual export from Spain to the UK. Co-operative Research on Environmental Problems in Europe (CREPE)

Simpson HJ, Herczeg AL (1991) Stable isotopes as an indicator of evaporation in the River Murray. Aust Water Resour Res 27(8):1925–1935

Tagma T, Hsissou Y, Bouchaou L, Bouragba L, Boutaleb S (2009) Groundwater NO_3^- pollution in Souss-Massa basin (south-west Morocco). Afr J Environ Sci Technol 3:301–309

Tang C, Chen J, Shindo S, Sakura Y, Zhang, W, Shen Y (2004) Assessment of groundwater contamination by nitrates associated with wastewater irrigation: a case study in Shijiazhuang region. China. Hydrol process 18(12):2303–2312

Wakida FT, Lerner DN (2005) Non-agricultural sources of groundwater nitrate: a review and case study. Water Res 39(1):3–16

Future Effects of Climate Change on the Dynamics of the Sierra Nevada Snowpack: Conclusions from Cellular Automata Models

E. Pardo-Igúzquiza, A. J. Collados-Lara and D. Pulido-Velazquez

1 Introduction

The management of water resources systems that include high mountain regions has to take into account the cryosphere, that is, the snow cover or snowpack. Any study of the spatiotemporal distribution of snow must cover the estimation of the snow cover area, the thickness of the snowpack, and the density of snow. The three variables are needed for estimating the snow water equivalent, although the estimation of each of them is an important problem on its own. In Alpine basins the snow may have an important influence on the yearly streamflow distribution due to its effects on both, surface water and groundwater flow balance components. In this work we study the estimation of the snow covered area (SCA). In normal conditions, the SCA can be estimated from satellite data, for example using MODIS (Hall and Riggs 2007). However the problem to be solved in this work is the estimation of the SCA when satellite data are unavailable. For example, periods previous to the launching of the satellite, periods in which the satellite was not operative, days covered by clouds or future predictions. In this sense cellular automata are pertinent because they are good simulators of the dynamics of natural phenomena, like the growth of crystals, snow avalanches or urban growth (Kumar et al. 2014) among others. The methodology of cellular automata is introduced in the next section.

E. Pardo-Igúzquiza (✉)
Instituto Geológico y Minero de España, Ríos Rosas, 23, 28003 Madrid, Spain
e-mail: e.pardo@igme.es

A. J. Collados-Lara · D. Pulido-Velazquez
Unidad del IGME en Granada, Urbanización Alcázar del Genil 4. Edificio Zulema Bajo, 18006 Granada, Spain
e-mail: ajcollados@gmail.com

D. Pulido-Velazquez
e-mail: d.pulido@igme.es

M. L. Calvache et al. (eds.), *Groundwater and Global Change in the Western Mediterranean Area*, Environmental Earth Sciences,
https://doi.org/10.1007/978-3-319-69356-9_31

2 Methodology

Cellular automata are discrete dynamical models that are able to simulate complex dynamics by using simple interaction rules (Wolfram 1984). In order to implement a cellular automata model, the study area is discretized in a grid of cells, and the cell size can be defined by the problem at hand or can be freely chosen. For example, in this work the cells are squares of around 460 m × 460 m because that is the spatial resolution of the MODIS images for the area of interest. Each cell (*i*, *j*) has an assigned state *S*(*i*, *j*) which can be one of two values: 1 if the cell is covered by snow, 0 if the cell is free of snow. Thus, only the binary state (0/1) is considered, while the covered snow fraction, that is, the percentage of the pixel that is covered by snow, has not been considered. The state of each cell for every time step *t*, depends on the state of the cell in the previous step, the states of the neighbor cells in the previous step and a given number of transition rules. In classic cellular automata, these transition rules and a function of the states of the cell and the neighbor cells in a previous time step. However, for simulating the snow cover in a realistic way it is necessary to introduce some driving variables that affect the transition rules. In the model proposed by Pardo-Igúzquiza et al. (2017), two climatic variables (precipitation and temperature) and a terrain variable (altitude) are used as the driving variables of the snowpack. A detailed description of the rules can be found in the previous reference, but basically they consists in deciding if precipitation and temperature are above a given threshold of precipitation and temperature respectively; also they are a function of the state of neighboring cells. In this way, there are five parameters that can be used to calibrate the model from experimental data. These parameters are: precipitation threshold, temperature threshold, number of neighbor cells with state 1 in the previous time step and the slope and intersection with the origin that define the snowline (Pardo-Igúzquiza et al. 2017). The hypothesis and variables that have been used have been sketched in Fig. 1.

3 Case Study

The study area is the Sierra Nevada Mountains, a Mediterranean mountain range in the South of Spain (Fig. 2). The area of interest has been restricted to a rectangular area represented in Fig. 2 and it includes the snowpack of Sierra Nevada. Occasionally there may be snow falls that covers the surface outside the area of interest but they have a duration of a few days and are not considered as part of the snowpack. In this way the study area has a surface of around 2000 km^2. Sierra Nevada is very important because of the water resources it stores, mainly because the accumulation of snow during winter. Additionally, it is a National Park that host a huge biodiversity, and where there is an important skiing resort. The aim of this work is to use the cellular automata models that were calibrated in Pardo-Igúzquiza

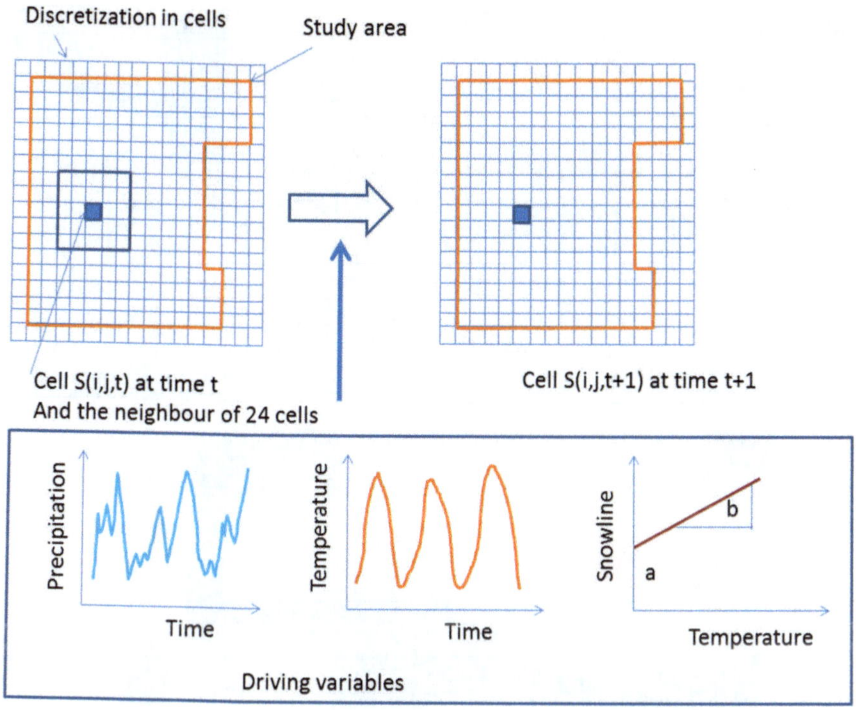

Fig. 1 Basis of the cellular automata models used in this work

et al. (2017) to study the response of the system if we introduce plausible scenarios of precipitation and temperature for this area, that were obtained from climate projection studies (Collados-Lara et al. 2017), as driving variables. To generate the future series of precipitation and temperature this study uses the results from the simulation of the CORDEX project under the emissions scenery RCP8.5 and the historical series from the Spain02 project (Herrera et al. 2016). The results are analyzed and discussed in the next section.

4 Results and Discussion

Using historical time series of precipitation and temperature, as well as control time series provided by the regional climate models and two approaches of correction (bias correction and delta change), using the projection of climate change and different perturbation techniques for the previous methodologies (first moment, second moment, quantile mapping), Collados-Lara et al. (2017) arrive to the conclusion that the models predict a reduction in precipitation and an increase in temperature for the Alto Genil basin. In particular, for the study area during the

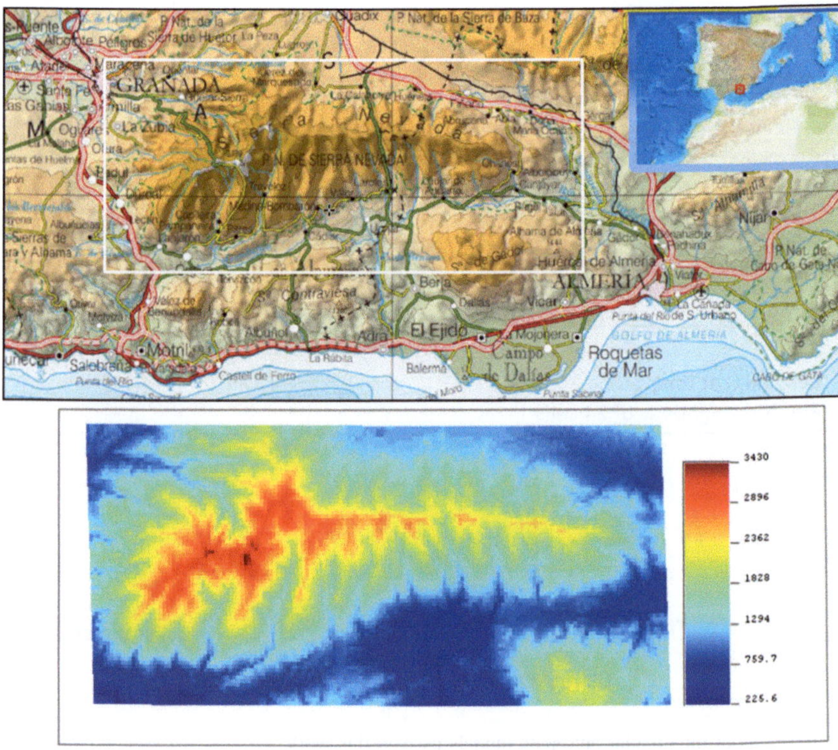

Fig. 2 Location of the study area. In the upper figure there is the geographical location of Sierra Nevada with the city of Granada at the west of the mountain range. In the bottom part there is a digital elevation model of Sierra Nevada

period 2071–2100 there is an estimated mean reduction in precipitation of around 28% while there is an expected increase of temperature of around 37%. There could be generated different climate series that fulfill the previous criteria. A simple possibility is to reduce the precipitation a 37% (Case A), another more sophisticated (Case B) is shown in Fig. 3.

In the second case (Fig. 3b) the time series has been simulated with a decrease of 28% in total precipitation, but with an increase in extreme events. It has been achieved by increasing the heavy precipitation events and eliminating low precipitation events at random. For the temperature, it has been considered the experimental temperature time series but with an increase of 37% of the temperature values. Both cases (A and B) have been introduced in the cellular automata models developed and calibrated in Pardo-Igúzquiza et al. (2017) and the time series with the evolution of the SCA shown in Fig. 4b, c respectively, have been obtained. Those figures can be compared with the experimental time series of the Fig. 4a. Basically, although the reduction of precipitation was of 28% and the increase of temperature of 37%, the SCA decreases by around 52 and 56% for the

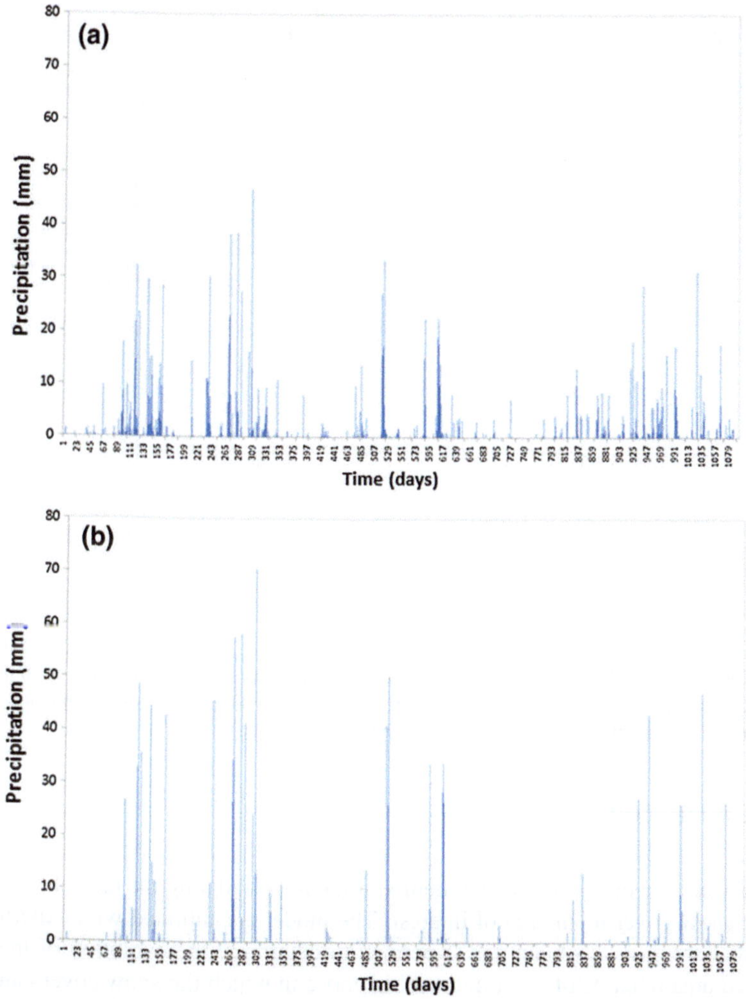

Fig. 3 **a** Experimental index of precipitation for a period of three years. **b** Simulated time series for three years with more extreme events and with a reduction of 28% in precipitation with respect to the series in (**a**)

cases A and B respectively. Also, the result is consistent no matter if the decrease of precipitation is homogeneous or if there is an increase in extreme events and a decrease of small events. Also there is an increase in the bimodality of the SCA evolution and even the first mode, coincidental with the first mode of precipitations in the wet season in Spain, can disappear.

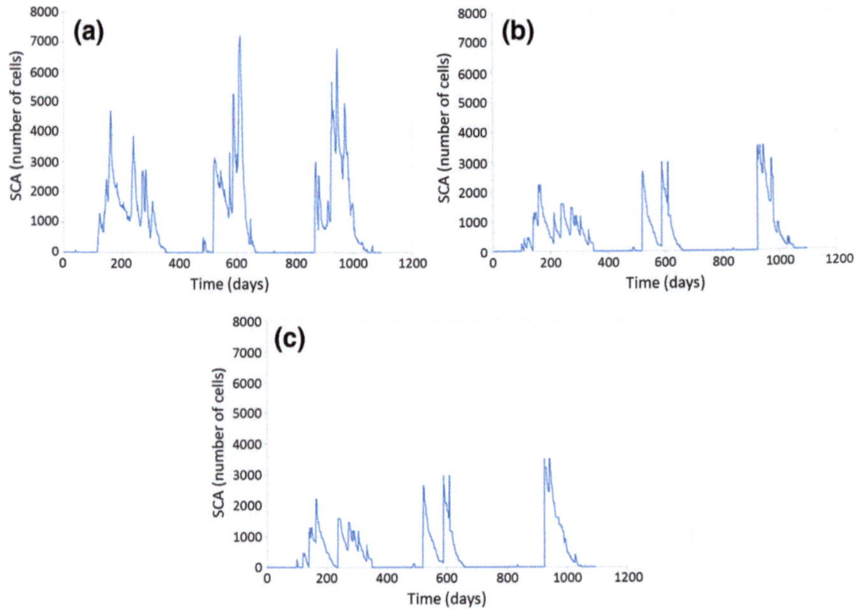

Fig. 4 **a** Experimental evolution of the SCA for three years (July 2003–July 2006). **b** Simulated evolution of the SCA according to a decrease in precipitation around 28% and an increase of temperature of 37%. **c** As the previous case but additionally to the decrease of precipitation there is an increase in extreme events as has been represented in Fig. 3b

5 Conclusions

In this work we have presented a methodology for simulating the discrete dynamics of the snow cover in an area of interest. The model is calibrated with experimental data that are easy to obtain and it permits to determine the evolution of the snow covered area in cases of great interest like those in which the snow cover cannot be obtained from satellite data. This is the case, by obvious reasons, for the prediction of the effects of climate change in the future. In this sense, the results of a study on the projection of climate change in the future, carried out by the same authors, has been used, as well as a cellular automata model that was calibrated with historical data. The task has been to predict its behavior under adverse conditions of increase in temperatures and decrease of precipitation. With respect to the results that have been shown, they have been obtained from simple formulations and require much more future work. Nevertheless, they point out to the big changes that are expected if there is not a reduction in CO_2 emissions into the atmosphere.

Acknowledgements This work was financed by the research projects CGL2013-48424-C2-2-R and CGL2015-71510-R (MINECO). We would also like to Spain02 project and the reviewers for providing a constructive criticism.

References

Collados-Lara AJ, Pulido-Velazquez D, Pardo-Iguzquiza E (2017) Which is the best method to generate future climate scenarios to analyse droughts in a catchment? Submitted

Hall DK, Riggs GA (2007) Accuracy assessment of the MODIS products. Hydrol Process 21:1534–1547

Herrera S, Fernández J, Gutiérrez JM (2016) Update of the Spain02 gridded observational dataset for Euro-CORDEX evaluation: assessing the effect of the interpolation methodology. Int J Climatol 36:900–908

Kumar U, Mukhopadhyay C, Ramachandra TV (2014) Cellular automata calibration model to capture urban growth. Boletín Geológico y Minero 125(3):285–299

Pardo-Igúzquiza E, Collados-Lara AJ, Pulido-Velazquez D (2017) Estimation of the spatiotemporal dynamics of snow covered area by using cellular automata models. J Hydrol 550:230–238

Wolfram S (1984) Cellular automata: a model of complexity. Nature 31:419–424

Numerical Modeling of Groundwater Age Distribution in Motril-Salobreña Coastal Aquifer (SE Spain)

J. P. Sánchez-Úbeda, M. L. Calvache, P. Engesgaard, C. Duque, M. López-Chicano and R. Purtschert

1 Introduction

Groundwater age dating has been used in several studies with different purposes such as identification of groundwater recharge sources (Woolfenden and Ginn 2009), calibration of groundwater models (Sanford 2011), description of flow distributions (Broers 2004), or to infer apparent ages (Cook and Böhlke 2000). The age distribution is also conditioned by both the actual heterogeneity and geometry of the aquifer system, as well as by the spatial variations in recharge and discharge (McCallum et al. 2015, 2017).

The present study was conducted in Motril-Salobreña coastal aquifer that is in a good hydrological state regarding both groundwater quantity and quality. However, its current conditions are changing due to the growing population and pressure from tourism, as well as the increase of the agricultural activities and changes in the irrigation systems, which could reduce significantly the recharge of the aquifer

---

J. P. Sánchez-Úbeda (✉) · M. López-Chicano
Departamento de Geodinámica, Facultad de Ciencias, Universidad de Granada,
Campus Fuentenueva s/n, 18071 Granada, Spain
e-mail: juampesu@ugr.es

P. Engesgaard
Department of Geosciences and Natural Resources Management, University of Copenhagen,
Øster Voldgade 10, Copenhagen, Denmark

C. Duque
Department of Geological Sciences, University of Delaware, Penny Hall,
255 Academy 9 Street, Newark, DE 19716, USA

R. Purtschert
Climate and Environmental Physics, University of Bern, Sidlerstrasse 5,
3012 Bern, Switzerland

M. L. Calvache
Department of Geodynamics, University of Granada, Granada, Spain

M. L. Calvache et al. (eds.), *Groundwater and Global Change in the Western Mediterranean Area*, Environmental Earth Sciences,
https://doi.org/10.1007/978-3-319-69356-9_32

(Duque et al. 2011). For this reason, water quality of the aquifer has been monitored during a prolonged period in order to assess the impact of the anthropogenic changes on groundwater resources. Recently, a survey for determining the age of groundwater has been completed providing information about the apparent groundwater age at different locations and depths within the aquifer. The age of groundwater and hydrogeological information can be combined in a model to complement information provided by the age dating that is punctual, with more general spatially-distributed conditions that can be reproduced in a groundwater flow model. The integration of information required to develop a method since the modelling of all the processes is computationally challenging. The aims of this study are:

- Establish a numerical model calibrated using the ages measured in the field, which can integrate the different sources of information as a result of the application of a specific modeling methodology.
- Calibrate the dispersion and diffusion parameters in a mass and age transport model, since there is not previous studies about them in this aquifer.

2 Hydrogeological Characterization

The Motril-Salobreña coastal aquifer is located on the south eastern Spanish Mediterranean coast (Fig. 1), constituting an alluvial-deltaic system, with an extension of about 42 km^2.

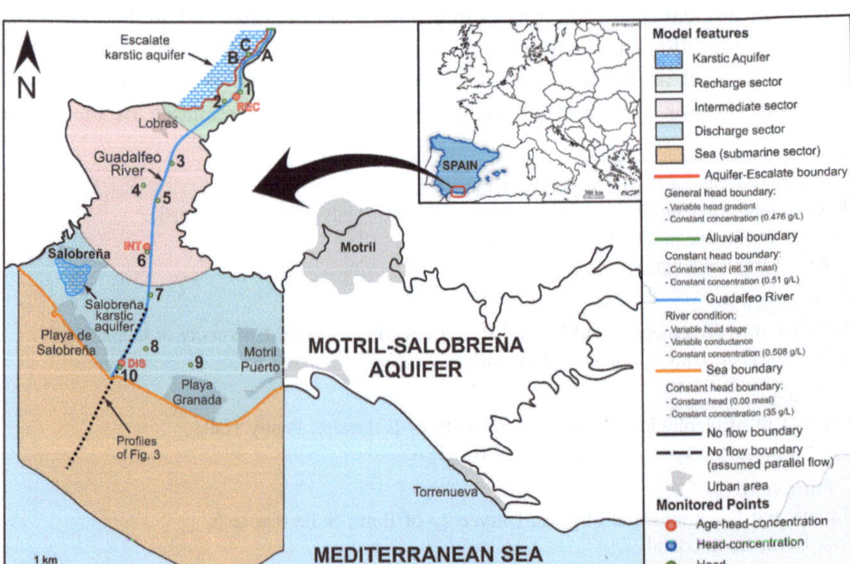

Fig. 1 Location of the Motril-Salobreña coastal aquifer, and the conceptual model approach with the type of boundary conditions imposed, location of sampled and monitored points

The aquifer is composed by coarse sediments layered, with a highly variable grain size. The aquifer thickness ranges from 30 to 50 m, in the northern sector of the fluvial sedimentary environment, to more than 250 m, in areas near the coastline in the deltaic sedimentary environment (Duque et al. 2008). The unconfined aquifer is constituted by the upper 140 m near the discharge zone, and comprises a heterogeneous succession of marine and continental units with deltaic-alluvial nature. It is in contact with the Mediterranean Sea along its South border, which constitutes the main output of the system.

The recharge sources of the aquifer are: the Guadalfeo River, which is a losing river throughout most of its course, the irrigation returns, since there are extensive agricultural activity with high water supplies, and the northern boundary, where the aquifer is connected with the alluvial aquifer of the Guadalfeo River and the Escalate carbonate aquifer. On all other borders, the aquifer is in contact with impermeable materials (Duque 2009).

3 Methodology

The present study is focused in the western part of Motril-Salobreña aquifer, where groundwater flow is higher. This part has been differentiated into zones named as recharge zone (REC), intermediate zone (INT) and discharge zone (DIS), according to distinct hydrological zones (Fig. 1). The model has been divided in 6 layers with different hydrogeological characteristics. Cells are 60 m wide by 60 m long, and their thickness are variable based on the lithological information.

The model was implemented in finite-difference using MODFLOW-2000 (Harbaugh et al. 2000) and MT3DMS (Zheng and Wang 1999), coupled transport model for density-dependent and age distribution simulation with SEAWAT 4 (Langevin et al. 2008). The boundary conditions were defined based on the hydrogeological information collected during the period 2012–2014 (Fig. 1). The method for the age distribution simulation was the direct age, which is based on the consideration of the age as a species in the mass transport model, applying the advective-dispersive equation in the following form (Voss and Wood 1994; Goode 1996; Post et al. 2013):

$$\delta\tau/\delta t = \nabla \cdot D \cdot \nabla\tau - \nabla \cdot (v\tau) + R \quad (1)$$

where τ [T] is the age distribution defined as the solute concentration with time, D [L^2/T] is the hydrodynamic dispersion tensor, v [L/T] is the specific discharge vector, and R [T/T] is the age production ratio, equal to unity in this case. This method allows simulating the groundwater age distribution in a direct way, taking into account the advective and dispersive processes.

The modeling process has been developed in several steps: (1) a groundwater flow and mass transport model to obtain the initial conditions in quasi-steady state for successive models, (2) a mass transport for 20 years of simulation (20 stress periods of one year), in which a quasi-steady state is achieved, (3) a mass transport for two hydrological years (24 stress periods of one month) with the aim of calibrating the hydraulic conductivity (K), storage coefficient (S) and porosity (ϕ), using the flow and transport solutions from the previous model, and (4) a mass transport model for 20 years for achieving a new quasi-steady state using the calibrated parameters. This model was run for 200 years (20 stress periods of 10 years) in transient state, and used then for calibrating longitudinal, transversal dispersion ($\propto_L$ and $\propto_T$, respectively) and molecular diffusion (D_M) coefficients.

The calibration was done based on observed groundwater heads and the concentration distribution (step 3) in the discharge zone (DIS points). The calibration of the age data was conducted with a previous sensitivity analysis. The sensibility analysis was conducted in the last step of the modeling process, and based on the transport parameters molecular diffusion (D_M), longitudinal dispersion ($\propto_L$) and transversal dispersion ($\propto_T$).

The values for these parameters in the analysis were chosen according with previous studies (Pool et al. 2015; Voss and Souza 1987; Post et al. 2013; Abarca et al. 2007). These parameters were then calibrated based on the comparison between the observed mean ages from the analysed samples and the simulated mean ages, trying to obtain a similar range of ages (from 0 to 200 years) than the tracer age data. The observed salinity distribution in the discharge zone was also used in the calibration process of the transport parameters.

The last model was run again with the calibrated D_M and $\propto$ for 220 years using the two species in the transport simulation to obtain a final quasi-steady state for the age and salinity distribution in the aquifer, fitting with the results from the analysed samples.

4 Results

The transient flow model had a maximum root mean square residual error of 0.75 m that was considered an acceptable model fit (Fig. 2). The calibration based on the observed concentration data focused on points located in the discharge zone (DIS points).

For the transport model, the salt concentration was represented in the DIS zone (Fig. 3a). At 300 m from the shoreline and 235 m depth, the mean concentration observed is 13.3 g/L, and the mean simulated value was 15.6 g/L, whereas at 40 m depth, the mean concentration observed was 0.6 g/L and the simulated is 0.43 g/L. The mean ages obtained in the transport model were in agreement with the dating estimated in the REC and INT zones (Fig. 3b), which have a high percentage of young water, and higher age at higher depth.

The differences in age distribution also responses to the heterogeneity of the system. The oldest ages obtained are related to the low permeability values of the deepest layer (from −150 m to the aquifer bottom). The SWI influence can be

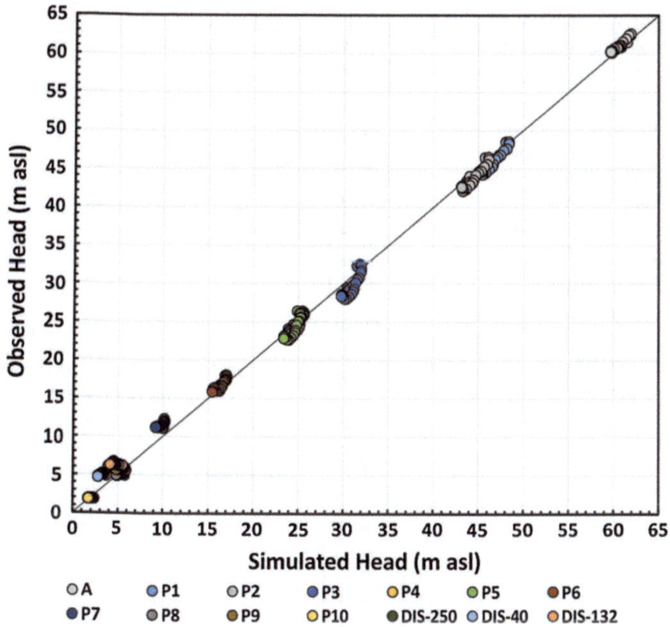

Fig. 2 Observed versus simulated head from the calibration. Each point represents the result for head in one stress period. Points labelled DIS are measured at the depth indicated in the labels (in m). See Fig. 1 for checking the position of the points

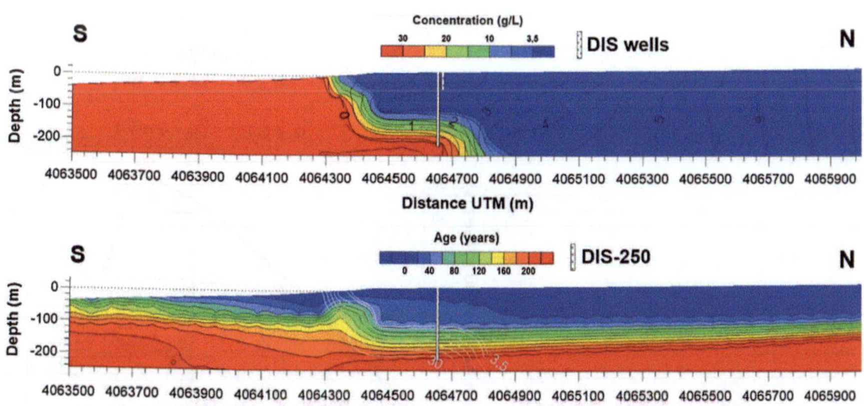

Fig. 3 **a** Chloride concentration distribution results of the transport model for the S-N profile in the DIS zone. Groundwater head (black lines) is 0.5 m spaced, and concentration contours are 5 g/L spaced except for the thick black line of 3.5 g/L (corresponding to 10% of salt water). **b** Results of the groundwater age distribution model for the S-N profile, which shows both age and concentration distribution result. Isoconcentration lines (white lines) are 0.5 g/L spaced, except for 3.5 g/L line

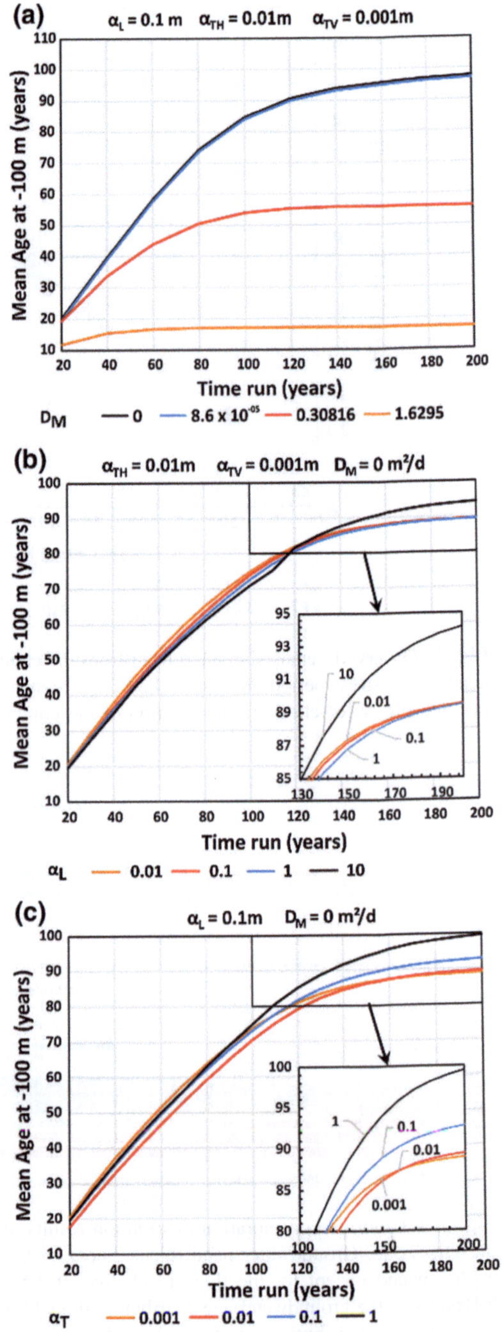

Fig. 4 Sensitivity analysis results for molecular diffusion and dispersion coefficients for the following cases **a** $\propto_L = 0.1$ m, $\propto_{TH} = 0.01$ m, $\propto_{TV} = 0.001$ m, and variable D_M, **b** $D_M = 0$, $\propto_{TH} = 0.01$ m, $\propto_{TV} = 0.001$ m, and variable $\propto_L$, and **c** $D_M = 0$, $\propto_L = 0.1$ m, and variables $\propto_{TH}$ and $\propto_{TV}$. The results show mean age simulated at 100 m depth and 300 m inland from shoreline (DIS zone)

observed in the discharge zone, it yields in an increment of the apparent ages towards the upper half of the aquifer.

The sensibility analysis carried out to determine the transport parameters (Fig. 4) shows that for high values of D_M—Age (molecular diffusion defined for the specie "age"), the maximum mean ages reached in the model are drastically reduced, and the maximum age values are achieved in short simulation times. However, for low or zero values of D_M, the maximum mean ages increase exponentially, and the quasi-steady state is reached for longer simulation times.

The influence of $\propto_L$ on the mean age is comparatively not affecting to the results, and only an increase of maximum mean age values are noticeable when $\propto_L$ is very high (95 years for a run time of 200 years). This does not happen in the case of $\propto_T$, where significant changes in the mean age distribution are found. These changes are higher with longer simulation times.

5 Discussion

The obtained mean ages in the model for the observation points are consistent with the dating results. The calculated values for DIS points are adjusted to the observed values of 24.5 and 167 years, respectively. The dating results for the rest of samples indicate a high percentage in young waters (less than 50 years). The model has shown that in the shallow part of the aquifer, the mixing of waters of different ages can provide different results, but always within a young mean age.

The sensitivity analysis show that the effects of the D_M and $\propto$ in the mean age distribution is higher than on the salt distribution. The age gradient increases with simulation times, from young waters to old waters. The effect of D_M on the age distribution is important in zones with low K, but the influence of $\propto_L$ turns out to be irrelevant. This fact is due to that dispersion tensor is diminished in the direction of flow in the age transport context (Goode 1996). Otherwise, $\propto_T$ changes led to significant changes in the age distribution.

In the present case of study, the effects of D_M were associated with the presence of the salt wedge and the low permeability, mainly in the deeper sector of the aquifer discharge zone. These effects led to an increase of the maximum simulated values of age (Goode 1996), and on the simulation time necessary to reach a quasi-steady state for the mean groundwater age distribution. The differences in the hydrogeological features between layer and transient conditions influence in a more significant way than the D_M on the age distribution, mainly in zones where the flow velocity is greater. However, in low flow velocity zones, D_M is a key factor, keeping constant the other conditions.

6 Conclusions

Two differentiated flow zones within the Motril-Salobreña aquifer were defined based on a flow model and age data, a shallow sector (0–50 m) with a faster groundwater flow, in which the presence of water younger than 50 years dominates, and another deep sector (50–150 m) where the groundwater flow is much slower and the older water is in a higher proportion, as an effect of the saline wedge presence in the discharge zone. A stepwise methodology for the implementation of the numerical model including flow, transport and variable density was required for the calibration of the D_M and $\propto$ parameters using the mean apparent age data obtained from field sampling. The impact of D_M on the simulated age distribution was investigated with the model, and the influence of both $\propto_L$ and $\propto_T$ especially when the groundwater flow velocity is relatively low. Simulating the diffusive-dispersive processes in coastal aquifer have a direct impact on the mass and age distribution, maximum age obtained and simulation time necessary to reach a quasi-steady state solution, especially in the discharge zone, where the saline wedge presence also implies changes in the groundwater flow and transport processes.

Acknowledgements This study was supported by project CGL2016-77503-R, which was funded by the Ministerio de Economía y Competitividad (Government of Spain), by the research group RNM-369 of the Junta de Andalucía and by the Marie Curie International Outgoing Fellowship (624496).

References

Abarca E, Carrera J, Sánchez-Vila X, Dentz M (2007) Anisotropic dispersive henry problem. Adv Water Resour 30:913–926

Broers HP (2004) The spatial distribution of groundwater age for different geohydrological situations in the Netherlands: implications for groundwater quality monitoring at the regional scale. J Hydrol 299(1):84–106

Cook PG, Böhlke JK (2000) Determining timescales for groundwater flow and solute transport. In: Environmental tracers in subsurface hydrology, pp 1–30

Duque C (2009) Influencia antrópica sobre la hidrogeología del acuífero Motril-Salobreña. Doctoral thesis, University of Granada, Granada, Spain

Duque C, Calvache ML, Pedrera A, Martín-Rosales W, López-Chicano M (2008) Combined time domain electromagnetic soundings and gravimetry to determine marine intrusion in a detrital coastal aquifer (southern Spain). J Hydrol 349:536–547

Duque C, López-Chicano M, Calvache ML, Martín-Rosales W, Gómez-Fontalva J, Crespo F (2011) Recharge sources and hydrogeological effects of irrigation and an influent river identified by stable isotopes in the Motril-Salobreña aquifer (southern Spain). Hydrol Process 25:2261–2274

Goode DJ (1996) Direct simulation of groundwater age. Water Resour Res 32:289–296

Harbaugh AW, Banta ER, Hill MC, McDonald MG (2000) MODFLOW-2000, the U. S. Geological Survey modular ground-water model. In: User guide to modularization concepts and the ground-water flow process. U. S. G. S. open-file Report

Langevin CD, Thorne Jr DT, Dausman AM, Sukop MC, Guo W (2008) SEAWAT-4: A computer program for simulation of multi-species solute and heat transport. U. S. G. S. Report

McCallum JL, Cook PG, Simmons CT (2015) Limitations of the use of environmental tracers to infer groundwater age. Groundwater 53:56–70

McCallum JL, Cook PG, Dogramaci S, Purtschert R, Simmons CT, Burk L (2017) Identifying modern and historic recharge events from tracer-derived groundwater age distributions. Water Resour Res 53:1039–1056

Post VE, Vandenbohede A, Werner AD, Teubner MD (2013) Groundwater ages in coastal aquifers. Adv Water Resour 57:1–11

Pool M, Post VE, Simmons CT (2015) Effects of tidal fluctuations and spatial heterogeneity on mixing and spreading in spatially heterogeneous coastal aquifers. Water Resour Res 51:1570–1585

Sanford W (2011) Calibration of models using groundwater age. Hydrogeol J 19:13–16

Voss CI, Souza WR (1987) Variable density flow and solute transport simulation of regional aquifers containing a narrow freshwater-saltwater transition zone. Water Resour Res 23: 1851–1866

Voss CI, Wood WW (1994) Synthesis of geochemical, isotopic and groundwater modelling analysis to explain regional flow in a coastal aquifer of southern Oahu, Hawaii. In: Proceedings on mathematical models and their applications to isotope studies in groundwater hydrology. International Atomic Energy Agency, Vienna, Austria, pp 147–178

Woolfenden LR, Ginn TR (2009) Modeled groundwater age distributions. Ground Water 47: 547–557

Zheng C, Wang PP (1999) MT3DMS: A modular three-dimensional multispecies transport model for simulation of advection, dispersion, and chemical reactions of contaminants in groundwater systems. In: Documentation and user's guide DTIC document

Groundwater Age Dating in Motril-Salobreña Coastal Aquifer with Environmental Tracers ($\delta^{18}O/\delta^{2}H$, $^{3}H/^{3}He$, ^{4}He, ^{85}Kr, and ^{39}Ar)

J. P. Sánchez-Úbeda, M. López-Chicano, M. L. Calvache, R. Purtschert, P. Engesgaard, C. Martín-Montañés, J. Sültenfuβ and C. Duque

1 Introduction

The Motril-Salobreña aquifer system is in a good state regarding both its quantity and quality due to high recharge and low pumping activity (Duque et al. 2011). However, its actual conditions are changing due to the growing population and pressure from tourism, as well as the increasing of the agricultural activities and higher demands in the irrigation systems, which could reduce significantly the water sources of the aquifer.

J. P. Sánchez-Úbeda (✉) · M. López-Chicano
Departamento de Geodinámica, Facultad de Ciencias, Universidad de Granada, Campus Fuentenueva s/n, 18071 Granada, Spain
e-mail: juampesu@ugr.es

R. Purtschert
Climate and Environmental Physics, University of Bern, Sidlerstrasse 5, 3012 Bern, Switzerland

P. Engesgaard
Department of Geosciences and Natural Resources Management, University of Copenhagen, Øster Voldgade 10, Copenhagen, Denmark

C. Martín-Montañés
Instituto Geológico y Minero de España (IGME), Alcázar del Genil 4, Edificio Zulema (bajo), 18006 Granada, Spain

J. Sültenfuβ
Institute of Environmental Physics, University of Bremen, Otto-Hahn-Allee, 28359 Bremen, Germany

C. Duque
Department of Geological Sciences, University of Delaware, Penny Hall, 255 Academy 9 Street, Newark, DE 19716, USA

M. L. Calvache
Department of Geodynamics, University of Granada, Granada, Spain

M. L. Calvache et al. (eds.), *Groundwater and Global Change in the Western Mediterranean Area*, Environmental Earth Sciences,
https://doi.org/10.1007/978-3-319-69356-9_33

The application of dating methods in this kind of coastal systems may provide knowledge about the residence times and differential flow paths, which is of great importance in the coastal water resources management. Due to the slow flow of groundwater, it can take a long time to detect hydrogeological changes and in order to predict and be prepared for the effects of climate change it is required a good knowledge of groundwater movement and residence time in aquifers.

Groundwater age dating has becomes a powerful technique in the study of groundwater recharge sources (Woolfenden and Ginn 2009), in the calibration of groundwater models (Sanford 2011), description of flow distributions (Broers 2004), understanding of the regional-scale groundwater dynamics (Sültenfuβ et al. 2011), or simply to infer apparent ages (Cook and Bohlke 2000). The applications also encompass the evaluation of contaminants transport and aquifer vulnerability (Åkesson et al. 2015), quantification of flow (Reilly et al. 1994), or groundwater mixing processes (Benettin et al. 2013).

Groundwater dating methods were applied in Motril-Salobreña coastal aquifer, where the main recharge comes from the river infiltration and irrigation returns and the main discharge take place as a submarine freshwater discharge towards the sea above the saline wedge. The aims of this study are:

- To describe the general distribution of groundwater age in Motril-Salobreña aquifer.
- To characterize flow circulation paths and residence times of groundwater considering the hydrogeological and geometrical characteristics of the aquifer.
- To determine the relation between groundwater age distribution and saltwater intrusion processes in the discharge zone.

2 Hydrogeological Characterization of the Study Site

The Motril-Salobreña coastal aquifer is located on the south-eastern Spanish Mediterranean coast (Fig. 1), constituting an alluvial-deltaic system, with an area of 42 km^2. Aquifer materials are mostly made up of coarse metapelitic sediments, with a highly variable grain size. Aquifer thickness ranges from 30 to 50 m in the northern sector of the fluvial sedimentary environment to more than 250 m in areas near the coastline in the deltaic sedimentary environment (Duque et al. 2008). The unconfined aquifer is constituted by the upper 140 m near the discharge zone, and comprises a heterogeneous succession of marine and continental units with deltaic-alluvial nature.

One of the main recharge sources of the Motril-Salobreña detrital aquifer is the Guadalfeo River, which is a losing stream throughout most of its course. The other main recharge source is the irrigation returns, since there is an extensive agricultural activity in the zone based on vegetables and subtropical tree species which have high water demands (Duque 2009). In the northern sector, there are other entries of groundwater to the aquifer, through the carbonated Escalate aquifer and through the

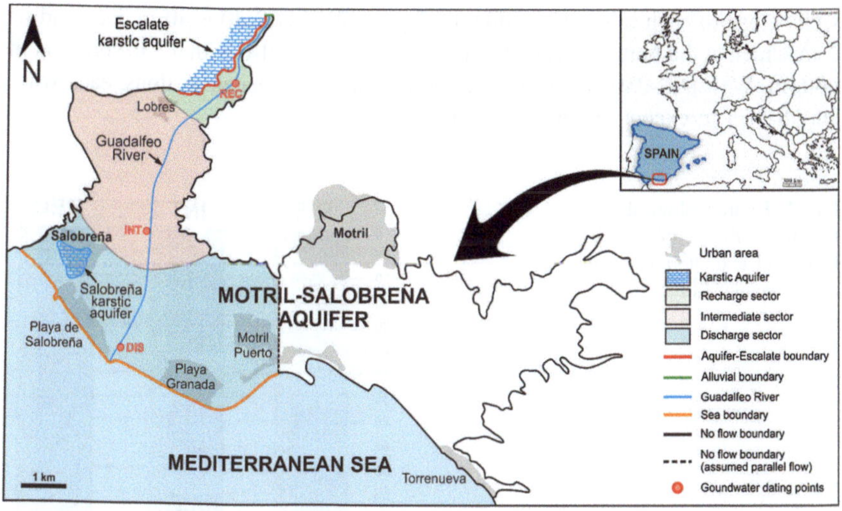

Fig. 1 Location of the Motril-Salobreña coastal aquifer, its boundaries with other systems, and the location of sampled points for groundwater dating. The sampled points are located in the western sector, since it constitutes the principal zone of groundwater flow

alluvium of the Guadalfeo River that continues upstream, with an average total contribution of 6 hm^3 per year (Duque 2009). The southern boundary is the Mediterranean Sea (Fig. 1), where most of the aquifer discharge takes place. On all other borders, the aquifer is in contact with schists and phyllites that also constitute the basement, which can be considered as impermeable.

3 Methodology

3.1 Aquifer Zonation and Sampling Procedure

The studied area was restricted to the western part of the aquifer, where the thickness and hydraulic conductivity are higher and where the Guadalfeo River is a significant recharge source. The sampling locations have been chosen according to distinct hydrological zones. These zones are the recharge (REC), the intermediate (INT) and the discharge zones (DIS) (Fig. 1). The samples for groundwater dating were taken at five different wells, one well in REC zone, one in INT zone and three in DIS zone, and at variable depths within the aquifer. The samples were named with the zone and well where they were sampled followed by the reference depth of sampling in each case.

The well REC was drilled until the bedrock, located at 62 m, and four slotted sections were installed. The bedrock in well INT was reached at 90 m depth, and the casing was installed with four screened sections. In the DIS zone there are three wells close to each other with different depths. Well DIS-40 is 40 m deep and has 3 m screen at the bottom. Well DIS-132 is 132 m deep also with a 3 m screen at the bottom. Well DIS-250 is 250 m deep, cased with 12 screened sections, each with a length of 3 m except one with 6 m (Fig. 2).

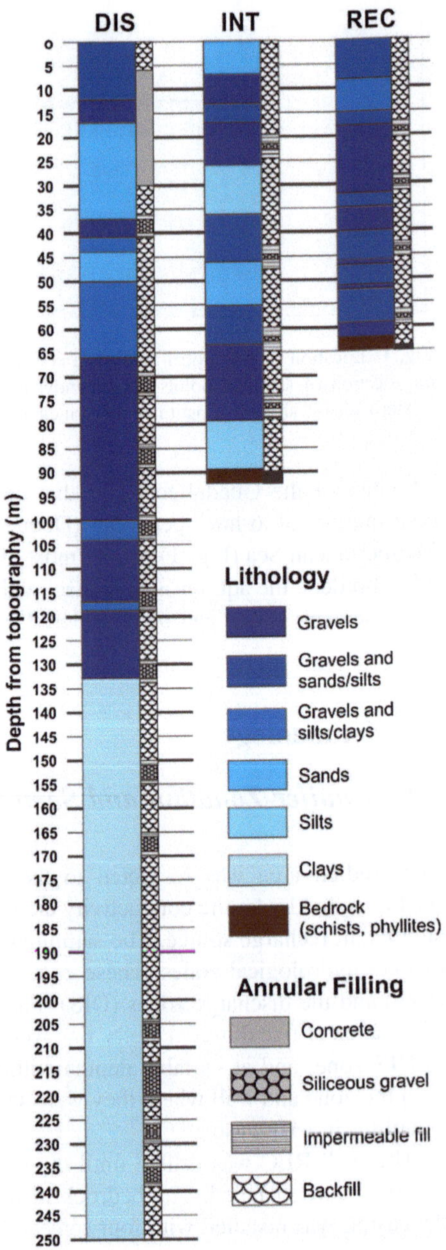

Fig. 2 Hydrogeological features of the sampled wells REC, INT and DIS and details of the annular filling

The sampling took place in May 2015, when the groundwater heads are usually close to the mean annual values of the aquifer in spring season, when the recharge is increased by the Guadalfeo River. In this area there are yearly oscillations of until several meters of the water due to dry periods in the summer and rainy conditions during winter and fall. This year was relatively drier than previous five years with precipitation values close to the mean or slightly higher.

The samples were taken at different depths isolating the screened sections with a packer system and pumping in the middle of the section to obtain a sample, whereas the samples from wells with only one screen were pumped directly after purging the water volume of the entire tube. Samples D-40, D-87, D-132, D-168 and D-mix were taken by free outflow due to its artesian nature introducing a tube at different depths.

3.2 *Environmental Tracers in Groundwater Dating*

The environmental tracers $\delta^{18}O/\delta^{2}H$, ^{3}H, $^{3}H/^{3}He$, ^{4}He, ^{85}Kr, and ^{39}Ar were used. Transient tracers like ^{3}H, ^{3}He and ^{85}Kr are sensitive for young groundwater components with residence times less than about 50 years (Corcho Alvarado et al. 2007). ^{39}Ar was used to date the oldest water components within the aquifer (50–1000 years), and stable isotopes for contrasting the youngest inputs (0.1–3 years), according with Suckow (2014). $^{4}He_{rad}$ serves here as a qualitative age indicator (Solomon 2000).

The analysis of stable isotopes, ^{85}Kr, and ^{39}Ar were analysed in the Climate and Environmental Physics Department (University of Bern, Switzerland) by low level proportional counting (Riedmann and Purtschert 2016; Loosli and Purtschert 2005), and ^{3}H–^{3}He were analysed in the Noble Gas Laboratory of the Department of Oceanography, Institute of Environmental Physics (University of Bremen, Germany), through Noble Gas Mass Spectrometry (NG-MS).

4 Results

Younger groundwater is found in almost the entire aquifer at shallow depths, both in the REC, INT and DIS zones, as indicated by the ^{3}H values closed to 2 TU in all samples (Table 1). Moreover, almost all samples have a value of ^{85}Kr between 70 and 80 dpm/cm^{3}Kr, close to the recent atmospheric level (Winger et al. 2005). However, an age gradient with depth in the DIS zone is observed, where a higher percentage of older groundwater in the analysed samples at greater depths are registered.

In this case, the obtained mean ages vary from 2.5 years to the maximum of 167 years, corresponding to the sample DIS 132, with an estimated 70% of old water (>50 years). In the DIS zone, the mean age estimates in intermediate depths

Table 1 Groundwater dating results for the used tracers

Sample ID	^{85}Kr (dpm/cm^3 Kr)	Err. (dpm/cm^3 Kr)	^{39}Ar (%modern)	Err. (%)	^{85}Kr age (years)	^{39}Ar age (years)	Err. (years)	^3H (TU)	Err. (TU)	$^3He_{tri}$ (TU)	Err. (TU)	$^3H/^3He$ age (years)	$^4He_{rad}$ (cm^{3}STP/kg)	ΔNe (%)
REC-57								2.30	0.1	0	0.5	0	0.000002	26
REC-mix	80	2.9			2.5			2.25	0.1					
INT-45	75.1	3.7			2.5			2.30	0.1	0	0.5	0	0.000002	26
INT-62	79.6	6.8			2.5			2.26	0.1	0	0.5	0	0.000002	28
INT-76	74.7	2.9	100	7	2.5	0.00	27.17	2.35	0.1					
INT-mix								2.42	0.1	0	0.5	0	0.000005	31
DIS-40								2.50	0.1	0.5	0.5	3.5	0.00001	44
DIS-87								2.20	0.1	6.5	0.5	24.5	0.00003	140
DIS-132	5.1	0.3	65	10	41	167	59.71	1.57	0.08	14.5	0.5	41.5	0.00003	42
DIS-168								1.40	0.07					
DIS-mix	47.2	2.3	94	6	12	24.01	24.77	2.14	0.1	1.8	0.5	10.5	0.00002	56

The concentrations for each tracer and the corresponding estimation of age are shown. The sampling work was conducted from 5th to 7th of May, 2015

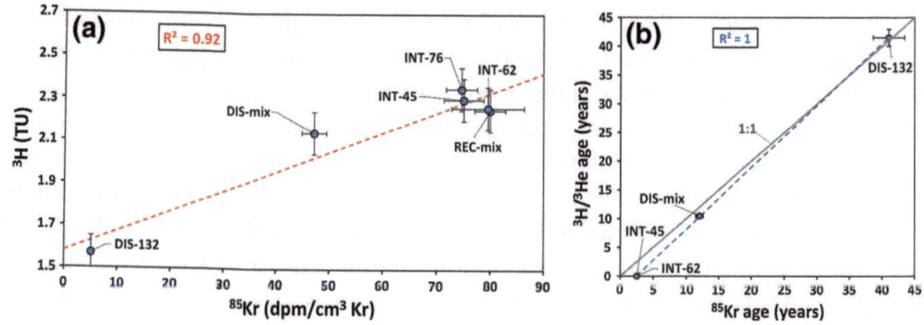

Fig. 3 **a** Linear correlation between concentrations of ^{3}H and ^{85}Kr and **b** between $^{3}H/^{3}He$ age ratio and ^{85}Kr age. The red dashed line in (**a**) and the blue dashed line in (**b**) denote the linear fit, and grey solid line in (**b**) denotes the 1:1 ratio

are in agreement with this tendency, and they also indicate that the increase in age is higher as depth increases. The sample taken at 168 m (DIS-168) for the ^{39}Ar analysis was lost during the laboratory procedure and its concentration could not be estimated, which might result in a higher maximum mean age for this depth as indicated by its low ^{3}H concentration.

Groundwater age data show a good agreement between different tracers. For example, the linear correlation between ^{3}H and ^{85}Kr is 0.92, which implies that the apparent ages obtained by $^{3}H/^{3}He$ and ^{85}Kr to be close each other. Although the linear correlation is close to the 1:1 ratio, there is a slight lack of fit in the lower ages (Fig. 3).

5 Discussion

In zones REC and INT, the direct recharge of the river and the irrigation returns enters young water in the aquifer system that, due to the low saturated thickness and high hydraulic conductivity, makes lower than 5 years the mean ages. Dating results from ^{3}H and ^{85}Kr indicate that the percentage of young water is elevated in almost all the aquifer. This indicates that the aquifer can be very sensitive to hydrogeological changes since surface water takes a short time to reach deep locations.

In the DIS zone, the results suggest a well-defined age gradient, with an apparent young age until 50 m depth and travel times ~25 m/year in the vertical direction. The mean age varies between the minimum assumed of 2.5 years and the maximum of 167 years (sample DIS-132). The presence of the salt wedge in the discharge zone forces the deep flow to ascend, producing a higher mixture of waters towards the surface with greater percentages of older water. This yields higher mean ages at shallower depths in DIS zone compared to the rest of the aquifer.

The groundwater of Motril-Salobreña aquifer has apparently a maximum residence times not much higher than 200 years. Groundwater younger than 5 years is present in almost all of the shallowest parts of the aquifer, which indicates a relatively fast groundwater flow recharge. Water younger than 50 years are present in greater or lesser proportions of almost in the whole aquifer indicating a very rapid circulation. Older water in the discharge zone are located in deeper parts of the aquifer, but are also related to the position of the salt wedge toe. The changes in the hydrodynamic circulation of groundwater in the aquifer generates a change in the age distribution in the aquifer. The presence of the saltwater wedge increases the mean ages in the discharge zone connected to the vertical flow component and also due to the slow flow near the toe of saltwater wedge toe.

There is faster flow in the shallow part of the aquifer, where a high percentage of groundwater gives transit times lower than 50 years (even less than 5 years in highly permeable areas associated with the Guadalfeo river alluvial), and a deep part, where the maximum average ages are obtained, indicating that the flow velocity is much smaller. However, the age results show some incongruences between ^{39}Ar and the other tracers, indicating that younger ages could be considered and faster groundwater infiltration and flow would be taking place into the aquifer.

6 Conclusions

The groundwater dating using environmental tracers indicates the presence of groundwater within the age range, from less than 5 to 170 years with variable percentages of young water in almost the entire aquifer. A steep gradient towards older water is observed in the deep part of the aquifer discharge zone. There are two well differentiated flow zones within the aquifer, a shallow part (0–50 m) with a faster groundwater flow, where the presence of groundwater younger than 50 years dominates, and a deeper part (50–150 m) where the groundwater flow is much slower and the older waters are in greater proportion. The mixing processes that affect the groundwater age distribution are determined by both low hydraulic conductivity and the flow field generated by the presence of the salt wedge. The groundwater age distribution shows the influence of changes in the hydraulic properties for the groundwater flow within the aquifer, and the implications of the freshwater-saltwater interface in the discharge zone of coastal aquifers.

Acknowledgements This study was supported by project CGL2016-77503-R, which was funded by the Ministerio de Economía y Competitividad (Government of Spain), by the research group RNM-369 of the Junta de Andalucía and by the Marie Curie International Outgoing Fellowship (624496).

References

Åkesson M, Suckow A, Visser A, Sültenfuβ J, Laier T, Purtschert R, Sparrenbom CJ (2015) Constraining age distributions of groundwater from public supply wells in diverse hydrogeological settings in Scania, Sweden. J Hydrol 528:217–229

Benettin P, Rinaldo A, Botter G (2013) Kinematics of age mixing in advection-dispersion models. Water Resour Res 49:8539–8551

Broers HP (2004) The spatial distribution of groundwater age for different geohydrological situations in the Netherlands: implications for groundwater quality monitoring at the regional scale. J Hydrol 299(1):84–106

Cook PG, Böhlke JK (2000) Determining timescales for groundwater flow and solute transport. In: Environmental tracers in subsurface hydrology, pp 1–30

Corcho-Alvarado JA, Purtschert R, Barbecot F, Chabault C, Rueedi J, Schneider V, Aeschbach-Hertig W, Kipfer R, Loosli H (2007) Constraining the age distribution of highly mixed groundwater using ^{39}Ar: A multiple environmental tracer ($^{3}H/^{3}He$, ^{85}Kr, ^{39}Ar, and ^{14}C) study in the semiconfined Fontainebleau sands aquifer (France). Water Resour Res 43(3):16 p

Duque C, Calvache ML, Pedrera A, Martín-Rosales W, López-Chicano M (2008) Combined time domain electromagnetic soundings and gravimetry to determine marine intrusion in a detrital coastal aquifer (southern Spain). J Hydrol 349:536–547

Duque C, López-Chicano M, Calvache M, Martín-Rosales W, Gómez-Fontalva J, Crespo F (2011) Recharge sources and hydrogeological effects of irrigation and an influent river identified by stable isotopes in the Motril-Salobreña aquifer (southern Spain). Hydrol Process 25:2261–2274

Duque C (2009) Influencia antrópica sobre la hidrogeología del acuífero Motril-Salobreña. Doctoral thesis, University of Granada, Granada, Spain, 196 pp

Loosli HH, Purtschert R (2005) Rare gases. In: Aggarwal P, Gat JR, Froehlich K (eds) Isotopes in the water cycle: past, present and future of a developing science. IAEA, Vienna, pp 91–95

Reilly TE, Plummer LN, Phillips PJ, Busenberg E (1994) The use of simulation and multiple environmental tracers to quantify groundwater flow in a shallow aquifer. Water Resour Res 30:421–433

Riedmann RA, Purtschert R (2016) Separation of argon from environmental samples for Ar-37 and Ar-39 analyses. Sep Purif Technol 170:217–223

Sanford W (2011) Calibration of models using groundwater age. Hydrogeol J 19:13–16

Solomon DK (2000) ^{4}He in groundwater. In: Environmental tracers in subsurface hydrology, pp 425–439

Suckow A (2014) The age of groundwater—definitions, models and why we do not need this term. Appl Geochem 50:222–230

Sültenfuβ J, Purtschert R, Führböter JF (2011) Age structure and recharge conditions of a coastal aquifer (Northern Germany) investigated with ^{39}Ar, ^{14}C, ^{3}H, He isotopes and Ne. Hydrogeol J 19:221–236

Winger K, Feichter J, Kalinowski MB, Sartorius H, Schlosser C (2005) A new compilation of the atmospheric ^{85}krypton inventories from 1945 to 2000 and its evaluation in a global transport model. J Environ Radioact 80:183–215

Woolfenden LR, Ginn TR (2009) Modeled groundwater age distributions. Ground Water 47:547–557

Multi-tracers Strategy to Define a Conceptual Model for the Coastal Aquifers of Mediterranean Islands, Case Study of the Bonifacio Aquifer (Corsica, France)

S. Santoni, E. Garel and F. Huneau

1 Introduction

A hydrochemical and isotope study was led to identify the flow paths, the recharge areas and the geochemical processes that control the evolution of groundwater in a Mediterranean carbonate coastal aquifer. The study is expected to propose a hydrogeological conceptual model based on environmental tracers to characterise and quantify the aquifer system of Bonifacio. The groundwater resource represents the unique drinking water resource of the southern Corsica and the region faces to high pressures during the touristic period (2,000,000 tourists each year). A well-documented description of the geology and structure of this basin was the starting point for a detailed hydrogeochemical and isotopic investigation at the aquifer scale (Orsini et al. 2010; Reynaud et al. 2012).

2 Study Site

The aquifer of Bonifacio (Fig. 1) is filling a Hercynian granitic depression and is composed of three Miocene formations: the continental Balistra formation (granitic alteration products), the marine Cala di Labra formation (calcarenites and sandy siltstones) and the marine Bonifacio formation (calcarenites and carbonates). The sandy siltstones level of the Cala di Labra formation materializes the separation between the upper and the lower aquifer. The study site is known for its strong winds blowing up to 300 days/year, which brings sea sprays over all this area.

S. Santoni · E. Garel · F. Huneau (✉)
Hydrogeology Department, University of Corsica, CNRS UMR 6134 SPE, Campus Grimaldi, BP 52, 20250 Corte, France
e-mail: huneau@univ-corse.fr

M. L. Calvache et al. (eds.), *Groundwater and Global Change in the Western Mediterranean Area*, Environmental Earth Sciences,
https://doi.org/10.1007/978-3-319-69356-9_34

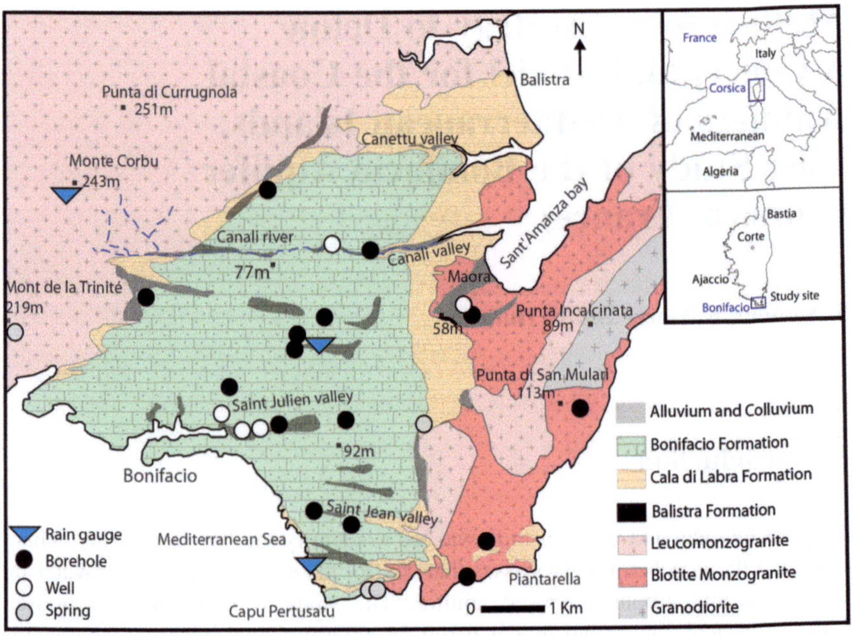

Fig. 1 Map of the study site and location of the sampling points (Santoni et al. 2016a)

3 Methodology

A monthly monitoring approach combining major ions and water stable isotopes in rain, runoff and ground waters has been carried out during two years. 15 boreholes (up to 270 m deep), 5 wells, 4 springs, 2 streams and 3 rain gauges located throughout the study site were sampled from January 2013 to February 2015 for physico-chemical parameters, major ions, ^{18}O and ^{2}H.

Punctual investigations were carried out for CFC-11, CFC-12, CFC-113, SF_6, trace elements, ^{3}H, ^{13}C, ^{222}Rn, $^{224,223}Ra$ and $^{87}Sr/^{86}Sr$ measurements.

4 Results and Discussion

4.1 Aquifer Recharge

The climate is Mediterranean with annual mean temperature and precipitation about 16.8 °C and 605 mm respectively. Annual potential evapotranspiration for the 2006–2014 period is estimated between 897 and 1258 mm following the method (Thornthwaite, Turc or Penman-Monteith). The ensuing infiltration rates are between 15 and 27%, with strong inter-annual variation. A Local Meteoric Water

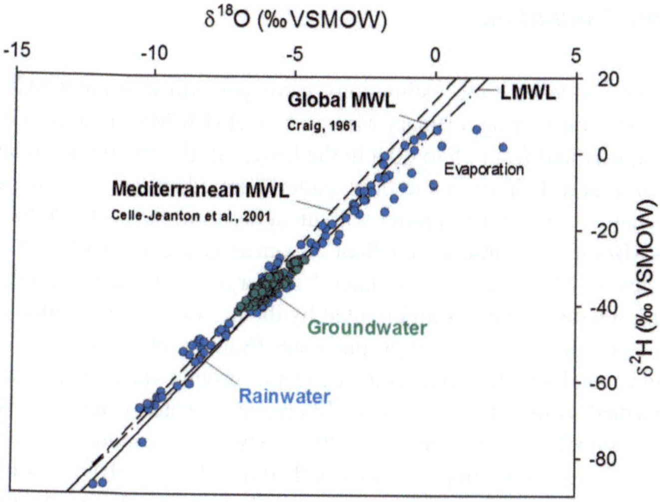

Fig. 2 Stable isotope content in rainwater and groundwater (Santoni et al. 2016b)

Line (LMWL, Fig. 2) has been established for Bonifacio ($n = 24$, $r^2 = 0.99$) and is defined as:

$$\text{LMWL}: \delta^2\text{H} = 7.39(\pm 0.19)\delta^{18}\text{O} + 6.03(\pm 0.53)$$

The LMWL is consistent with the meteoric water line in Sicily defined as $\delta^2H = 6.75\ \delta^{18}O + 8.2$. The weighted means about −6.84‰ for $\delta^{18}O$ and about −44.2‰ for δ^2H, are in good agreement with the neighbouring data from Ajaccio, Corsica (−6.89‰; −44.0‰) in Huneau et al. (2015). The rainwater collected upland in Monte Corbu and in the centre of the plateau displays more depleted stable isotopes values than near the shores. $\delta^{18}O$ and δ^2H mean in precipitation evolve along the LMWL from −7.90‰ and −50.8‰ in winter to −3.01‰ and −16.3‰ in summer. Same seasonal variations with lower intensities are observed in groundwater.

Groundwater from the surrounding granites and the autochthonous recharge area display temporal variability up to ±1.27‰ for $\delta^{18}O$ and ±8.6‰ for δ^2H. Conversely, groundwater from the lower aquifer displays variability only about ±0.11‰ for $\delta^{18}O$ and ±0.5‰ for δ^2H. Since uncertainties in stable isotopes analyses are about 0.1 for δ^2H and 0.01 for $\delta^{18}O$, the changes are thus little but significant, implying the presence of important mixing with older groundwater.

4.2 Flow Conditions

Groundwater residence time estimations were performed using CFCs and SF_6. Weighted ages considering a Binary Mixing Model (BMM) range from 0 to 50 a in the upper aquifer and from 35 to 60 a in the lower aquifer implying a relatively low dynamic of renewal. Two main flow pattern are identified: (1) autochthonous vertical recharge of the upper aquifer with an aging from the centre of the plateau to the shores (Bonifacio harbour and Sant'Amanza) and consistent with the potentiometric heads, (2) allochthonous lateral recharge of the lower aquifer via the surrounding fractured granites highlighted by the presence of terrigenous SF_6 in the granite leads to an SF_6 excess in groundwater (Santoni et al. 2016e).

Hydraulic conductivities (K) from using the groundwater residence times has been performed considering the time difference a water parcel needs to travel between two sampling points (Suckow, 2014). Hydraulic conductivities range from 1.4 to 2.2 10^{-6} m/s in the upper aquifer and from 2.1 to 4.9 10^{-6} m/s in the lower aquifer. These values are in good agreement with the hydraulic conductivities from a previous geophysical survey (Dörfliger et al. 2002).

An approach using $^{87}Sr/^{86}Sr$ allowed to validate and compute the mixing rates between the granites, the upper and the lower aquifer (Fig. 3). The recharge provided by the neighbouring fractured granites represents up to 20% of the total

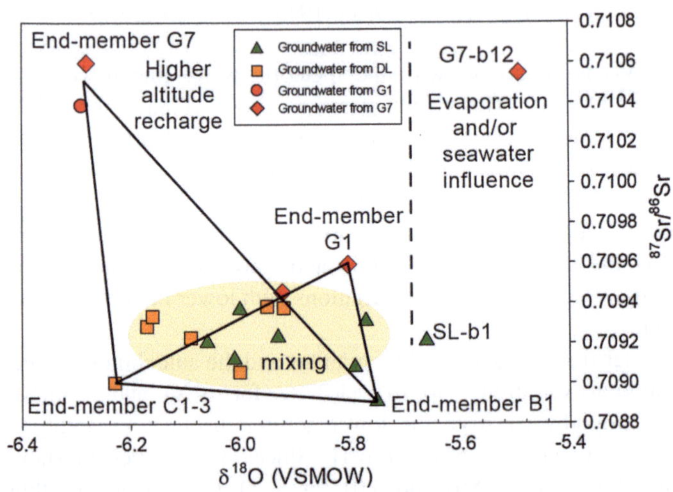

Fig. 3 Groundwater mixing determination using $\delta^{18}O$ and $^{87}Sr/^{86}Sr$ (Santoni et al. 2016d). Green symbols are for groundwater from the shallow level (SL) of the upper aquifer, orange symbols are for groundwater from the deep level (DL) of the lower aquifer and red symbols are for groundwater from granites of both sides of the plateau (G1 and G7)

recharge. This additional natural recharge rises consequently the recharge area of the aquifer (+32% in surface), extended to the whole topographic catchment boundaries (Santoni et al. 2016d).

4.3 Groundwater Types

Environmental tracers allowed assessing mineralization processes within this carbonate coastal aquifer. Three groundwater types are observed: (1) Ca^{2+}–HCO_3^- water type in the upper aquifer conditioned by carbonate dissolution, (2) Na^+–Cl^- water type in the granites conditioned by sea sprays and enhanced by the winds and (3) intermediate Ca^{2+}–Na^+–Cl^- water type in the lower aquifer conditioned by the mixing of groundwater from the upper aquifer and granites.

4.4 Aquifer Discharge

Two outlets are possible for groundwater: (1) anthropogenic, with up to 180,000 m^3/year of groundwater taken for drinking water demand, mainly for a public supply and (2) natural, through the Submarine Groundwater Discharge (SGD). $^{224,223}Ra$ and ^{222}Rn measurements along the coastline display low activities. The main discharge points were found in the Bonifacio harbor (SGD1) and in Sant'Amanza (SGD2), in agreement with the flow scheme defined previously and with the low temperatures observed using thermal infrared images (Fig. 4). The low radon activities reflect thus low discharge conditions (Garel et al. 2015).

4.5 Anthropogenic Influence

Some groundwater samples show noticeable NO_3 concentrations (up to 40 mg/L) highlighting the vulnerability of the aquifer facing human activities. Stable isotopes in groundwater have not yet demonstrated any seawater intrusion for the investigated sampling points. In addition, low SGD flows detected in the Bonifacio harbour (close to the drinking water supply wells) imply that groundwater flow conditions do not appear influenced by over pumping suggesting the possibility to plan a sustainable management of the aquifer (Santoni et al. 2016c).

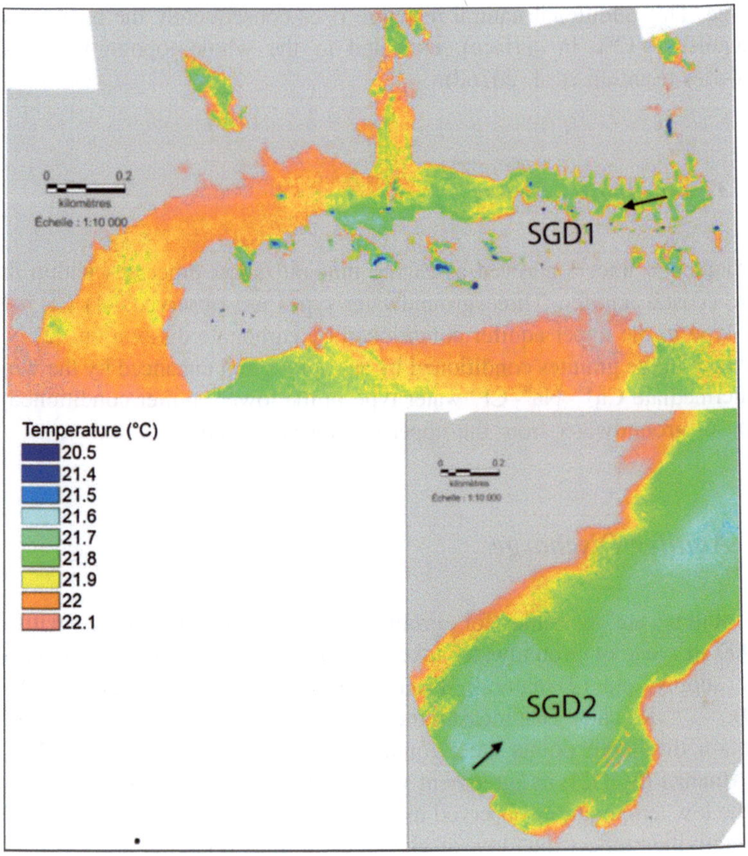

Fig. 4 Submarine groundwater discharge evidenced by thermal infrared images (Garel et al. 2015)

5 Conclusion

This study highlights the relevance of a multi-tracer approach to define conceptual models (Fig. 5) for coastal aquifers under Mediterranean climate. This approach reveals a relatively undisturbed hydrosystem in spite of increasing tourism and drinking water demand. The outputs of the study will help promoting the establishment of a proactive management strategy, in clear opposition with the widespread curative management operating in many coastal aquifers in the Mediterranean area.

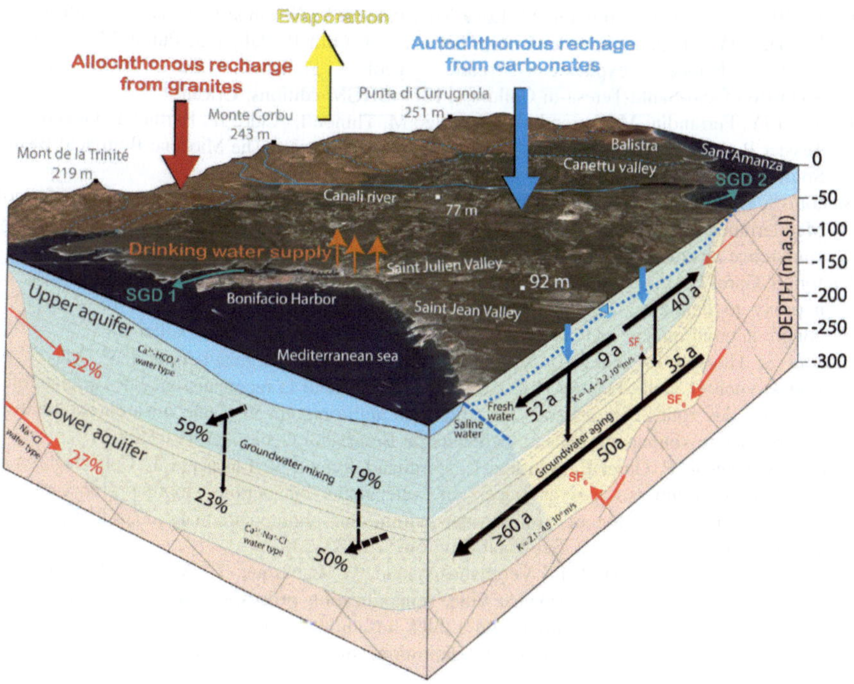

Fig. 5 Conceptual model of the coastal Bonifacio aquifer (Santoni et al. 2016a)

Acknowledgements The Ph.D. grant of Sébastien Santoni was financially supported by the regional council of Corsica (Collectivité Territoriale de Corse). The authors would like to thank the municipality of Bonifacio for its support and contribution to the implementation of the project. The authors would also like to thank the Bonifacio inhabitants for access to their wells and boreholes during the field sampling.

References

Celle-Jeanton H, Travi Y, Blavoux B (2001) Isotopic typology of the precipitation in the Western Mediterranean Region at three different time scales. Geophys Res Lett 28(7):1215–1218

Craig H (1961) Isotopic variations in meteoric waters. Science 133:1702–1703

Dörfliger N, Ferrandini J, Ferrandini M, Mathieu F (2002) Caractérisation géométrique et hydrodynamique du causse de Bonifacio (Corse) à partir d'une synthèse des connaissances géologiques et hydrogéologiques et par méthodes géophysiques. BRGM/RP-51860-FR

Garel E, Santoni S, Huneau F, Mayer A, Travi Y, Radakovitch O, Cockenpot S, Claude C (2015) A radium and radon tracing experiment to estimate low submarine groundwater discharge (aquifer of Bonifacio, Corsica, France). In: International symposium on isotope hydrology: revisiting foundations and exploring frontiers, IAEA, Vienna, Austria

Huneau F, Garel E, Santoni S, Jaunat J, Celle-Jeanton H (2015) A new network for isotopes in precipitation of Corsica (France). In: International symposium on isotope hydrology, revisiting foundations and exploring frontiers, IAEA, Vienna, Austria

Orsini JB, Ferrandini J, Ferrandini M, Loÿe MD, Pluquet F, Guennoc P, Thinon I, Santiago M, Reynaud JY, Oggiano G, Cherchi G, Gattaccaca J, Orrù P, Puliga G, Pintus M, Ulzega A (2010) Notice explicative, carte géol. France (1/50 000), feuille Sotta-Bonifacio-Santa-Teresa-di-Gallura (1127). BRGM éditions, Orléans

Reynaud JY, Ferrandini M, Ferrandini J, Santiago M, Thinon I, André JP, Barthet Y, Guennoc P, Tessier B (2012) From non-tidal shelf to tide-dominated strait: The Miocene Bonifacio Basin, Southern Corsica. Sedimentology 60:599–623

Santoni S, Garel E, Huneau F (2016a) Defining a conceptual model for the coastal aquifers of Mediterranean islands, an example from Corsica (France). In: EGU general assembly conference abstracts, vol 18. Vienna, Austria, p 9268

Santoni S, Garel E, Huneau F, Celle-Jeanton H (2016b) Stable isotope composition of rainwater as a proxy to delineate recharge processes to Mediterranean coastal aquifers. In: 43rd IAH congress, environment and society: 60 years of IAH, Le Corum, Montpellier

Santoni S, Huneau F, Garel E, Aquilina L, Labasque T, Vergnaud-Ayraud V, Mayer A, Radakovitch O, Celle-Jeanton H (2016c) Quels outils pour la reconnaissance des potentialités aquifères des zones côtières sous forte pression anthropique? Vers une mise en exploitation durable des ressources du plateau carbonaté de Bonifacio. Géologues 191:58–62

Santoni S, Huneau F, Garel E, Aquilina L, Vergnaud-Ayraud V, Labasque T, Celle-Jeanton H (2016d) Strontium isotopes as tracers of water-rocks interactions, mixing processes and residence time indicator of groundwater within the granite-carbonate coastal aquifer of Bonifacio (Corsica, France). Sci Total Environ 573:233–246

Santoni S, Huneau F, Garel E, Vergnaud-Ayraud V, Labasque T, Aquilina L, Jaunat J, Celle-Jeanton H (2016e) Residence time, mineralization processes and groundwater origin within a carbonate coastal aquifer with a thick unsaturated zone. J Hydrol 540:50–63

Suckow A (2014) The age of groundwater—definitions, models and why we do not need this term. Appl Geochem 50:222–230

Assessment of Artificial Recharge Efficiency Against Groundwater Stress in the El Khairat Aquifer

M. Zammouri, N. Brini and F. Jarraya Horriche

1 Introduction

In semi-arid regions, aquifers are mainly recharged by infiltration of flood water through beds of ephemeral streams (wadis). A good knowledge of aquifer recharge is recommended for the efficient and sustainable groundwater resources.

Today, in Tunisia, most aquifers show signs of stress manifesting by a continuous decline of the water table and a drying up of natural outlets, due to an excessive increase of groundwater extraction. To remedy this, Tunisian manager policy is oriented towards the development of surface water mobilization works (dams, benches etc.). Many upland dams are used for artificial aquifer recharge by releasing the reservoir water. Direct and indirect measurements of artificial aquifer recharge are expensive. Modeling can be a useful tool to assess the artificial groundwater recharge effectiveness.

2 Objectives

The El Khairat aquifer, which is located in the eastern center of Tunisia, is affected by the groundwater extraction. The construction of El Khairat dam upstream of the plain has harmed the aquifer recharge as no artificial recharge operations were carried out during the first three years following the dam construction. Since 2000, many artificial recharge operations were carried out by Tunisian Water Authorities

M. Zammouri (✉) · N. Brini
Faculty of Sciences of Tunis, University of Tunis El Manar, 2092 Tunis, Tunisia
e-mail: z_mounira@yahoo.fr

F. Jarraya Horriche
Georesources Laboratory, Centre of Research and Water Technologies,
Technopole Borj Cedria, BP273, 8020 Soliman, Tunisia

M. L. Calvache et al. (eds.), *Groundwater and Global Change in the Western Mediterranean Area*, Environmental Earth Sciences,
https://doi.org/10.1007/978-3-319-69356-9_35

to re-establish the natural recharge. They consisted of releasing water from the El Khairat reservoir to the natural downstream bed of the wadi channel.

This paper aims to assess the efficiency of the artificial groundwater recharge campaigns carried out in the El Khairat alluvial plain. Data from the Tunisian agencies responsible for the management of water resources (DGRE and CRDA of Sousse) have been collected and explored. A conceptual model is used to evaluate the artificial recharge of El Khairat aquifer. The model findings are analyzed to determine the factors and constraints influencing the yield of this technique of artificial recharge and to compare the groundwater recharge corresponding to artificialized regime to the recharge occurring in natural conditions.

3 Materials and Methods

3.1 Study Area Description

The El Khairat plain, extending over an area of about 63 km^2 (Fig. 1), is characterized by a semiarid climate with an aridity index of 11.5 mm/°C, a yearly potential evapotranspiration reaching 1,730 mm and erratic annual rainfall ranging from 39 to 744 mm over the period 1972–2015. The plain contains an alluvial aquifer which is important for the economic activity of the region. The geology of the alluvial plain is dominated by tertiary and quaternary deposits. The alluvial deposits are constituted by pebbles, gravel, sand and clayey sand. The alluvial thickness is variable and can exceed 100 m in the centre of the plain. The bedrock of the alluvial deposits consists of late Eocene marls or Mio-Pliocene clays.

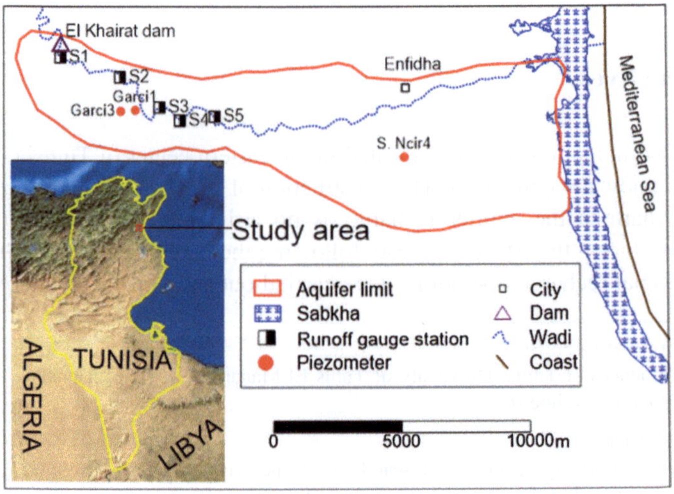

Fig. 1 Location and overview of the El Khairat plain

The aquifer is recharged mainly by infiltration of water from flood events. The El Khairat wadi is the most important stream. In 1999, El Khairat wadi was dammed upstream of the plain in order to protect the Enfidha town against flooding. Between 2002 and 2015, part of the stored reservoir water was released to the natural downstream bed of the wadi channel for artificial aquifer recharge.

3.2 Conceptual Modelling of the Groundwater Recharge

The groundwater recharge is one of the most hard water balance terms to assess, often poorly known. It involves open-channel flow, infiltration into the unsaturated zone and saturated groundwater flow. Using direct or indirect measurements to assess the groundwater recharge (Dassi et al. 2005; Sophocleous 1991; Xi et al. 2008) are expensive and difficult to implement at the regional level.

Conceptual modelling is a good alternative approach to model the recharge mechanism (Hernandez 2000; Hughes 1995; Zammouri and Feki 2005). This approach is used in modelling of artificial groundwater recharge by water releases from the El Khairat reservoir. The water releases are transformed to runoff by sequence of reservoirs. The wadi bed is divided into several serial reaches. Each reach considered to be an entity with homogeneous characteristics flows into the downstream one. The model input consists of the daily water volume released from the dam. For each reach, the production function is represented by a soil storage zone and four transfer reservoirs (Fig. 2). The conceptual model structure was described in details in a previous paper (Zammouri and Feki 2005).

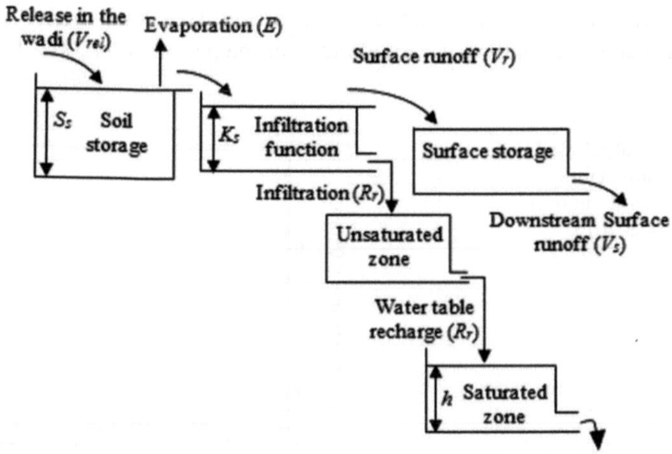

Fig. 2 Structure of the conceptual model showing the linkage between the different reservoirs

Daily released water (V_{rel}) from the El Khairat reservoir enters the soil storage zone, which retains all moisture until the storage (S_s) is full. The storage is emptied by evapotranspiration (E). Actual evapotranspiration occurs at the potential rate. The first transfer reservoir having a finite capacity (K_s) controls the amount of moisture that infiltrates into the soil (R_r). Overflow from this reservoir constitutes the surface runoff (V_r). The second transfer reservoir is a surface storage zone. It is aimed to the runoff routing through the wadi course according to an exponential depletion and allows calculating the effective runoff (V_e). The third transfer reservoir represents the unsaturated zone. It allows defining the infiltration function delay. The fourth transfer reservoir represents the saturated zone. It is characterized by the water table level (h), which is calculated by applying the water balance for the groundwater reservoir under unsteady conditions.

Since 2002 onwards, several artificial recharge operations were carried out by CRDA of Sousse. The annual released water ranged from 1.2×10^6 m^3 in 2003 to 10.7×10^6 m^3 in 2002, with daily volume varying between 4000 and 79,500 m^3. The recharge operation may last from one month to one year. For the dry years such as 2010, there was no artificial recharge operation (Fig. 3).

Measurements of the surface runoff are available only for the first recharge campaign carried out over the period 2/28/2002–6/8/2002. During this campaign, flood routing was controlled by five runoff gauging stations (Fig. 1). Accordingly, the El Khairat wadi bed was divided into four reaches, in relation to the runoff gauging location, with a total length of 9.7 km, downstream the El Khairat reservoir.

The calibration of the model consists of reproducing the daily surface runoff volume observed at the runoff gauging stations as well as the water table level observed at the piezometers located in close proximity of the wadi course. The reference period for the calibration of the surface runoff is 2/28/2002–4/18/2002 while the validation is carried out over the period 4/19/2002–6/8/2002. For the groundwater flow, calibration is carried out over the period 2/28/2002–11/20/2003.

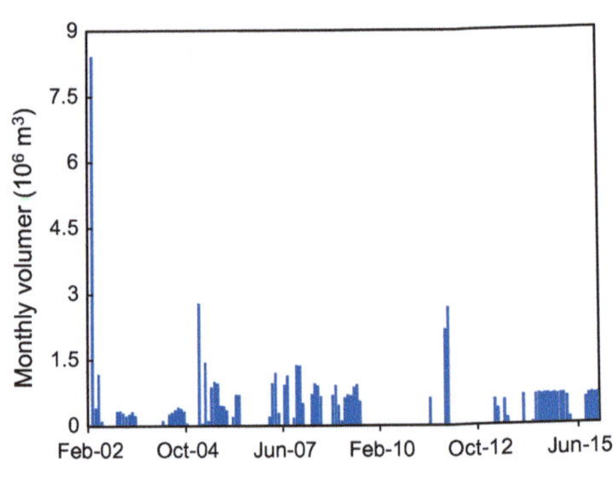

Fig. 3 Historical released water in the artificial campaigns carried out over the period 2002–2015

4 Results and Discussion

The calibration process indicates the high model sensitivity to the infiltration capacity (K_s) and the depletion parameter of the superficial storage reservoir, which describe the runoff routing technique. The calibration results are of good to medium quality. The correlation coefficient varies between 0.5 and 0.93 and the coefficient of efficiency ranges from 0.21 to 0.8. Figure 4 shows model calibration results at the runoff gauging station S2 as well as the observation well Garci3.

The simulated water balance of the artificial groundwater recharge operations over the period 2/28/2002–11/20/2003 indicates an infiltration coefficient ranging between 40 and 80% of the released water. Evaporation losses are insignificant, in general lower than 1%. Losses by surface runoff beyond the permeable part of the wadi course can be important. They are strongly related to the releases discharge rate, which varied on average between 0.25 and 0.5 m^3/s. High discharge rates of flood waves reaching sometimes 0.9 m^3/s involved significant losses by runoff of up to 30%.

The calibrated model is used to simulate the artificial recharge operations carried out over the period 2002–2015. The yearly groundwater artificial recharge over this period varied between 10^6 m^3 and 5.3×10^6 m^3 (Fig. 5), with an average value of

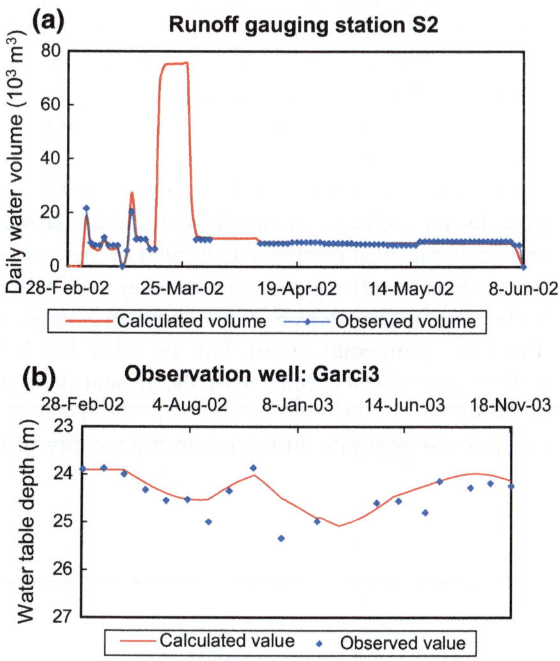

Fig. 4 Model calibration and validation results: **a** comparison between the observed and simulated streamflow volumes; **b** temporal reproduction of water table level

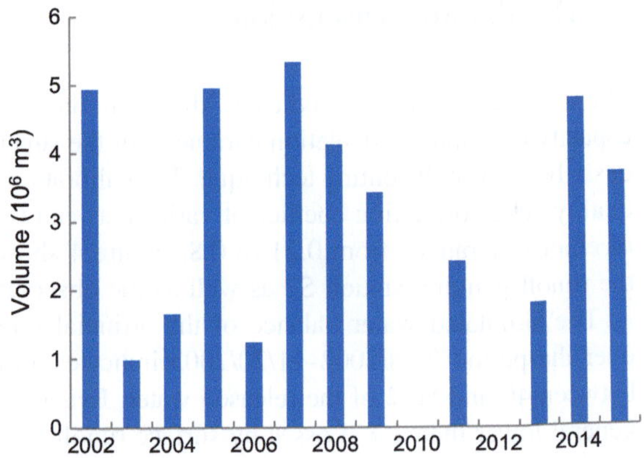

Fig. 5 Calculated artificial groundwater recharge

2.8×10^6 m^3/year. It was null in the years 2010 and 2012 because of the absence of recharge operations. These values are still below the mean natural recharge rate estimated between 6×10^6 and 10.5×10^6 m^3/year (Manaa et al. 1996; Brini and Zammouri 2016).

The important deficit in recharge is due to the absence of artificial recharge operation in 2010 and 2012 and to low released water in several years. The analysis of the historical artificial recharge operations carried out over the period 2002–2015 indicate a total released water of 59.4×10^6 m^3, with a mean yearly value of 4.2×10^6 m^3 and a standard variation of 3.2×10^6 m^3. This is inconsistent with one of the objectives of El Khairat wadi damming that consists of mobilizing large flood waters for groundwater artificial recharge. The maximum released volume of 10.7×10^6 m^3 in 2002 doesn't reach the capacity storage of the El Khairat reservoir equal to 13×10^6 m^3.

The released water quantities seem to be insufficient to compensate the groundwater recharge in natural conditions. However, as shown in Fig. 6, the piezometric level at piezometers located in close proximity of the potential recharge area such as Garci1 shows rather a rising trend while piezometers located far from the recharge area such as S. Ncir4 show, in general, a groundwater level decreasing (Brini and Zammouri 2016). On the other hand, the analysis of the impact of artificial groundwater recharge on the groundwater quality showed a slight decrease of the salinity in the zone located near the recharge site (Ketata et al. 2011). This confirms the high rate infiltration in the vicinity of the wadi.

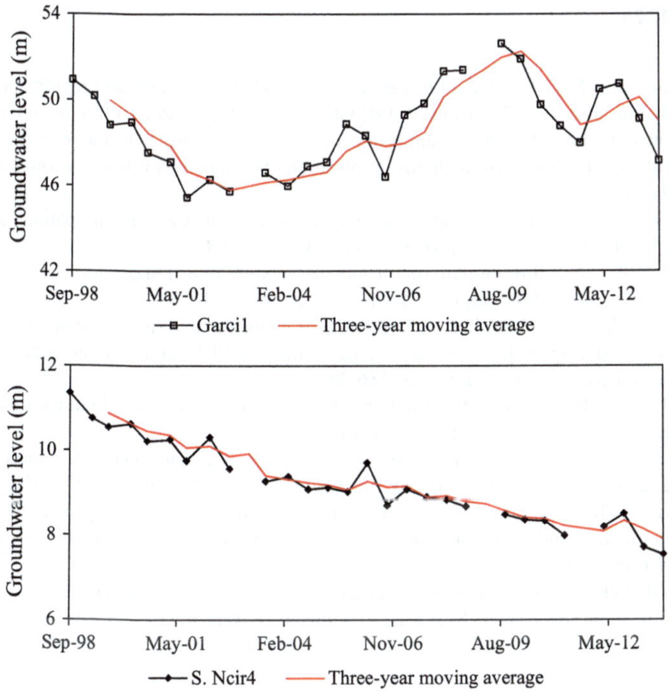

Fig. 6 Observed groundwater level (above sea level) at selected observation wells

5 Conclusion

The results of the artificial groundwater recharge modelling indicated high infiltration rates and insignificant evaporation losses. Losses by surface runoff beyond the permeable part of El Khairat wadi were large when the releases discharge rates were high. To avoid this, releasing water at discharge rate lower than 0.3 m^3/s is recommended. The calculated mean groundwater artificial recharge over the period 2002–2015 is 2.8×10^6 m^3/year. It is low compared to the mean value corresponding to the groundwater recharge in natural conditions. At present, the El Khairat reservoir is used for artificial recharge and cereal irrigation. The artificial recharge yield can be improved if the reservoir will be exclusively used for the groundwater artificial recharge. The artificial recharge effect is mainly located along the permeable course El Khairat wadi, in the upstream of the plain. It is advisable to extend the technique of artificial groundwater recharge in remote areas affected by groundwater extraction activity in the aquifer.

References

Brini N, Zammouri M (2016) Groundwater recharge modelling in semi-arid regions; a case study of El Khairat alluvial plain (Tunisia). QJEGH 49(3):229–236. doi:10.1144/qjegh2015-084

Dassi L, Zouari K, Faye S (2005) Identifying sources of groundwater recharge in the Merguellil basin (Tunisia) using isotopic methods: implication of dam reservoir water accounting. Environ Geol 49:114–123

Hernandez M (2000) Modeling runoff response to land cover and rainfall spatial variability in semi-arid watersheds. Environ Monitor Assess 64:285–298

Hughes DA (1995) Monthly rainfall–runoff models applied to arid and semi arid catchments for water resource estimation purposes. Hydrol Sci J 40(6):751–769

Ketata M, Gueddari M, Bouhlila R (2011) Hydrodynamic and hydrochemical evolution of groundwaters during artificial recharge: a case study of El Khairat aquifer system in Enfidha (Tunisian Sahel). J Appl Sci Res 7(3):286–293

Manaa M, Chaieb H, Amri R (1996) Etude par modèle hydrogéologique de la nappe alluviale de l'oued El Khairat. DGRE, Ministry of Agriculture, Tunisia

Sophocleous MA (1991) Combining the soil water balance and water-level fluctuation methods to estimate natural groundwater recharge: practical aspects. J Hydrol 124:229–241

Xi C, Zhang Z, Zhang X, Chen Y, Qian M, Peng S (2008) Estimation of groundwater recharge from precipitation and evapotranspiration by lysimeter measurement and soil moisture model. J Hydrol Eng 13(50):333–340

Zammouri M, Feki H (2005) Managing releases from small upland reservoirs for downstream recharge in semi-arid basins (Northeast of Tunisia). J Hydrol 314:125–138

MIX
Papier aus verantwortungsvollen Quellen
Paper from responsible sources
FSC® C105338

If you have any concerns about our products,
you can contact us on
ProductSafety@springernature.com

In case Publisher is established outside the EU,
the EU authorized representative is:
Springer Nature Customer Service Center GmbH
Europaplatz 3, 69115 Heidelberg, Germany

Printed by Libri Plureos GmbH
in Hamburg, Germany